D0803161

Introduction to Mathematical Physics

PRENTICE-HALL PHYSICS SERIES

Consulting Editors
Francis M. Pipkin
George A. Snow

Introduction to Mathematical Physics

CHARLIE HARPER

Department of Physics
California State University, Hayward

Prentice-Hall, Inc., Englewood Cliffs, New Jersey

Library of Congress Cataloging in Publication Data

HARPER, CHARLIE (date)
Introduction to mathematical physics.

Bibliography: p. 289
Includes index.
1. Mathematical physics. I. Title.
QC20.H38 530.1'5 75-28333
ISBN 0-13-487538-9

Printed in the United States of America

10 9 8 7 6 5 4 3 2 1

PRENTICE-HALL INTERNATIONAL, INC., *London*
PRENTICE-HALL OF AUSTRALIA, PTY. LIMITED, *Sydney*
PRENTICE-HALL OF CANADA, LTD., *Toronto*
PRENTICE-HALL OF INDIA PRIVATE LIMITED, *New Delhi*
PRENTICE-HALL OF JAPAN, INC., *Tokyo*
PRENTICE-HALL OF SOUTHEAST ASIA PTE. LTD., *Singapore*

Dedicated to the memory of

Sis

Contents

4 Calculus of Residues 101

5 Differential Equations 127

6 Special Functions of Mathematical Physics 163

7 Fourier Series 207

8 Fourier Transforms 227

9 Tensor Analysis 255

Preface

The chief aim of this book is to provide undergraduate students, who have a working knowledge of differential and integral calculus, with most of the mathematical prerequisites required for the study of (1) classical and quantum mechanics, (2) electromagnetism, (3) statistical thermodynamics, and (4) special and general relativity as well as other areas of physics, chemistry, applied mathematics, and engineering. The selected topics are based on my estimation of what is essential for undergraduate students in these areas and on the frequency with which these topics occur in physical applications.

While every effort has been made to present the material correctly, no attempt has been made to be absolutely rigorous. The included proofs are mainly (from a mathematician's point of view) plausibility arguments. It is assumed that students who plan to pursue advanced work in the natural sciences or applied mathematics will study mathematics and mathematical physics at a more advanced level.

The included illustrative examples are selected from general physics or developed from first principles. The problems at the end of each chapter are considered to be an integral part of the chapters and are occasionally used to convey new information in a self-contained manner.

I am grateful to Professor R. K. Cooper for the helpful suggestions and for enlightening discussions of several included topics. I am also indebted to Professors R. K. Cooper, R. H. Good, and J. C. Giles for proofreading the galleys.

Hayward, California CHARLIE HARPER

Introduction to Mathematical Physics

1

Vector Analysis

1.1 INTRODUCTION

1.1.1 Definition of Terms

In meaningful expressions and equations involving physical quantities, the dimension† (the power or powers to which the fundamental quantities appear) of each term must be the same. The fundamental physical quantities in mechanics are defined as length, L, mass, M, and time, T. For example, 3 mile/min + 7 cm/sec is a meaningful expression since the dimension of each term is length/time, L/T. It is also important to know that all physical quantities may be classified as a tensor.

A tensor quantity which can be completely specified by its magnitude is called a **scalar** (tensor of rank zero). Here magnitude means a number and a unit. (For dimensionless quantities, only a number is required.) The quantity "6 feet" is a magnitude since 6 is a number and feet is a unit. The following quantities are examples of scalars: mass, volume, density, energy, and tem-

†Certain physical quantities such as the coefficient of friction are dimensionless. Angles are geometrical constructions and are not physical quantities.

perature. The valid algebraic operations for scalars are the same as those for ordinary numbers.

A tensor quantity which can be completely specified by its magnitude and one direction is called a **vector** (tensor of rank one). The quantity "6 feet west" is a vector since it has magnitude, 6 feet, and one direction, west. A more general definition of a scalar and a vector in terms of transformation properties will be given in Chapter 9. We will restrict this chapter to the study of tensors of rank one (vector analysis). The following quantities are examples of vectors: displacement, velocity, acceleration, force, and torque. It is necessary for us to develop the laws of vector algebra and vector calculus since they are not, in general, the same as those for scalars. The subject of vector analysis was developed by J. Willard Gibbs (1839–1903) during the years 1880–1882.

1.1.2 Fundamental Concepts and Notations

Boldface print is used in most textbooks to designate a vector quantity. In writing, one of the following traditional notations for a vector quantity, $\vec{A}$, $\hat{A}$, or $\underset{\sim}{A}$, should be adopted. Note that the magnitude of a vector, $|\mathbf{A}| = A$, is a scalar. A vector with zero magnitude is called a **null vector,** and a vector whose magnitude is unity, $|\hat{\mathbf{B}}| = 1$, is called a **unit vector.** Throughout this chapter, the symbol ^ over a letter will denote a unit vector. For example, $\hat{\mathbf{B}}$ means that the quantity represented by $\hat{\mathbf{B}}$ is of unit magnitude, $|\hat{\mathbf{B}}| = 1$.

In Fig. 1.1 an arrow is used to represent a vector quantity. The direction of the quantity is indicated by the head of the arrow, and the magnitude of the quantity is characterized by the length of the arrow.

The vectors (Fig. 1.1) **A** and **B** are said to be equal, $\mathbf{A} = \mathbf{B}$, since they have the same length and direction; but they may not be equivalent. An understanding of the concept of vector equivalence is required in the study of mechanics. To be equivalent, vector quantities must produce identical mechanical effects.

The vector $-\mathbf{A}$ is equal in magnitude to vector **A** but opposite in direc-

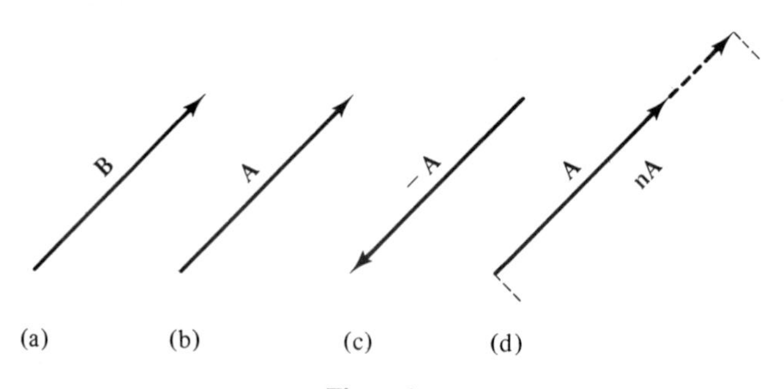

Figure 1.1

tion. The multiplication of a vector by a scalar, n, such that $n\mathbf{A} = \mathbf{A}n$ is shown in Fig. 1.1(d).

1.1.3 Vector Addition without Coordinates

The geometric addition of vectors is achieved by use of the following rule: (1) Place the vectors heads to tails. (2) Draw the vector from the tail of the first vector to the head of the last vector. The **resultant vector**, the **sum**, is the vector in part 2 of the rule. This procedure is illustrated in Figs. 1.2–1.4.

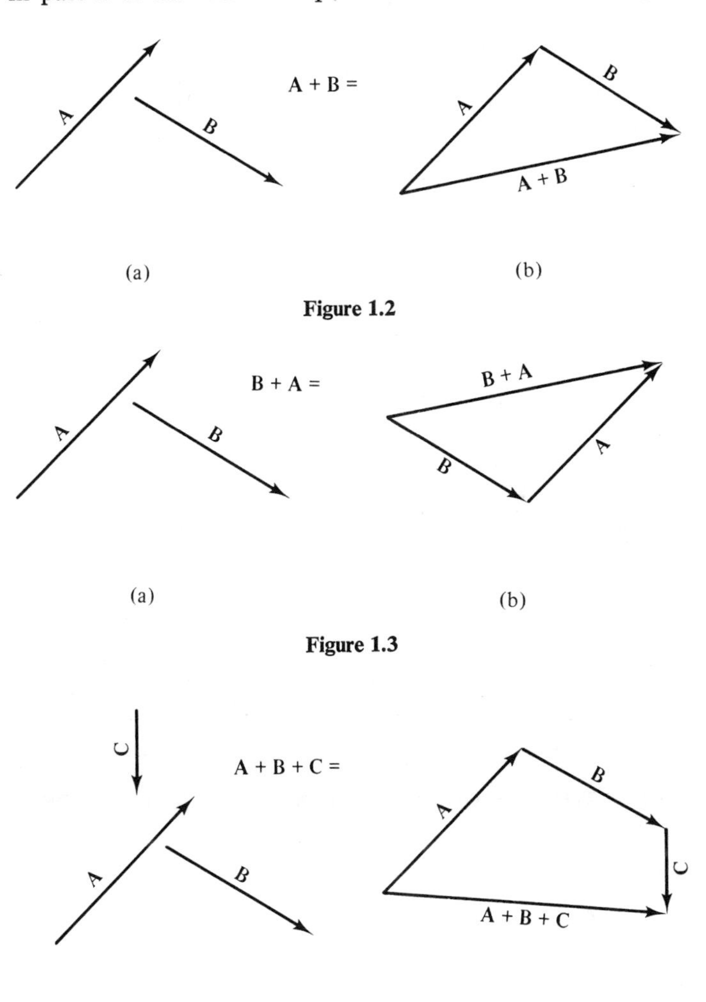

Figure 1.2

Figure 1.3

Figure 1.4

The extension to an arbitrary number of vectors is obvious, and the process of subtraction is trivial since $\mathbf{A} - \mathbf{B} = \mathbf{A} + (-\mathbf{B})$.

The operation of subtraction is illustrated in Fig. 1.5, and the commutative, $\mathbf{A} + \mathbf{B} = \mathbf{B} + \mathbf{A}$, and associative, $(\mathbf{C} + \mathbf{D}) + \mathbf{E} = \mathbf{C} + (\mathbf{D} + \mathbf{E})$, laws of addition are shown in Fig. 1.6.

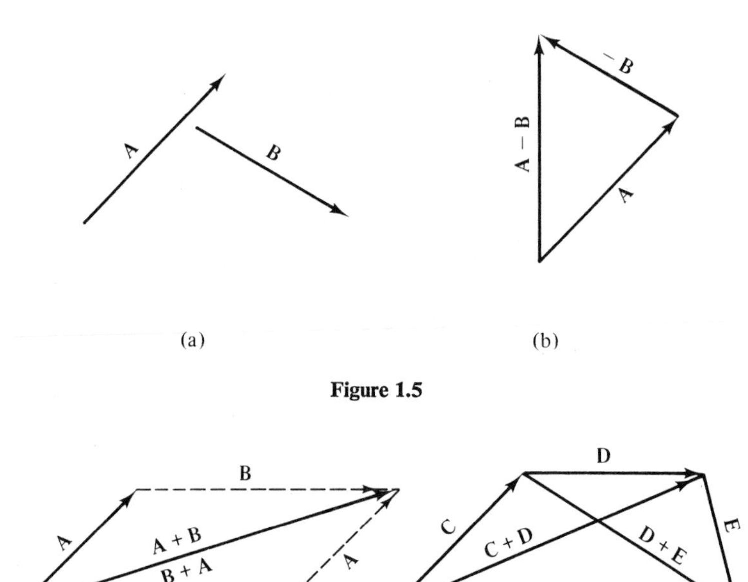

Figure 1.5

Figure 1.6

1.2 THE CARTESIAN SYSTEM OF BASE VECTORS

1.2.1 An Orthonormal Basis

An orthogonal basis for a vector space (see Chapter 2) of three dimensions consists of a set of three mutually perpendicular vectors. In Fig. 1.7, the **i**, **j**, and **k** set of vectors forms a normal orthogonal basis (orthonormal basis) since **i**, **j**, and **k** are mutually perpendicular unit vectors. This system is called a **Cartesian** (Descartes 1596–1650) **coordinate system.**

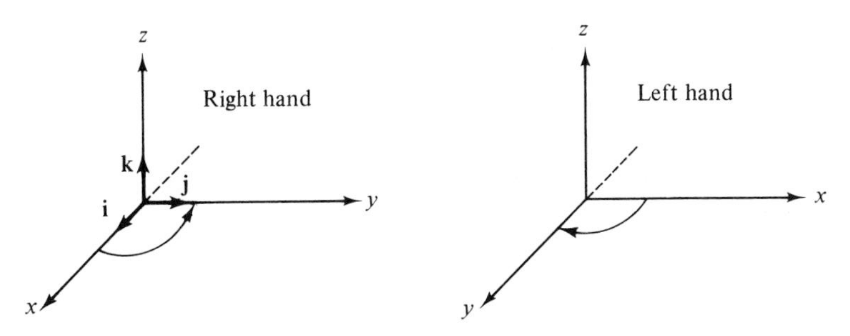

Figure 1.7

1.2.2 Position Vector (Radius Vector)

The **position** of a physical object is completely specified by its position vector (sometimes called **radius vector**). The position vector is a vector drawn from the origin of a coordinate system to the object in question. The orientation, with respect to a Cartesian coordinate system, may be in a right- or left-hand coordinate system as indicated in Fig. 1.7. However, we will use a right-hand system throughout this chapter.

1.2.3 Rectangular Resolution of a Vector

The position vector, **R**, of an object located at $P(x, y, z)$ is shown in Fig. 1.8. A small **r** will also be used to represent a position vector.

$$\overline{\mathbf{OP}} = \mathbf{R} = \overline{\mathbf{OA}} + \overline{\mathbf{AB}} + \overline{\mathbf{BP}}$$
$$= x\mathbf{i} + y\mathbf{j} + z\mathbf{k} \tag{1.1}$$

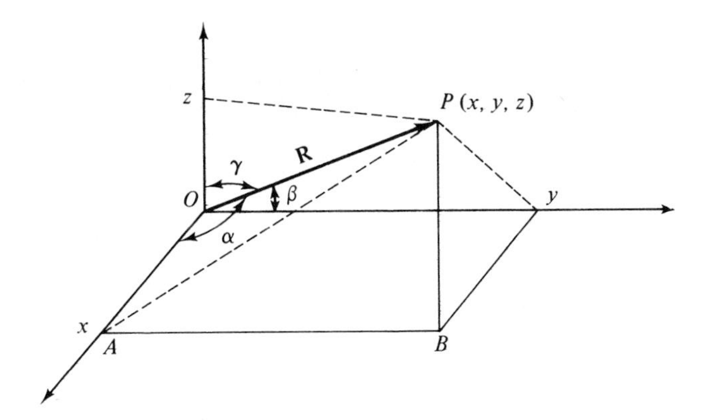

Figure 1.8

The vectors $x\mathbf{i}$, $y\mathbf{j}$, and $z\mathbf{k}$ are the three components of $\mathbf{R}$; they are the vector representations of the projections of $\mathbf{R}$ on the three coordinate axes, respectively. The quantities x, y, and z are the magnitudes of the vector components in the three respective directions.

On using the Pythagorean (Pythagoras, about 572–497 B.C.) theorem in Fig. 1.8, we find that

$$(\overline{\mathrm{OP}})^2 = |\mathbf{R}|^2 = (\overline{\mathrm{OA}})^2 + (\overline{\mathrm{AB}})^2 + (\overline{\mathrm{BP}})^2$$
$$= x^2 + y^2 + z^2 \tag{1.2}$$

where $|\mathbf{R}|$ is the magnitude (absolute value) of $\mathbf{R}$.

If the projections of an arbitrary vector $\mathbf{A}$ along the three axes of a Cartesian system are A_x, A_y, and A_z, the vector $\mathbf{A}$ in terms of these three components may be written as

$$\mathbf{A} = A_x\mathbf{i} + A_y\mathbf{j} + A_z\mathbf{k}. \tag{1.3}$$

The magnitude of $\mathbf{A}$ is given by

$$|\mathbf{A}| = (A_x^2 + A_y^2 + A_z^2)^{1/2}. \tag{1.4}$$

In terms of components, the equality $\mathbf{A} = \mathbf{B}$ means that

$$A_x = B_x, \qquad A_y = B_y, \quad \text{and} \quad A_z = B_z. \tag{1.5}$$

1.2.4 Direction Cosines

The angles α, β, and γ are the angles that $\overline{\mathbf{OP}}$ makes with the three coordinate axes (see Fig. 1.8). Here we have

$$x = Rl, \qquad y = Rm, \quad \text{and} \quad z = Rn \tag{1.6}$$

where $l = \cos\alpha$, $m = \cos\beta$, and $n = \cos\gamma$. The quantities l, m, and n are called **direction cosines.** Combining Eqs. (1.2) and (1.6), we find that

$$1 = l^2 + m^2 + n^2.$$

For an arbitrary vector $\mathbf{A}$, we may write

$$\frac{\mathbf{A}}{|\mathbf{A}|} = \mathbf{i}\cos\alpha + \mathbf{j}\cos\beta + \mathbf{k}\cos\gamma. \tag{1.7}$$

Clearly, the quantity $\mathbf{A}/|\mathbf{A}|$ is a unit vector since its magnitude is unity.

EXAMPLE 1.1 What are the direction cosines of the directed line segment from $P_1(-1, -4, 5)$ to $P_2(3, -2, 2)$?

Solution In general, we have

$$l = \cos\alpha = \frac{\Delta x}{d}, \qquad m = \cos\beta = \frac{\Delta y}{d}, \quad \text{and} \quad n = \cos\gamma = \frac{\Delta z}{d}.$$

The distance from P_1 to P_2 is

$$\begin{aligned} d &= \sqrt{(\Delta x)^2 + (\Delta y)^2 + (\Delta z)^2} \\ &= \sqrt{(3+1)^2 + (-2+4)^2 + (2-5)^2} \\ &= 5.39 \text{ units.} \end{aligned}$$

We therefore find that

$$l = \frac{4}{5.39}, \qquad m = \frac{2}{5.39}, \quad \text{and} \quad n = -\frac{3}{5.39}.$$

1.2.5 Vector Algebra with Coordinates

A. Addition The subject of vector algebra involves developing laws for the following two operations: (1) addition (subtraction) and (2) multiplication. The operation of addition is very simple and straightforward. To find the sum of **A** and **B**, we add like components together, that is,

$$\begin{aligned} \mathbf{A} + \mathbf{B} &= (A_x\mathbf{i} + A_y\mathbf{j} + A_z\mathbf{k}) + (B_x\mathbf{i} + B_y\mathbf{j} + B_z\mathbf{k}) \\ &= (A_x + B_x)\mathbf{i} + (A_y + B_y)\mathbf{j} + (A_z + B_z)\mathbf{k}. \end{aligned} \tag{1.8}$$

B. Scalar Product (Dot Product) There are two kinds of products defined for vectors: they are called **scalar (dot) product** and **vector (cross) product.**

The **scalar product** of two vectors **A** and **B** is a scalar and is defined as follows:

$$\begin{aligned} \mathbf{A}\cdot\mathbf{B} &\equiv |\mathbf{A}||\mathbf{B}|\cos\theta \\ &= A_xB_x + A_yB_y + A_zB_z \end{aligned} \tag{1.9}$$

where θ is the smaller angle between **A** and **B** when they are placed tail to tail, $\mathbf{i}\cdot\mathbf{i} = \mathbf{j}\cdot\mathbf{j} = \mathbf{k}\cdot\mathbf{k} = 1$, and $\mathbf{i}\cdot\mathbf{j} = \mathbf{i}\cdot\mathbf{k} = \mathbf{k}\cdot\mathbf{j} = 0$. The quantity $|\mathbf{B}| \cos\theta$ is the component of **B** along **A** (see Fig. 1.9). Hence the scalar product $\mathbf{A}\cdot\mathbf{B}$ equals the product of $|\mathbf{A}|$ and the component of **B** along **A**.

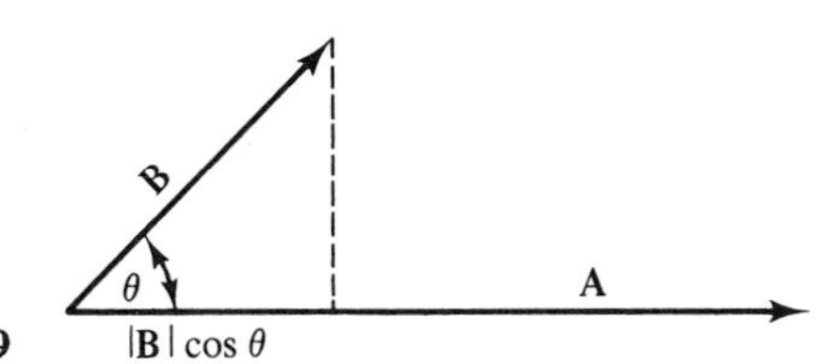

Figure 1.9

From the definition in Eq. (1.9), it is clear that the scalar product is commutative, and we may therefore write

$$\mathbf{A}\cdot\mathbf{B} = \mathbf{B}\cdot\mathbf{A}. \tag{1.10}$$

From Fig. (1.10), we note that

$$|\mathbf{A}|\,(b + c) = |\mathbf{A}|\,b + |\mathbf{A}|\,c \qquad \text{(scalar)}$$

or

$$\mathbf{A}\cdot(\mathbf{B} + \mathbf{C}) = \mathbf{A}\cdot\mathbf{B} + \mathbf{A}\cdot\mathbf{C} \tag{1.11}$$

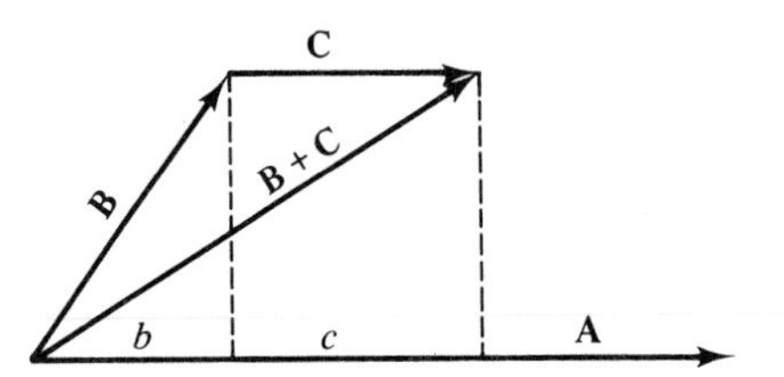

Figure 1.10

where $b + c$ is the component of $\mathbf{B} + \mathbf{C}$ along $\mathbf{A}$, b is the component of $\mathbf{B}$ along $\mathbf{A}$, and c is the component of $\mathbf{C}$ along $\mathbf{A}$.

The **work** done by a constant force, $\mathbf{F}$, is defined by means of the scalar product. In equation form, this definition is

$$W \equiv \mathbf{F}\cdot\mathbf{D} \tag{1.12}$$

where $\mathbf{D}$ is the displacement of the object.

C. Vector Product (Cross Product) The **vector product** of $\mathbf{A}$ and $\mathbf{B}$ is defined by

$$\mathbf{A} \times \mathbf{B} \equiv |\mathbf{A}|\,|\mathbf{B}|\sin\theta\,\hat{\mathbf{N}} \tag{1.13}$$

where θ is the smaller angle between $\mathbf{A}$ and $\mathbf{B}$ when they are placed tail to tail. The unit vector $\hat{\mathbf{N}}$ is perpendicular to the plane of $\mathbf{A}$ and $\mathbf{B}$ and is in the direction advanced by a right-hand woodscrew when it is turned from $\mathbf{A}$ to $\mathbf{B}$ (see Fig. 1.11). Note that

$$\mathbf{A} \times \mathbf{B} = -\mathbf{B} \times \mathbf{A} \tag{1.14}$$

because of the direction of $\hat{\mathbf{N}}$.

Since $\sin 0° = 0$,

$$\mathbf{i} \times \mathbf{i} = \mathbf{j} \times \mathbf{j} = \mathbf{k} \times \mathbf{k} = 0. \tag{1.15}$$

We also have (see Fig. 1.12)

$$\mathbf{i} \times \mathbf{j} = \mathbf{k}, \qquad \mathbf{j} \times \mathbf{k} = \mathbf{i}, \quad \text{and} \quad \mathbf{k} \times \mathbf{i} = \mathbf{j}. \tag{1.16}$$

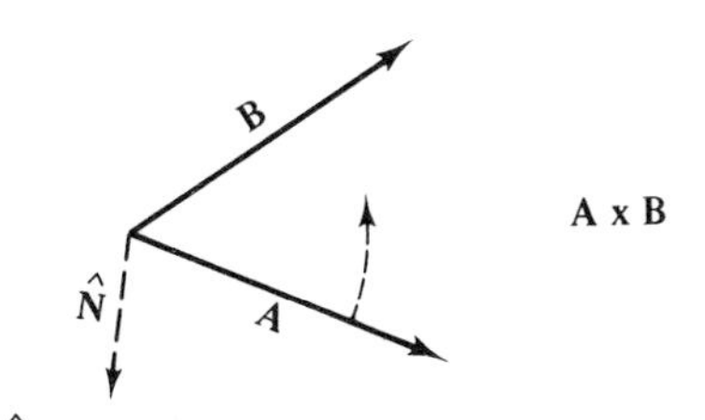

Figure 1.11 $\hat{\mathbf{N}}$ (out of the page) is perpendicular to the page and both **A** and **B** are on the page.

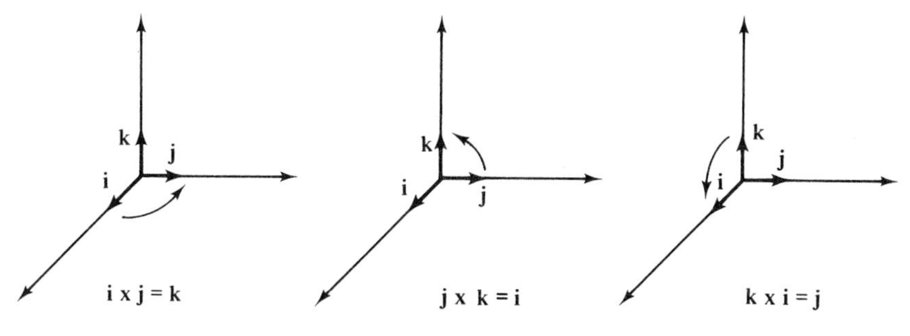

Figure 1.12

In terms of the components of **A** and **B**, the cross product $\mathbf{A} \times \mathbf{B}$ may be written as

$$\begin{aligned}\mathbf{A} \times \mathbf{B} &= (A_x\mathbf{i} + A_y\mathbf{j} + A_z\mathbf{k}) \times (B_x\mathbf{i} + B_y\mathbf{j} + B_z\mathbf{k}) \\ &= (A_yB_z - A_zB_y)\mathbf{i} + (A_zB_x - A_xB_z)\mathbf{j} + (A_xB_y - A_yB_x)\mathbf{k} \\ &= \begin{vmatrix} \mathbf{i} & \mathbf{j} & \mathbf{k} \\ A_x & A_y & A_z \\ B_x & B_y & B_z \end{vmatrix}. \end{aligned} \tag{1.17}$$

Equations (1.15) and (1.16) were used to obtain Eq. (1.17). (A discussion of determinants is given in the appendix of Chapter 2.)

In mechanics, **torque** (the moment of a force) is defined by use of the vector product (see Fig. 1.13), that is,

$$\boldsymbol{\tau}(\text{torque}) = \mathbf{R} \times \mathbf{F} \tag{1.18}$$

D. Division by a Vector Since division is normally thought of as the inverse process of multiplication, one would naturally assume that two different kinds of division processes (corresponding to each of the two kinds of multiplication processes) exist for vectors. However, there exists no unique definition for the division by a vector. To understand the difficulty involved in

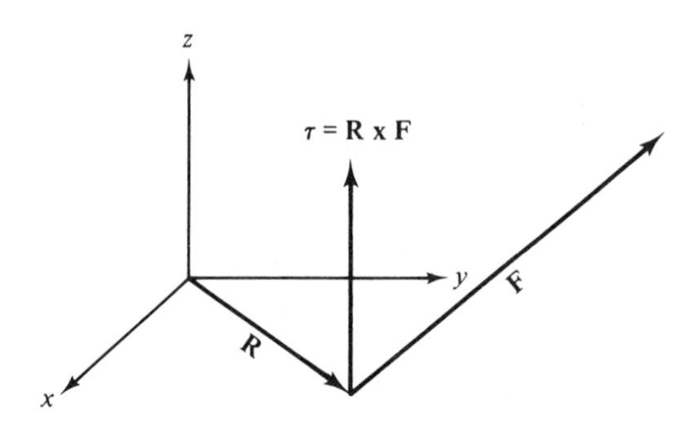

Figure 1.13

developing a unique definition for the division by a vector, consider the case of scalar multiplication. Here we have for two vectors **A** and **B**

$$\alpha = \mathbf{A} \cdot \mathbf{B}$$

where α is a scalar. The product on the right-hand side of the above equation would normally result from the division of α by **A**, $\alpha/\mathbf{A}$, giving a quotient **B**. However, **B** is not a unique quotient when obtained in this manner since we may write

$$\alpha = \mathbf{A} \cdot (\mathbf{B} + \mathbf{D})$$

where **D** is an arbitrary vector perpendicular to **A**. The quotient, $\mathbf{B} + \mathbf{D}$, for $\alpha/\mathbf{A}$ is therefore not unique since $\mathbf{B} + \mathbf{D}$ is an arbitrary vector. Similarly, we can show that the quotient of $\mathbf{A}/\mathbf{B}$ is not unique if $\mathbf{A} = \mathbf{B} \times \mathbf{C}$.

EXAMPLE 1.2 Given $\mathbf{A} = \mathbf{i} + 2\mathbf{j} + 3\mathbf{k}$ and $\mathbf{B} = 3\mathbf{i} + 2\mathbf{j} + \mathbf{k}$. Find

(a) $|\mathbf{A}|$, (b) $\mathbf{A} + \mathbf{B}$,
(c) $\mathbf{A} - \mathbf{B}$, (d) $\mathbf{A} \cdot \mathbf{B}$,
(e) $\mathbf{A} \times \mathbf{B}$,
(f) a unit vector in the direction of **A**,
(g) the angle between **A** and **B**, and
(h) a unit vector perpendicular to both **A** and **B**.

Solution

(a)
$$|\mathbf{A}| = \sqrt{1^2 + 2^2 + 3^2}$$
$$= 3.74$$

(b)
$$\mathbf{A} + \mathbf{B} = (\mathbf{i} + 2\mathbf{j} + 3\mathbf{k}) + (3\mathbf{i} + 2\mathbf{j} + \mathbf{k})$$
$$= 4\mathbf{i} + 4\mathbf{j} + 4\mathbf{k}$$

(c)
$$\mathbf{A} - \mathbf{B} = (\mathbf{i} + 2\mathbf{j} + 3\mathbf{k}) - (3\mathbf{i} + 2\mathbf{j} + \mathbf{k})$$
$$= -2\mathbf{i} + 2\mathbf{k}$$

(d)
$$\mathbf{A}\cdot\mathbf{B} = (\mathbf{i} + 2\mathbf{j} + 3\mathbf{k})\cdot(3\mathbf{i} + 2\mathbf{j} + \mathbf{k})$$
$$= 3 + 4 + 3$$
$$= 10$$

(e)
$$\mathbf{A}\times\mathbf{B} = \begin{vmatrix} \mathbf{i} & \mathbf{j} & \mathbf{k} \\ 1 & 2 & 3 \\ 3 & 2 & 1 \end{vmatrix}$$
$$= -4\mathbf{i} + 8\mathbf{j} - 4\mathbf{k}$$

(f)
$$\hat{A} = \frac{\mathbf{A}}{|\mathbf{A}|}$$
$$= \frac{\mathbf{i} + 2\mathbf{j} + 3\mathbf{k}}{3.74}$$

where $\hat{A}$ is a unit vector in the direction of **A**.

(g)
$$\mathbf{A}\cdot\mathbf{B} = |\mathbf{A}|\,|\mathbf{B}|\cos\theta$$
$$= 14\cos\theta$$
$$= 10$$

or

$$\cos\theta = \frac{10}{14} = 0.714$$

therefore $\theta \approx 44.4°$.

(h) By definition, **A** × **B** is perpendicular to both **A** and **B**.

$$\hat{\mathbf{N}}_{AB} = \frac{\mathbf{A}\times\mathbf{B}}{|\mathbf{A}\times\mathbf{B}|}$$
$$= \frac{-4\mathbf{i} + 8\mathbf{j} - 4\mathbf{k}}{9.8}$$

EXAMPLE 1.3 *Cauchy's inequality* Note that

$$(\mathbf{A}\cdot\mathbf{B})(\mathbf{A}\cdot\mathbf{B}) = |\mathbf{A}|^2\,|\mathbf{B}|^2\cos^2\theta$$
$$\leq |\mathbf{A}|^2\,|\mathbf{B}|^2$$

or

$$(A_1B_1 + A_2B_2 + A_3B_3)^2 \leq (A_1^2 + A_2^2 + A_3^2)(B_1^2 + B_2^2 + B_3^2)$$

or

$$\sum_{\alpha=1}^{3} A_\alpha B_\alpha \leq \Big(\sum_{\alpha=1}^{3} A_\alpha^2\Big)^{1/2}\Big(\sum_{\alpha=1}^{3} B_\alpha^2\Big)^{1/2}. \tag{1.19}$$

The above inequality is called **Cauchy's inequality.**

EXAMPLE 1.4 Calculate the moment of the force $\mathbf{F} = (5\mathbf{i} + 2\mathbf{j})$ lb about the origin if the force acts at the point (2, 1) ft.

Solution Here we have

$$\mathbf{R} = (2\mathbf{i} + \mathbf{j}) \text{ ft}$$

and

$$\mathbf{F} = (5\mathbf{i} + 2\mathbf{j}) \text{ lb}$$

The moment in this case becomes

$$\begin{aligned} \boldsymbol{\tau} &= \mathbf{R} \times \mathbf{F} \\ &= \begin{vmatrix} \mathbf{i} & \mathbf{j} & \mathbf{k} \\ 2 & 1 & 0 \\ 5 & 2 & 0 \end{vmatrix} = -\mathbf{k} \text{ lb}\cdot\text{ft} \end{aligned}$$

1.3 DIFFERENTIATION OF VECTOR FUNCTIONS

1.3.1 The Derivative of a Vector

Consider the vector $\mathbf{F}(s)$ where s is a scalar variable. Here $\mathbf{F}$ is a function of s; for each value of s, there is a corresponding value of $\mathbf{F}$ (see Fig. 1.14). The increment in $\mathbf{F}(s)$ when s changes to $s + \Delta s$ is given by

$$\begin{aligned} \Delta\mathbf{F} &= \mathbf{F}(s + \Delta s) - \mathbf{F}(s) \\ &= \overline{\mathbf{PQ}}. \end{aligned} \tag{1.20}$$

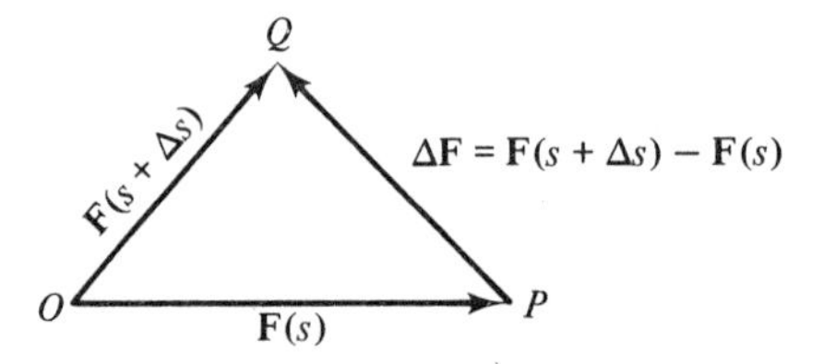

Figure 1.14

Dividing both sides of Eq. (1.20) by the scalar Δs, we obtain

$$\frac{\Delta\mathbf{F}}{\Delta s} = \frac{\overline{\mathbf{PQ}}}{\Delta s}$$

which is a vector along $\overline{\mathbf{PQ}}$. If $\Delta\mathbf{F}/\Delta s$ approaches a limit as Δs approaches zero, this limit is called the **derivative** of $\mathbf{F}$ with respect to s, that is,

$$\begin{aligned}\frac{d\mathbf{F}}{ds} &= \lim_{\Delta s \to 0}\left(\frac{\Delta \mathbf{F}}{\Delta s}\right)\\ &= \lim_{\Delta s \to 0}\left[\frac{\mathbf{F}(s+\Delta s)-\mathbf{F}(s)}{\Delta s}\right]\\ &= \mathbf{i}\frac{dF_x}{ds}+\mathbf{j}\frac{dF_y}{ds}+\mathbf{k}\frac{dF_z}{ds}. \end{aligned}\tag{1.21}$$

The derivative of $\mathbf{A}+\mathbf{B}$ is obtained in the following manner:

$$\Delta\mathbf{C} = \Delta\mathbf{A}+\Delta\mathbf{B}\tag{1.22}$$

where $\mathbf{C}=\mathbf{A}+\mathbf{B}$ and $\mathbf{C}+\Delta\mathbf{C}=\mathbf{A}+\Delta\mathbf{A}+\mathbf{B}+\Delta\mathbf{B}$. On dividing both sides of Eq. (1.22) by Δs and taking the limit as $\Delta s \longrightarrow 0$, we obtain

$$\begin{aligned}\frac{d\mathbf{C}}{ds} &= \lim_{\Delta s \to 0}\left(\frac{\Delta \mathbf{C}}{\Delta s}\right)\\ &= \frac{d}{ds}(\mathbf{A}+\mathbf{B})\\ &= \frac{d\mathbf{A}}{ds}+\frac{d\mathbf{B}}{ds}.\end{aligned}\tag{1.23}$$

Now consider the derivative of the scalar product where $C=\mathbf{A}\cdot\mathbf{B}$. Here we obtain

$$\begin{aligned}\frac{dC}{ds} &= \lim_{\Delta s \to 0}\left(\frac{\Delta C}{\Delta s}\right)\\ &= \frac{d}{ds}(\mathbf{A}\cdot\mathbf{B})\\ &= \mathbf{A}\cdot\frac{d\mathbf{B}}{ds}+\frac{d\mathbf{A}}{ds}\cdot\mathbf{B}\end{aligned}\tag{1.24}$$

where

$$\Delta C = \mathbf{A}\cdot\Delta\mathbf{B}+\Delta\mathbf{A}\cdot\mathbf{B}+\Delta\mathbf{A}\cdot\Delta\mathbf{B}$$

and

$$\lim_{\Delta s \to 0}\left(\Delta\mathbf{A}\cdot\frac{\Delta\mathbf{B}}{\Delta s}\right) = \lim_{\Delta s \to 0}\Delta\mathbf{A}\cdot\lim_{\Delta s \to 0}\frac{\Delta\mathbf{B}}{\Delta s} = 0$$

since

$$\lim_{\Delta s \to 0}\Delta\mathbf{A} = \lim_{\Delta s \to 0}[\mathbf{A}(s+\Delta s)-\mathbf{A}(s)] = 0.$$

The following relations can also be proved:

$$\frac{d}{ds}(\mathbf{A} \times \mathbf{B}) = \mathbf{A} \times \frac{d\mathbf{B}}{ds} + \frac{d\mathbf{A}}{ds} \times \mathbf{B} \tag{1.25a}$$

$$\frac{d\mathbf{C}}{ds} = 0 \qquad (\mathbf{C} \text{ is a constant vector}) \tag{1.25b}$$

$$\frac{d}{ds}(a\mathbf{A}) = a\frac{d\mathbf{A}}{ds} \qquad (a \text{ is a constant scalar}) \tag{1.25c}$$

$$\frac{d\mathbf{A}}{dt} = \frac{d\mathbf{A}}{ds}\frac{ds}{dt} \qquad [\text{where} \quad s = s(t)] \tag{1.25d}$$

$$\frac{d}{ds}(U\mathbf{V}) = U\frac{d\mathbf{V}}{ds} + \mathbf{V}\frac{dU}{ds} \qquad [\mathbf{V} = \mathbf{V}(s) \quad \text{and} \quad U = U(s)]. \tag{1.25e}$$

The quantities

$$\mathbf{v} = \frac{d\mathbf{R}}{dt} \equiv \dot{\mathbf{R}} \tag{1.26a}$$

and

$$\mathbf{a} = \frac{d\mathbf{v}}{dt} \equiv \dot{\mathbf{v}} \tag{1.26b}$$

where t is time, are called **velocity** and **acceleration,** respectively.

The **del (nabla) operator,** ∇, is a differential operator and is of immense importance in physics. It is defined by

$$\nabla \equiv \mathbf{i}\frac{\partial}{\partial x} + \mathbf{j}\frac{\partial}{\partial y} + \mathbf{k}\frac{\partial}{\partial z}. \tag{1.27}$$

EXAMPLE 1.5 Evaluate $\nabla \times \mathbf{i}\phi(x, y, z)$

Solution In this case, we have

$$\begin{aligned}
\nabla \times \mathbf{i}\phi &= \begin{vmatrix} \mathbf{i} & \mathbf{j} & \mathbf{k} \\ \frac{\partial}{\partial x} & \frac{\partial}{\partial y} & \frac{\partial}{\partial z} \\ \phi & 0 & 0 \end{vmatrix} \\
&= \mathbf{i}(0) - \mathbf{j}\left(-\frac{\partial \phi}{\partial z}\right) + \mathbf{k}\left(-\frac{\partial \phi}{\partial y}\right) \\
&= \mathbf{j}\frac{\partial \phi}{\partial z} - \mathbf{k}\frac{\partial \phi}{\partial y} \\
&= \left(\mathbf{j}\frac{\partial}{\partial z} - \mathbf{k}\frac{\partial}{\partial y}\right)\phi.
\end{aligned}$$

1.3.2 The Concept of a Gradient

Let $\phi(x, y, z)$ be a single-valued scalar function with continuous first derivatives in a certain region of space. If $\mathbf{R}$ is the position vector of an object located at $P(x, y, z)$, we have $\mathbf{R} = x\mathbf{i} + y\mathbf{j} + z\mathbf{k}$ and $d\mathbf{R} = dx\mathbf{i} + dy\mathbf{j} + dz\mathbf{k}$ (see Fig. 1.15a). The total differential of $\phi(x, y, z)$ is

$$\begin{aligned} d\phi &= \frac{\partial \phi}{\partial x}\, dx + \frac{\partial \phi}{\partial y}\, dy + \frac{\partial \phi}{\partial z}\, dz \\ &= \nabla\phi \cdot d\mathbf{R} \end{aligned} \tag{1.28}$$

The vector $\nabla\phi$ is called the **gradient** of ϕ and is often written as grad ϕ. Differentiating both sides of Eq. (1.28) with respect to R, we obtain

$$\frac{d\phi}{dR} = \nabla\phi \cdot \frac{d\mathbf{R}}{dR} \qquad \text{(directional derivative).} \tag{1.29}$$

By use of the definition of the scalar product (see Section 1.2.5 of this chapter), we see that the right-hand side of Eq. (1.29) is just the scalar component of $\nabla\phi$ in the direction of $d\mathbf{R}$. Hence the **gradient** of ϕ is a vector whose component in any direction, $d\mathbf{R}$, is the derivative of ϕ with respect to R. Note that when $\nabla\phi$ is parallel to $d\mathbf{R}$, $d\phi$ has its maximum value (see Fig. 1.15b).

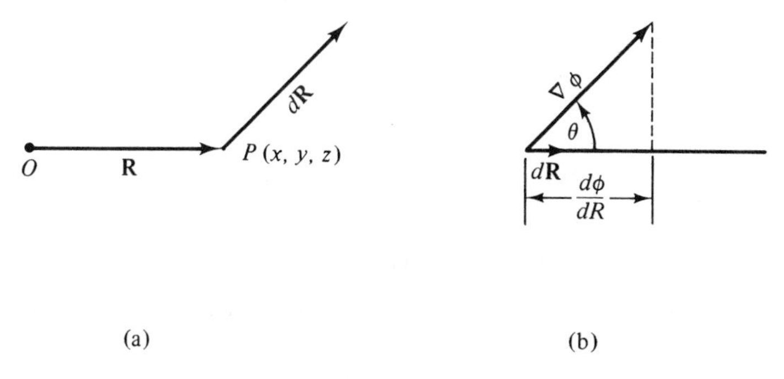

Figure 1.15

EXAMPLE 1.6 Find a unit vector normal to the surface $x^2 + y^2 - z = 1$ at $P(1, 1, 1)$.

Solution Here we have $\phi(x, y, z) = x^2 + y^2 - z$; therefore

$$\begin{aligned} \nabla\phi &= 2x\mathbf{i} + 2y\mathbf{j} - \mathbf{k} \\ &= 2\mathbf{i} + 2\mathbf{j} - \mathbf{k} \qquad [\text{at } \; P(1, 1, 1)] \end{aligned}$$

$$\hat{\mathbf{N}} = \frac{\nabla\phi}{|\nabla\phi|} = \frac{2\mathbf{i} + 2\mathbf{j} - \mathbf{k}}{3}$$

where $\hat{\mathbf{N}}$ is a unit vector normal to ϕ at $P(1, 1, 1)$.

Multiplication involving the ∇ operator is extremely useful in physics. The two products denoted by $\nabla\cdot\mathbf{F} \equiv \text{div}\,\mathbf{F}$ and $\nabla\times\mathbf{F} \equiv \text{curl}\,\mathbf{F}$ are called the **divergence** of **F** and the **curl** of **F**, respectively. Their connections with physical problems will be revealed in later sections. The scalar operator $\nabla\cdot\nabla = \nabla^2$ is also useful in physics; it is called the **Laplacian** (Laplace, 1749–1827) **operator.** In equation form, it is

$$\nabla^2 = \frac{\partial^2}{\partial x^2} + \frac{\partial^2}{\partial y^2} + \frac{\partial^2}{\partial z^2} \tag{1.30}$$

since $\mathbf{i}\cdot\mathbf{i} = \mathbf{j}\cdot\mathbf{j} = \mathbf{k}\cdot\mathbf{k} = 1$ and $\mathbf{i}\cdot\mathbf{j} = \mathbf{i}\cdot\mathbf{k} = \mathbf{j}\cdot\mathbf{k} = 0$.

The divergence and curl of **F** are given, respectively, by

$$\nabla\cdot\mathbf{F} = \frac{\partial F_x}{\partial x} + \frac{\partial F_y}{\partial y} + \frac{\partial F_z}{\partial z} \tag{1.31}$$

and

$$\nabla\times\mathbf{F} = \begin{vmatrix} \mathbf{i} & \mathbf{j} & \mathbf{k} \\ \frac{\partial}{\partial x} & \frac{\partial}{\partial y} & \frac{\partial}{\partial z} \\ F_x & F_y & F_z \end{vmatrix}. \tag{1.32}$$

If at some point P

$$\nabla\cdot\mathbf{V}\begin{cases} > 0, & \text{then } \mathbf{V} \text{ has a source at } P \\ < 0, & \text{then } \mathbf{V} \text{ has a sink at } P \\ = 0, & \text{then } \mathbf{V} \text{ is said to be solenoidal} \end{cases} \tag{1.33}$$
$$\nabla\times\mathbf{V} = 0, \quad \text{then } \mathbf{V} \text{ is classified as irrotational.}$$

EXAMPLE 1.7 Establish a physical meaning for the divergence of a vector by use of an illustration from hydrodynamics.

Solution Consider the flow indicated in Fig. 1.16 and let $\mathbf{A} = \rho\mathbf{v}$, that is, **A** represents the mass of fluid flowing through a unit area normal to side $ABOC$ per unit time. The density of the fluid is denoted by ρ, and $\mathbf{v}$ is its velocity.

The y-component of **A** through the area $ABOC$ indicated in Fig. 1.16 per unit time is given by

$$A_y\,dx\,dz.$$

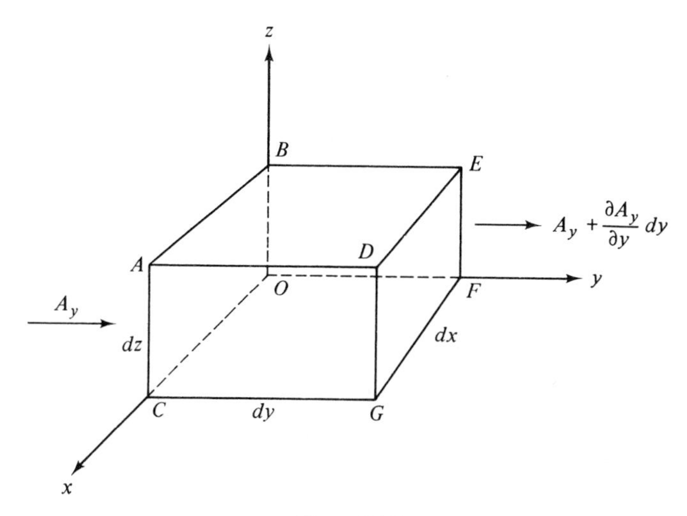

Figure 1.16

The flow through the area *DEFG* per unit time may be represented by the following Taylor's series:

$$A_y(y + dy)\, dx\, dz = \left[A_y(y) + \frac{\partial A_y}{\partial y} dy + \cdots\right] dx\, dz.$$

We will neglect higher-order terms in the above expansion. The net increase in the mass of the fluid inside the volume element $d\tau = dx\, dy\, dz$ per unit time due to the flow through the two faces is

$$A_y\, dx\, dz - \left(A_y + \frac{\partial A_y}{\partial y} dy\right) dx\, dz = -\frac{\partial A_y}{\partial y} d\tau.$$

Similarly, the net increase in the mass of fluid per unit time due to the flow through *BEFO* and *ADGC* is

$$-\frac{\partial A_x}{\partial x} d\tau$$

and that through *FGCO* and *EDAB* is

$$-\frac{\partial A_z}{\partial z} d\tau.$$

The total increase in the mass of fluid per unit volume per unit time due to the excess of inward flow over the outward flow is

$$\frac{\left(-\dfrac{\partial A_x}{\partial x} - \dfrac{\partial A_y}{\partial y} - \dfrac{\partial A_z}{\partial z}\right) d\tau}{d\tau} = -\boldsymbol{\nabla} \cdot \mathbf{A}$$

which is just the rate of increase of the density of the fluid inside of the volume element $d\tau$. That is to say,

$$\frac{\partial \rho}{\partial t} = -\nabla \cdot \mathbf{A}. \tag{1.34}$$

The above equation is called the **continuity equation.** For an incompressible fluid, we have

$$\frac{\partial \rho}{\partial t} = 0 \quad \text{or} \quad \nabla \cdot \mathbf{A} = 0 \qquad \text{(solenoidal).}$$

In this case, the excess of outward flow over inward flow is zero.

The physical interpretation of the curl of a vector is connected with the rotation (or circulation) of a vector field; its full meaning will be made clear by its use in later sections.

1.4 INTEGRATION OF VECTOR FUNCTIONS

Here we consider the question of the integration of vector functions and three important related theorems.

1.4.1 Line Integrals

The **line integral (tangential line integral),** I_λ, of a vector **F** along a curve C from A to B (see Fig. 1.17) is defined as the definite integral of the scalar component of **F** in the direction of the tangent to the curve at $P(x, y, z)$. Thus

$$I_\lambda = \int_A^B \mathbf{F} \cdot d\mathbf{R} \tag{1.35}$$

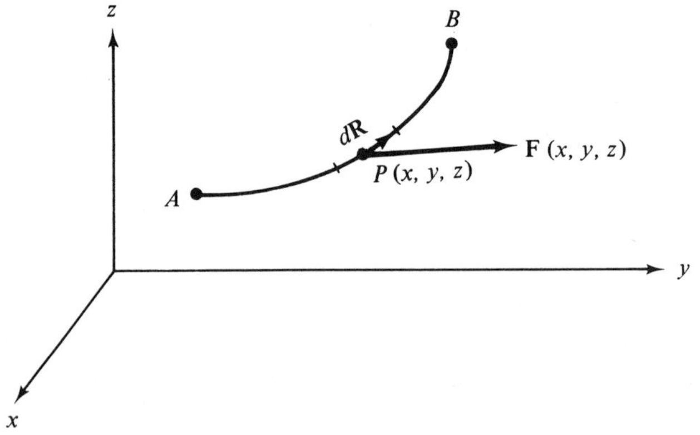

Figure 1.17

where $d\mathbf{R}$ is an element of displacement at $P(x, y, z)$. Note that the line integral from A to B is the negative of that from B to A.

The **work** done by a variable force $\mathbf{F}(x, y, z)$ in moving an object from A to B is defined as

$$\text{Work} = \int_A^B \mathbf{F} \cdot d\mathbf{R} \tag{1.36}$$

EXAMPLE 1.8 Calculate the work done by the force $\mathbf{F} = 2y\mathbf{i} + xy\mathbf{j}$ lb in moving an object along a straight line from $A(0, 0, 0)$ to $B(2, 1, 0)$ ft.

Solution

$$\begin{aligned}\text{Work} &= \int_A^B \mathbf{F} \cdot d\mathbf{R} \\ &= \int (2y\mathbf{i} + xy\mathbf{j}) \cdot (dx\mathbf{i} + dy\mathbf{j} + dz\mathbf{k}) \\ &= \int (2y\,dx + xy\,dy) \\ &= \int_0^2 x\,dx + 2\int_0^1 y^2\,dy \\ &= 2.67 \text{ ft-lb}\end{aligned}$$

since the equation for the path is $x = 2y$.

EXAMPLE 1.9 Show that the work done on an object by a resultant force $\bar{\mathbf{F}}$ during a displacement from A to B is equal to the change in the kinetic energy ($\frac{1}{2}mv^2$) of the object.

Solution The work is given by

$$\text{Work} = \int_A^B \bar{\mathbf{F}} \cdot d\mathbf{R}$$

where

$$\bar{\mathbf{F}} = m\mathbf{a} = m\frac{d\mathbf{v}}{dt} \qquad \text{(Newton's second law)}$$

and

$$d\mathbf{R} = \frac{d\mathbf{R}}{dt}\,dt = \mathbf{v}\,dt.$$

The expression for the work can therefore be put in the form

$$\begin{aligned}\text{Work} &= \int_A^B \bar{\mathbf{F}} \cdot d\mathbf{R} \\ &= m\int_{t_A}^{t_B} \frac{d\mathbf{v}}{dt} \cdot \mathbf{v}\,dt\end{aligned}$$

$$= m \int_{v_A}^{v_B} \mathbf{v} \cdot d\mathbf{v}$$

$$= \frac{m}{2} \int_{v_A}^{v_B} d(\mathbf{v} \cdot \mathbf{v})$$

$$= \frac{m}{2}(v_B^2 - v_A^2) = \text{change in the kinetic energy} \tag{1.37}$$

(the **work-energy theorem** of mechanics)

since

$$\frac{d}{dt}(\mathbf{v} \cdot \mathbf{v}) = \frac{d\mathbf{v}}{dt} \cdot \mathbf{v} + \mathbf{v} \cdot \frac{d\mathbf{v}}{dt} = 2\mathbf{v} \cdot \frac{d\mathbf{v}}{dt}.$$

If the force **F** is the gradient of a single-valued scalar function $\phi(x, y, z)$ and the path of integration is closed, we obtain

$$\begin{aligned} \text{Work} &= \oint \mathbf{F} \cdot d\mathbf{R} \\ &= \oint \nabla\phi \cdot d\mathbf{R} \\ &= \oint d\phi \\ &= 0. \end{aligned}$$

The symbol $\oint$ means integration around a closed path. In this case, we see that the work done is zero. From the result of Example 1.9 (the work-energy theorem), we note that zero work means that the final kinetic energy of the object is equal to the initial kinetic energy. Hence energy is conserved; such a force, $\mathbf{F} = \nabla\phi$, is said to be a **conservative force.**

At position $P(x, y, z)$, an object subjected to a conservative force **F** is said to possess a potential energy relative to some fixed point, B. This **potential energy** is defined by

$$\text{Potential energy} = \int_P^B \mathbf{F} \cdot d\mathbf{R}. \tag{1.38}$$

1.4.2 The Divergence Theorem Due to Gauss (1777–1855)

Before discussing the divergence theorem, we explain some useful concepts related to vector integration. The **normal surface integral** of $\mathbf{F}(x, y, z)$ over a closed boundary is defined as the surface (double) integral of the scalar component of **F** in the direction of a normal to the surface. This is written

$$I_\sigma = \oiint_\sigma \mathbf{F} \cdot \hat{\mathbf{n}}\, d\sigma = \oiint_\sigma \mathbf{F} \cdot d\boldsymbol{\sigma} \tag{1.39}$$

where $d\sigma$ is the area of an element of the boundary surface and $\hat{\mathbf{n}}$ is a unit outward normal to this element of area. The symbol $\oint\!\!\oint_\sigma$ means integration around a closed surface.

The sign convention for the unit normal is as follows: (1) For a closed surface (a surface which encloses a volume), the outward normal is called positive. (2) For an open surface, the right-hand screw rule is used; the direction of rotation is the same as that in which the periphery is traversed.

We will now delineate a very important theorem due to Gauss. The **divergence theorem** is as follows: The normal surface integral of a function **F** over the boundary of a closed surface of arbitrary shape is equal to the volume (triple) integral of the divergence of **F** taken throughout the enclosed volume. In equation form, we write

$$\oint\!\!\oint_\sigma \mathbf{F}\cdot d\boldsymbol{\sigma} = \iiint_\tau \nabla\cdot\mathbf{F}\, d\tau. \tag{1.40}$$

To prove this theorem, we first expand the right-hand side of Eq. (1.40) and obtain

$$\iiint_\tau \nabla\cdot\mathbf{F}\, d\tau = \iiint_\tau \left[\frac{\partial F_x}{\partial x} + \frac{\partial F_y}{\partial y} + \frac{\partial F_z}{\partial z}\right] dx\, dy\, dz. \tag{1.41}$$

Although the theorem is valid for an arbitrarily shaped closed surface, we choose the volume in Fig. 1.18 for convenience. In the figure, we have

(bottom)	$d\boldsymbol{\sigma}_1 = -\mathbf{k}\, dx\, dy;$	(back)	$d\boldsymbol{\sigma}_4 = -\mathbf{i}\, dy\, dz$
(front)	$d\boldsymbol{\sigma}_2 = \mathbf{i}\, dy\, dz;$	(top)	$d\boldsymbol{\sigma}_5 = \mathbf{k}\, dx\, dy$
(left)	$d\boldsymbol{\sigma}_3 = -\mathbf{j}\, dx\, dz;$	(right)	$d\boldsymbol{\sigma}_6 = \mathbf{j}\, dx\, dz.$

Integrating the last term on the right-hand side of Eq. (1.41) with respect to z from z' to z'', we obtain

$$\iiint_{z'}^{z''} \frac{\partial F_z}{\partial z}\, dx\, dy\, dz = \iint_\sigma [F_z(x, y, z'') - F_z(x, y, z')]\, dx\, dy$$
$$= \iint_{\sigma_5} F_z(x, y, z'')\, dx\, dy - \iint_{\sigma_1} F_z(x, y, z')\, dx\, dy. \tag{1.42}$$

Note that

$$\begin{aligned} \hat{\mathbf{n}}_1\cdot\mathbf{k}\, d\sigma_1 &= -dx\, dy \qquad \text{(bottom)} \\ \hat{\mathbf{n}}_5\cdot\mathbf{k}\, d\sigma_5 &= dx\, dy \qquad \text{(top).} \end{aligned} \tag{1.43}$$

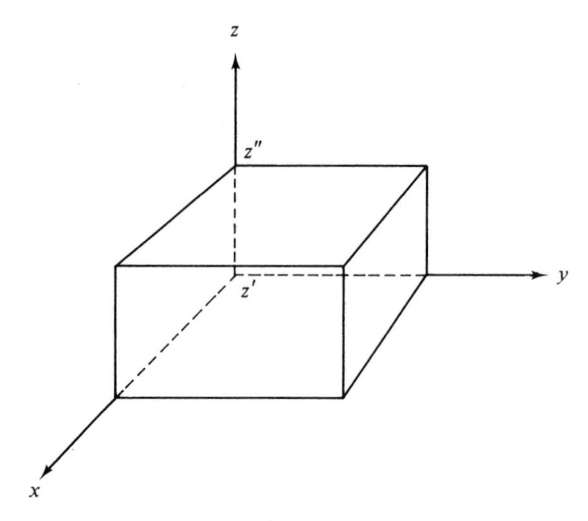

Figure 1.18

Using Eq. (1.43) in Eq. (1.42), we write

$$\iiint_{z'}^{z''} \frac{\partial F_z}{\partial z}\, dx\, dy\, dz = \iint_{\sigma_5} F_z(x, y, z'')\hat{\mathbf{n}}_5 \cdot \mathbf{k}\, d\sigma_5 + \iint_{\sigma_1} F_z(x, y, z')\hat{\mathbf{n}}_1 \cdot \mathbf{k}\, d\sigma_1$$

$$= \oiint_{\sigma} F_z \mathbf{k} \cdot \hat{\mathbf{n}}\, d\sigma \tag{1.44}$$

where

$$\iint_{\text{sides}} (\cdots) = 0$$

since $\mathbf{k}$ is perpendicular to $d\boldsymbol{\sigma}_2$, $d\boldsymbol{\sigma}_3$, $d\boldsymbol{\sigma}_4$, and $d\boldsymbol{\sigma}_6$. Similarly, we can show that

$$\iiint_{\tau} \frac{\partial F_x}{\partial x}\, d\tau = \oiint_{\sigma} F_x \mathbf{i} \cdot \hat{\mathbf{n}}\, d\sigma \tag{1.45}$$

and

$$\iiint_{\tau} \frac{\partial F_y}{\partial y}\, d\tau = \oiint_{\sigma} F_y \mathbf{j} \cdot \hat{\mathbf{n}}\, d\sigma. \tag{1.46}$$

Combining Eqs. (1.44), (1.45), and (1.46), we obtain

$$\iiint_{\tau} \nabla \cdot \mathbf{F}\, d\tau = \oiint_{\sigma} \mathbf{F} \cdot d\boldsymbol{\sigma}. \tag{1.40$'$}$$

Hence the theorem is proved.

EXAMPLE 1.10 *An electrostatics application of the divergence theorem.*

Solution Let $\mathbf{F} = \epsilon_0 \mathbf{E}$, where $\mathbf{E}$ is the electric field intensity and ϵ_0 is the permittivity of free space. Note that $\nabla \cdot \mathbf{E} = \rho/\epsilon_0$ (Maxwell's first equation), where ρ is the volume charge density. In this example, the divergence theorem becomes

$$\oiint_\sigma \mathbf{E} \cdot d\boldsymbol{\sigma} = \iiint_\tau \nabla \cdot \mathbf{E} \, d\tau$$

$$= \frac{1}{\epsilon_0} \iiint_\tau \rho \, d\tau. \tag{1.47}$$

In electrostatics, the above equation is called **Gauss' law,** and the arbitrarily shaped hypothetical surface σ is known as a **Gaussian surface.**

Gauss' law may be used to find the electric field at a distance r from a charge Q (see Fig. 1.19). For convenience, we let σ be a spherical surface enclosing the total charge Q. We then obtain

$$E \oiint_\sigma d\sigma = E 4\pi r^2$$

$$= \frac{1}{\epsilon_0} \iiint_\tau \rho \, d\tau$$

$$= \frac{Q}{\epsilon_0}$$

or

$$E = \frac{Q}{4\pi\epsilon_0 r^2}$$

which is just the usual expression for the field of a point charge a distance r from the charge Q.

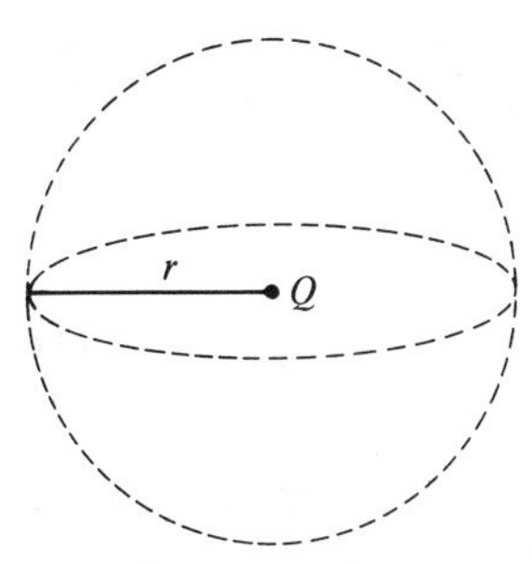

Figure 1.19 Gaussian surface

EXAMPLE 1.11 *An application of the divergence theorem to the problem of heat conduction* Here we consider the problem of the transfer of heat through a body by means of conduction. We assume that (1) the temperature, T, within the body is a finite and continuous point function $T(x, y, z, t)$, (2) through any point, a surface on which the temperature is everywhere the same (isothermal surface) may be drawn, (3) the temperature gradient, ∇T, and the direction of heat flow are normal to the isothermal surface at the point in question, (4) ∇T is a finite and continuous function and is in the direction of increasing T, and (5) the rate of heat flow per unit area across the isothermal surface, $\mathbf{F}$, is given by $\mathbf{F} = -k\,\nabla T$ (**Fourier's law**), where k is called the **thermal conductivity** of the substance under investigation and the negative sign indicates that the heat flows from points at higher temperatures to points at lower temperatures.

Solution If σ is an arbitrary closed surface drawn entirely within a certain region of the substance under investigation, then the total amount of heat flowing out of σ in time Δt is given by

$$Q = -\Delta t \oint\!\!\oint_{\sigma} k\,\nabla T \cdot \hat{\mathbf{n}}\, d\sigma$$

$$= -\Delta t \iiint_{\tau} \nabla \cdot (k\,\nabla T)\, d\tau$$

We have used the divergence theorem to obtain the last equation.

The quantity

$$Q = -\Delta t \iiint_{\tau} c\rho \frac{\partial T}{\partial t}\, d\tau$$

also represents the amount of heat flowing out of σ in time Δt provided c is the specific heat of the substance, ρ is the density, and $\partial T/\partial t$ is the rate of increase in temperature within σ. Equating the two expressions for the heat flow, we obtain

$$\iiint_{\tau} \left[c\rho \frac{\partial T}{\partial t} - \nabla \cdot (k\,\nabla T)\right] d\tau = 0$$

or

$$c\rho \frac{\partial T}{\partial t} = \nabla \cdot (k\,\nabla T) \tag{1.48}$$

since the volume is arbitrary. Equation (1.48) is called the **heat conduction equation** and was first developed by Fourier in 1822.

1.4.3 Green's (1793–1841) Theorem

Green's theorem is an important corollary of the divergence theorem, and it has numerous applications in various branches of physics.

Let ψ and ϕ be two scalar functions of position with continuous derivatives within a certain region bounded by a closed surface σ. On applying the divergence theorem to the vector $\psi \nabla\phi$ in this region, we obtain

$$\oiint_\sigma \psi \nabla\phi \cdot \hat{\mathbf{n}}\, d\sigma = \iiint_\tau \nabla\cdot(\psi \nabla\phi)\, d\tau$$

$$= \iiint_\tau [\psi \nabla\cdot\nabla\phi + \nabla\psi\cdot\nabla\phi]\, d\tau$$

$$= \iiint_\tau [\psi \nabla^2\phi + \nabla\psi\cdot\nabla\phi]\, d\tau$$

or

$$\oiint_\sigma \psi \frac{\partial\phi}{\partial n}\, d\sigma = \iiint_\tau [\psi \nabla^2\phi + \nabla\psi\cdot\nabla\phi]\, d\tau \tag{1.49}$$

where

$$\nabla\cdot\psi \nabla\phi = \psi \nabla\cdot\nabla\phi + \nabla\psi\cdot\nabla\phi$$

and

$$\nabla\phi\cdot\hat{\mathbf{n}} = \frac{\partial\phi}{\partial n} \qquad \text{(directional derivative).} \tag{1.50}$$

Equation (1.49) is known as **Green's theorem in the first form.**

A second form of Green's theorem is obtained if we consider the following two equations:

$$\nabla\cdot(\psi \nabla\phi) = \psi \nabla\cdot\nabla\phi + \nabla\psi\cdot\nabla\phi \tag{1.51}$$

and

$$\nabla\cdot(\phi \nabla\psi) = \phi \nabla\cdot\nabla\psi + \nabla\phi\cdot\nabla\psi. \tag{1.52}$$

On subtracting Eq. (1.52) from Eq. (1.51) and integrating over an arbitrary volume, we obtain

$$\iiint_\tau \nabla\cdot(\psi \nabla\phi - \phi \nabla\psi)\, d\tau = \iiint_\tau [\psi \nabla^2\phi - \phi \nabla^2\psi]\, d\tau \tag{1.53}$$

Converting the left-hand side of Eq. (1.53) to a surface integral by use of the divergence theorem, we obtain

$$\oiint_\sigma \left[\psi \frac{\partial \phi}{\partial n} - \phi \frac{\partial \psi}{\partial n}\right] d\sigma = \iiint_\tau [\psi \nabla^2 \phi - \phi \nabla^2 \psi] \, d\tau. \quad (1.54)$$

Equation (1.54) is the **second form of Green's theorem.**

1.4.4 The Curl Theorem Due to Stokes (1819–1903)

A third and equally important theorem is due to Stokes. The **curl theorem** is as follows: If **F** and its first derivatives are continuous, the line integral of **F** around a closed curve λ is equal to the normal surface integral of curl **F** over an open surface bounded by λ. In other words, the surface integral of curl **F** taken over any open surface σ is equal to the line integral of **F** around the periphery λ of the surface. In equation form, we write

$$\oint_\lambda \mathbf{F} \cdot d\boldsymbol{\lambda} = \iint_\sigma \nabla \times \mathbf{F} \cdot d\boldsymbol{\sigma}. \quad (1.55)$$

To prove the theorem, we first expand the right-hand side of Eq. (1.55); it becomes

$$\iint_\sigma \nabla \times \mathbf{F} \cdot \hat{\mathbf{n}} \, d\sigma = \iint_\sigma \hat{\mathbf{n}} \cdot [\nabla \times \mathbf{i} F_x + \nabla \times \mathbf{j} F_y + \nabla \times \mathbf{k} F_z] \, d\sigma. \quad (1.56)$$

The first integral on the right-hand side of Eq. (1.56) reduces to

$$\iint_\sigma \hat{\mathbf{n}} \cdot (\nabla \times \mathbf{i} F_x) \, d\sigma = \iint_\sigma \left[\hat{\mathbf{n}} \cdot \mathbf{j} \frac{\partial F_x}{\partial z} - \hat{\mathbf{n}} \cdot \mathbf{k} \frac{\partial F_x}{\partial y}\right] d\sigma. \quad (1.57)$$

Note that the projection of $d\boldsymbol{\sigma}$ onto the xy-plane (see Fig. 1.20) leads to

$$\hat{\mathbf{n}} \cdot \mathbf{k} \, d\sigma = dx \, dy \quad (1.58)$$

We now let the line segment $\overline{P_1 P_2}$ be the intersection of the surface σ with a plane that is parallel to the yz-plane at a distance x from the origin (see Fig. 1.20). Along the strip $\overline{P_1 P_2}$, we find that

$$dF_x = \frac{\partial F_x}{\partial y} dy + \frac{\partial F_x}{\partial z} dz \quad (1.59)$$

and

$$d\mathbf{R} = dy \, \mathbf{j} + dz \, \mathbf{k}.$$

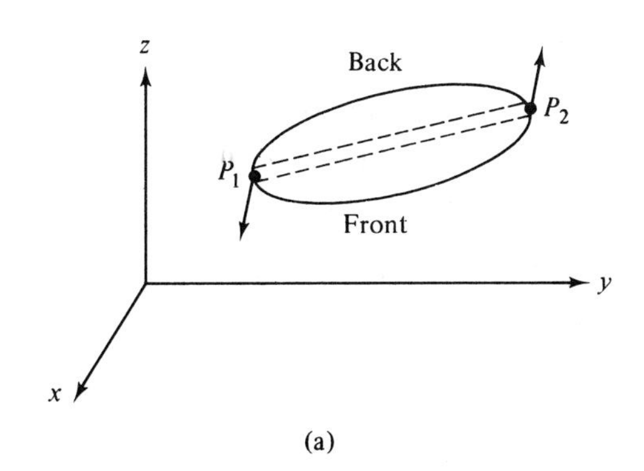

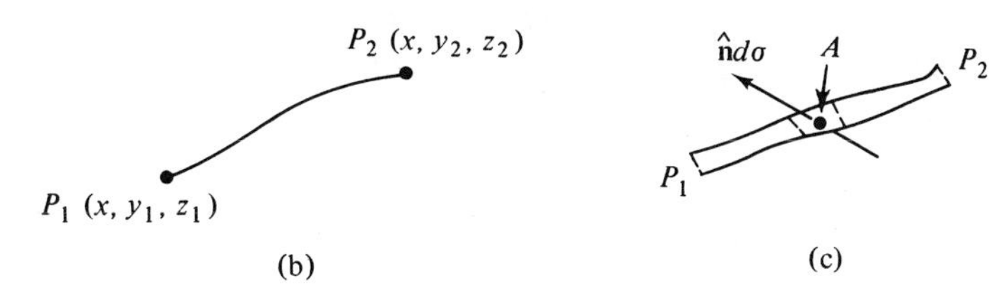

Figure 1.20

The vector $d\mathbf{R}$ is tangent to $\overline{P_1P_2}$ at A and perpendicular to $\hat{\mathbf{n}}$. We may therefore write

$$d\mathbf{R}\cdot\hat{\mathbf{n}} = 0 = dy\,\hat{\mathbf{n}}\cdot\mathbf{j} + dz\,\hat{\mathbf{n}}\cdot\mathbf{k}$$

or

$$\hat{\mathbf{n}}\cdot\mathbf{j} = -\frac{dz}{dy}\,\hat{\mathbf{n}}\cdot\mathbf{k} = -\frac{dz}{dy}\left(\frac{dx\,dy}{d\sigma}\right)$$

or

$$\hat{\mathbf{n}}\cdot\mathbf{j}\,d\sigma = -dx\,dz. \tag{1.60}$$

On substituting Eqs. (1.58), (1.59), and (1.60) into Eq. (1.57), we obtain

$$\begin{aligned}\iint_\sigma \hat{\mathbf{n}}\cdot(\nabla\times\mathbf{i}F_x)\,d\sigma &= -\iint_\sigma\left[\frac{\partial F_x}{\partial z}\,dz + \frac{\partial F_x}{\partial y}\,dy\right]dx \\ &= -\int dx\int dF_x \\ &= -\int [F_x(x, y_2, z_2) - F_x(x, y_1, z_1)]\,dx. \end{aligned}\tag{1.61}$$

The sense of the periphery at P_1 is positive, $dx = d\lambda_x$, and it is negative,

$dx = -d\lambda_x$, at P_2 (see Fig. 1.20). Hence Eq. (1.61) becomes

$$\iint_\sigma \hat{\mathbf{n}}\cdot(\nabla \times \mathbf{i}F_x)\, d\sigma = \underbrace{\int F_x(x, y_2, z_2)\, d\lambda_x}_{\text{back part}} + \underbrace{\int F_x(x, y_1, z_1)\, d\lambda_x}_{\text{front part}}$$

$$= \oint_\lambda F_x\, d\lambda_x \tag{1.62}$$

Similarly, we find that

$$\iint_\sigma \hat{\mathbf{n}}\cdot(\nabla \times \mathbf{j})F_y\, d\sigma = \oint_\lambda F_y\, d\lambda_y \tag{1.63}$$

and

$$\iint_\sigma \hat{\mathbf{n}}\cdot(\nabla \times \mathbf{k})F_z\, d\sigma = \oint_\lambda F_z\, d\lambda_z \tag{1.64}$$

Combining Eqs. (1.62), (1.63), and (1.64), we obtain

$$\iint_\sigma \hat{\mathbf{n}}\cdot(\nabla \times \mathbf{F})\, d\sigma = \oint_\lambda \mathbf{F}\cdot d\boldsymbol{\lambda} \tag{1.55$'$}$$

and the theorem is proved. This theorem is extremely useful in potential theory and in other areas of mathematical physics.

1.4.5 Two Useful Integral Relations

In this section, we present the proof of two useful integral relations.

RELATION 1

$$\iiint_\tau \nabla\phi\, d\tau = \oiint_\sigma \phi\hat{\mathbf{n}}\, d\sigma. \tag{1.65}$$

Proof Let $\mathbf{F} = \phi\mathbf{C}$, where $\mathbf{C}$ is a constant vector. On substituting this form for $\mathbf{F}$ into the divergence theorem, we obtain

$$\iiint_\tau \nabla\cdot(\phi\mathbf{C})\, d\tau = \oiint_\sigma \phi\mathbf{C}\cdot\hat{\mathbf{n}}\, d\sigma. \tag{1.66}$$

Note that

$$\nabla\cdot(\phi\mathbf{C}) = \mathbf{C}\cdot\nabla\phi \tag{1.67}$$

since $\nabla\cdot\mathbf{C} = 0$. By use of Eq. (1.67), we may write Eq. (1.66) in the form

$$\iiint_\tau \mathbf{C}\cdot\nabla\phi\, d\tau = \oiint_\sigma \mathbf{C}\cdot(\phi\hat{\mathbf{n}})\, d\sigma$$

or

$$\mathbf{C}\cdot\left[\iiint_{\tau} \nabla\phi \, d\tau - \oiint_{\sigma} \phi\hat{\mathbf{n}} \, d\sigma\right] = 0$$

or

$$\iiint_{\tau} \nabla\phi \, d\tau = \oiint_{\sigma} \phi\hat{\mathbf{n}} \, d\sigma \tag{1.65$'$}$$

since $\mathbf{C}$ is an arbitrary constant vector. Therefore the relation in Eq. (1.65) has been established.

EXAMPLE 1.12 If σ is a closed surface which encloses a volume τ, prove that $\oiint_{\sigma} \hat{\mathbf{n}} \, d\sigma = 0$.

Solution In Eq. (1. 65), let $\phi = 1$. We therefore obtain

$$\iiint_{\tau} \nabla\phi \, d\tau = 0 = \oiint_{\sigma} \hat{\mathbf{n}} \, d\sigma$$

and the result is proved.

RELATION 2

$$\iiint_{\tau} \nabla \times \mathbf{A} \, d\tau = \oiint_{\sigma} \hat{\mathbf{n}} \times \mathbf{A} \, d\sigma. \tag{1.68}$$

Proof Let $\mathbf{F} = \mathbf{A} \times \mathbf{B}$ where $\mathbf{B}$ is an arbitrary constant vector. Substituting this form for $\mathbf{F}$ into the divergence theorem, we obtain

$$\iiint_{\tau} \nabla\cdot(\mathbf{A} \times \mathbf{B}) \, d\tau = \oiint_{\sigma} (\mathbf{A} \times \mathbf{B})\cdot\hat{\mathbf{n}} \, d\sigma. \tag{1.69}$$

The integrand on the left-hand side of Eq. (1.69) may be written as

$$\begin{aligned} \nabla\cdot(\mathbf{A} \times \mathbf{B}) &= \mathbf{B}\cdot\nabla \times \mathbf{A} - \mathbf{A}\cdot\nabla \times \mathbf{B} \\ &= \mathbf{B}\cdot\nabla \times \mathbf{A} \end{aligned} \tag{1.70}$$

since $\mathbf{B}$ is a constant vector. Also note that

$$\mathbf{A} \times \mathbf{B}\cdot\hat{\mathbf{n}} = \mathbf{B}\cdot(\hat{\mathbf{n}} \times \mathbf{A}). \tag{1.71}$$

If we substitute Eqs. (1.70) and (1.71) into Eq. (1.69), Eq. (1.69) becomes

$$\iiint_{\tau} \mathbf{B}\cdot\nabla \times \mathbf{A} \, d\tau = \oiint_{\sigma} \mathbf{B}\cdot(\hat{\mathbf{n}} \times \mathbf{A}) \, d\sigma$$

or

$$\mathbf{B}\cdot\left[\iiint_\tau \nabla \times \mathbf{A}\, d\tau - \oiint_\sigma (\hat{\mathbf{n}} \times \mathbf{A})\, d\sigma\right] = 0$$

which yields

$$\iiint_\tau \nabla \times \mathbf{A}\, d\tau = \oiint_\sigma \hat{\mathbf{n}} \times \mathbf{A}\, d\sigma \tag{1.68'}$$

since $\mathbf{B}$ is arbitrary. Hence integral relation 2 is proved.

EXAMPLE 1.13 If σ is a closed surface which encloses a volume τ, prove that $\oiint_\sigma \mathbf{r} \times d\boldsymbol{\sigma} = 0$.

Solution In Eq. (1.68), let $\mathbf{A} = \mathbf{r}$ and $\hat{\mathbf{n}}\, d\sigma = d\boldsymbol{\sigma}$. We therefore obtain

$$-\oiint_\sigma \mathbf{r} \times d\boldsymbol{\sigma} = \iiint_\tau \nabla \times \mathbf{r}\, d\tau$$
$$= 0$$

since $\nabla \times \mathbf{r} = 0$.

1.5 SOME USEFUL RELATIONS INVOLVING VECTORS

1.5.1 Relations Involving the ∇ Operator

1. $\nabla(\phi + \psi) = \nabla\phi + \nabla\psi$
2. $\nabla(\phi\psi) = \phi\,\nabla\psi + \psi\,\nabla\phi$
3. $\nabla\cdot(\mathbf{A} + \mathbf{B}) = \nabla\cdot\mathbf{A} + \nabla\cdot\mathbf{B}$
4. $\nabla \times (\mathbf{A} + \mathbf{B}) = \nabla \times \mathbf{A} + \nabla \times \mathbf{B}$
5. $\nabla\cdot(\phi\mathbf{A}) = \mathbf{A}\cdot\nabla\phi + \phi\,\nabla\cdot\mathbf{A}$
6. $\nabla \times (\phi\mathbf{A}) = \phi\,\nabla \times \mathbf{A} - \mathbf{A} \times \nabla\phi$
7. $\nabla\cdot(\mathbf{A} \times \mathbf{B}) = \mathbf{B}\cdot\nabla \times \mathbf{A} - \mathbf{A}\cdot\nabla \times \mathbf{B}$
8. $\nabla \times (\mathbf{A} \times \mathbf{B}) = (\mathbf{B}\cdot\nabla)\mathbf{A} - \mathbf{B}\,\nabla\cdot\mathbf{A} - (\mathbf{A}\cdot\nabla)\mathbf{B} + \mathbf{A}\,\nabla\cdot\mathbf{B}$
9. $\nabla(\mathbf{A}\cdot\mathbf{B}) = (\mathbf{B}\cdot\nabla)\mathbf{A} + (\mathbf{A}\cdot\nabla)\mathbf{B} + \mathbf{B} \times \nabla \times \mathbf{A} + \mathbf{A} \times \nabla \times \mathbf{B}$
10. $\nabla \times \nabla\phi = 0$
11. $\nabla\cdot\nabla \times \mathbf{A} = 0$
12. $\nabla \times \nabla \times \mathbf{A} = \nabla\,\nabla\cdot\mathbf{A} - \nabla^2\mathbf{A}$

1.5.2 Other Vector Relations

1. $\mathbf{A} \times (\mathbf{B} + \mathbf{C}) = \mathbf{A} \times \mathbf{B} + \mathbf{A} \times \mathbf{C}$
2. $(\mathbf{B} + \mathbf{C}) \times \mathbf{A} = \mathbf{B} \times \mathbf{A} + \mathbf{C} \times \mathbf{A}$

3. $\mathbf{A} \times \mathbf{B} \cdot \mathbf{C} = \mathbf{A} \cdot \mathbf{B} \times \mathbf{C}$
4. $\mathbf{A} \times (\mathbf{B} \times \mathbf{C}) = \mathbf{B}(\mathbf{A} \cdot \mathbf{C}) - \mathbf{C}(\mathbf{A} \cdot \mathbf{B})$
5. $\nabla \cdot \mathbf{R} = 3$
6. $\nabla \times \mathbf{R} = 0$
7. $\nabla R^n = nR^{n-2}\mathbf{R}$
8. $\nabla^2 R^n = n(n+1)R^{n-2}$
9. $\mathbf{A} \times \mathbf{B} = \begin{vmatrix} \mathbf{i} & \mathbf{j} & \mathbf{k} \\ A_x & A_y & A_z \\ B_x & B_y & B_z \end{vmatrix}$
10. $\mathbf{A} \cdot \mathbf{B} \times \mathbf{C} = \begin{vmatrix} A_x & A_y & A_z \\ B_x & B_y & B_z \\ C_x & C_y & C_z \end{vmatrix}$

1.5.3 Some Important Equations in Physics

1. net $\mathbf{F} = \sum \mathbf{F} = m\mathbf{a}$ (Newton's second law)
2. $\mathbf{F} = q(\mathbf{E} + \mathbf{v} \times \mathbf{B})$ (Lorentz's (1853–1928) force law)
3. $$\left.\begin{aligned} &\nabla \cdot \mathbf{D} = \rho \qquad \nabla \times \mathbf{E} = -\frac{\partial \mathbf{B}}{\partial t} \\ &\nabla \cdot \mathbf{B} = 0 \qquad \nabla \times \mathbf{H} = \frac{\partial \mathbf{D}}{\partial t} + \mathbf{J} \\ &\text{where} \quad \mathbf{D} = \epsilon_0 \mathbf{E} \quad \text{and} \quad \mathbf{B} = \mu_0 \mathbf{H} \end{aligned}\right\}$$ (Maxwell's equations for a vacuum)
4. $$\nabla^2 \phi = \begin{cases} -\dfrac{\rho}{\epsilon_0} & \text{(Poisson's (1781–1840) equation)} \\ 0 & \text{(Laplace's equation)} \end{cases}$$
5. $\nabla^2 \phi = \dfrac{1}{v^2} \dfrac{\partial^2 \phi}{\partial t^2}$ (the wave equation)
6. $\nabla^2 \phi = \dfrac{1}{\sigma} \dfrac{\partial \phi}{\partial t}$ [the diffusion (or heat) equation]
7. $\nabla^2 \phi + k^2 \phi = 0$ (Helmholtz's equation)
8. $\dfrac{\partial \rho}{\partial t} + \nabla \cdot (\rho \mathbf{v}) = 0$ (the continuity equation)

In the above equations, each symbol has its usual meaning.

1.6 GENERALIZED COORDINATES

Thus far, our presentation has made exclusive use of the right-hand Cartesian coordinate system. Often the solution of a physical problem is made easier by first selecting an appropriate coordinate system. In general, it is easier to solve a problem involving spherical geometry by use of spherical coordinates than by use of Cartesian coordinates. It is therefore desirable to develop a procedure for making the transformation from Cartesian coordinates to other coordinate systems. Such a procedure is developed in this section.

1.6.1 General Curvilinear Coordinates

Let the position of a point P in space be completely described by $P(u^1, u^2, u^3)$, where u^1, u^2, and u^3 are three single-valued functions of position. For $u^1 =$ constant, $u^2 =$ constant, and $u^3 =$ constant, the u's are three surfaces which intersect at P. These surfaces are called **coordinate surfaces,** and the three curves of intersection are called **coordinate lines.** Tangents to the coordinate lines at P are called **coordinate axes.** If the relative orientation of the coordinate surfaces changes from point to point, the u^i $(i = 1, 2, 3)$ are called **general curvilinear coordinates.** If the three surfaces are everywhere mutually perpendicular, the u^i are called **orthogonal curvilinear coordinates.**

It is assumed that the following transformations exist:

$$\begin{aligned} x = x^1 &= f_1(u^1, u^2, u^3) \\ y = x^2 &= f_2(u^1, u^2, u^3) \\ z = x^3 &= f_3(u^1, u^2, u^3) \end{aligned}$$

and

$$\begin{aligned} u^1 &= F_1(x^1, x^2, x^3) \\ u^2 &= F_2(x^1, x^2, x^3) \\ u^3 &= F_3(x^1, x^2, x^3). \end{aligned}$$

The position vector of the point P is a function of the u^i, that is, $\mathbf{r} = \mathbf{r}(u^1, u^2, u^3)$ (see Fig. 1.21). An element of displacement is given by

$$\begin{aligned} d\mathbf{r} &= \frac{\partial \mathbf{r}}{\partial u^1} du^1 + \frac{\partial \mathbf{r}}{\partial u^2} du^2 + \frac{\partial \mathbf{r}}{\partial u^3} du^3 \\ &= \sum_{i=1}^{3} \frac{\partial \mathbf{r}}{\partial u^i} du^i \\ &\equiv d\mathbf{s} \end{aligned} \tag{1.72}$$

The square of the length of this displacement is given by

$$\begin{aligned} ds^2 &= d\mathbf{r} \cdot d\mathbf{r} \\ &= \sum_{i=1}^{3} \sum_{j=1}^{3} \frac{\partial \mathbf{r}}{\partial u^i} \cdot \frac{\partial \mathbf{r}}{\partial u^j} du^i \, du^j. \end{aligned} \tag{1.73}$$

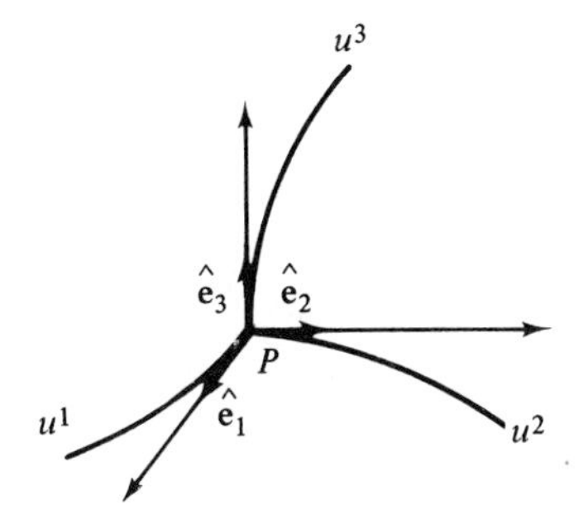

Figure 1.21 $\hat{\mathbf{e}}_1 = \dfrac{\mathbf{a}_1}{|\mathbf{a}_1|} \quad \hat{\mathbf{e}}_2 = \dfrac{\mathbf{a}_2}{|\mathbf{a}_2|} \quad \hat{\mathbf{e}}_3 = \dfrac{\mathbf{a}_3}{|\mathbf{a}_3|}$

Note that the $\partial\mathbf{r}/\partial u^i$ are tangent to the u^i respectively, since $\partial\mathbf{r}/\partial u^1$ means u^2 and u^3 are fixed and $\mathbf{r} = \mathbf{r}(u^1)$ is constrained to move along the u^1 surface. Let the three coordinate axes be represented by $\mathbf{a}_i$, where $\mathbf{a}_i = \partial\mathbf{r}/\partial u^i$ (see Fig. 1.21). Hence Eq. (1.73) becomes

$$\begin{aligned} ds^2 &= \sum_{i=1}^{3} \sum_{j=1}^{3} \mathbf{a}_i \cdot \mathbf{a}_j \, du^i \, du^j \\ &= \sum_{i=1}^{3} \sum_{j=1}^{3} g_{ij} \, du^i \, du^j \end{aligned} \tag{1.74}$$

where $g_{ij} = \mathbf{a}_i \cdot \mathbf{a}_j$. Since $\mathbf{a}_i \cdot \mathbf{a}_j = \mathbf{a}_j \cdot \mathbf{a}_i$, $g_{ij} = g_{ji}$. The quantities g_{ij} are called **metric coefficients** and characterize the relative nature of the space. Note that $\mathbf{a}_i$ and $\mathbf{a}_j$ may vary in direction and magnitude from point to point when general curvilinear coordinates are under consideration.

1.6.2 Orthogonal Curvilinear Coordinates

In an orthogonal curvilinear coordinate system (see Fig. 1.21),

$$\mathbf{a}_i \cdot \mathbf{a}_j = 0 \qquad (\text{for } i \neq j). \tag{1.75}$$

Hence the square of the element of length becomes

$$ds^2 = g_{11}(du^1)^2 + g_{22}(du^2)^2 + g_{33}(du^3)^2. \tag{1.76}$$

Note that $du^2 = du^3 = 0$ when the element of length ds is along u^1. We may therefore write

$$\begin{aligned} ds_1 &= \sqrt{g_{11}}\, du^1 \\ &= h_1 \, du^1 \\ ds_2 &= \sqrt{g_{22}}\, du^2 \\ &= h_2 \, du^2 \end{aligned}$$

and

$$\begin{aligned} ds_3 &= \sqrt{g_{33}}\, du^3 \\ &= h_3\, du^3 \end{aligned}$$

where $h_1 = \sqrt{g_{11}}$, $h_2 = \sqrt{g_{22}}$, and $h_3 = \sqrt{g_{33}}$. The h's are called **scale factors.**

We now develop a scheme for determining g_{ii} (or h_i) when the x^i are known. In Cartesian coordinates, $g_{11} = g_{22} = g_{33} = 1$ and

$$\begin{aligned} ds^2 &= \sum_{k=1}^{3} dx^k\, dx^k \\ &= \sum_{k=1}^{3} \left[\left(\sum_{i=1}^{3} \frac{\partial x^k}{\partial u^i} du^i \right) \left(\sum_{j=1}^{3} \frac{\partial x^k}{\partial u^j} du^j \right) \right] \\ &= \sum_{i=1}^{3} \sum_{j=1}^{3} \left(\sum_{k=1}^{3} \frac{\partial x^k}{\partial u^i} \frac{\partial x^k}{\partial u^j} \right) du^i\, du^j. \end{aligned} \tag{1.77}$$

Comparing Eq. (1.74) with Eq. (1.77), we see that

$$g_{ij} = \sum_{k=1}^{3} \frac{\partial x^k}{\partial u^i} \frac{\partial x^k}{\partial u^j}$$

or

$$g_{ii} = \sum_{k=1}^{3} \left(\frac{\partial x^k}{\partial u^i} \right)^2. \tag{1.78}$$

The area and volume elements are given by

$$\begin{aligned} d\sigma_{ij} &= h_i h_j\, du^i\, du^j \\ d\tau &= h_1 h_2 h_3\, du^1\, du^2\, du^3. \end{aligned} \tag{1.79}$$

Expressions for the gradient, divergence, curl, and the Laplacian in orthogonal curvilinear coordinates will now be developed.

1.6.3 The Gradient in Orthogonal Curvilinear Coordinates

Since $\partial\psi/\partial s$ is the component of $\nabla\psi$ in the $d\mathbf{s}$ direction (see Section 1.3.2 of this chapter), $\nabla\psi$ is given by

$$\begin{aligned} \nabla\psi &= \hat{\mathbf{e}}_1 \frac{\partial\psi}{\partial s_1} + \hat{\mathbf{e}}_2 \frac{\partial\psi}{\partial s_2} + \hat{\mathbf{e}}_3 \frac{\partial\psi}{\partial s_3} \\ &= \frac{\hat{\mathbf{e}}_1}{h_1} \frac{\partial\psi}{\partial u^1} + \frac{\hat{\mathbf{e}}_2}{h_2} \frac{\partial\psi}{\partial u^2} + \frac{\hat{\mathbf{e}}_3}{h_3} \frac{\partial\psi}{\partial u^3} \end{aligned} \tag{1.80}$$

where $\hat{\mathbf{e}}_1$, $\hat{\mathbf{e}}_2$, and $\hat{\mathbf{e}}_3$ are unit vectors along $d\mathbf{s}_1$, $d\mathbf{s}_2$, and $d\mathbf{s}_3$, respectively.

1.6.4 The Divergence and Curl in Orthogonal Curvilinear Coordinates

The arbitrary vector $\mathbf{V}$ in terms of components can be written as

$$\begin{aligned}\mathbf{V} &= \hat{\mathbf{e}}_1 V_1 + \hat{\mathbf{e}}_2 V_2 + \hat{\mathbf{e}}_3 V_3 \\ &= \frac{\hat{\mathbf{e}}_1}{h_2 h_3}(h_2 h_3 V_1) + \frac{\hat{\mathbf{e}}_2}{h_1 h_3}(h_1 h_3 V_2) + \frac{\hat{\mathbf{e}}_3}{h_1 h_2}(h_1 h_2 V_3).\end{aligned} \tag{1.81}$$

The divergence of $\mathbf{V}$ is given by

$$\begin{aligned}\nabla\cdot\mathbf{V} &= \frac{\hat{\mathbf{e}}_1}{h_2 h_3}\cdot\nabla(h_2 h_3 V_1) + \frac{\hat{\mathbf{e}}_2}{h_1 h_3}\cdot\nabla(h_1 h_3 V_2) + \frac{\hat{\mathbf{e}}_3}{h_1 h_2}\cdot\nabla(h_1 h_2 V_3) \\ &= \frac{1}{h_1 h_2 h_3}\left[\frac{\partial}{\partial u^1}(h_2 h_3 V_1) + \frac{\partial}{\partial u^2}(h_1 h_3 V_2) + \frac{\partial}{\partial u^3}(h_1 h_2 V_3)\right]\end{aligned} \tag{1.82}$$

since

$$\nabla\cdot(\phi\mathbf{A}) = \mathbf{A}\cdot\nabla\phi + \phi\nabla\cdot\mathbf{A}$$

and (see Problem 1.50)

$$\nabla\cdot\left(\frac{\hat{\mathbf{e}}_1}{h_2 h_3}\right) = \nabla\cdot\left(\frac{\hat{\mathbf{e}}_2}{h_1 h_3}\right) = \nabla\cdot\left(\frac{\hat{\mathbf{e}}_3}{h_1 h_2}\right) = 0. \tag{1.83}$$

In a similar manner, we find that

$$\nabla\times\mathbf{V} = \frac{1}{h_1 h_2 h_3}\begin{vmatrix} h_1\hat{\mathbf{e}}_1 & h_2\hat{\mathbf{e}}_2 & h_3\hat{\mathbf{e}}_3 \\ \dfrac{\partial}{\partial u^1} & \dfrac{\partial}{\partial u^2} & \dfrac{\partial}{\partial u^3} \\ h_1 V_1 & h_2 V_2 & h_3 V_3 \end{vmatrix} \tag{1.84}$$

since

$$\mathbf{V} = \frac{\hat{\mathbf{e}}_1}{h_1}(h_1 V_1) + \frac{\hat{\mathbf{e}}_2}{h_2}(h_2 V_2) + \frac{\hat{\mathbf{e}}_3}{h_3}(h_3 V_3)$$

and (see Problem 1.50)

$$\nabla\times\left(\frac{\hat{\mathbf{e}}_1}{h_1}\right) = \nabla\times\left(\frac{\hat{\mathbf{e}}_2}{h_2}\right) = \nabla\times\left(\frac{\hat{\mathbf{e}}_3}{h_3}\right) = 0. \tag{1.85}$$

1.6.5 The Laplacian in Orthogonal Curvilinear Coordinates

The Laplacian in orthogonal curvilinear coordinates is given by

$$\begin{aligned}\nabla^2\psi &= \nabla\cdot\nabla\psi \\ &= \nabla\cdot\left[\frac{\hat{\mathbf{e}}_1}{h_1}\frac{\partial\psi}{\partial u^1} + \frac{\hat{\mathbf{e}}_2}{h_2}\frac{\partial\psi}{\partial u^2} + \frac{\hat{\mathbf{e}}_3}{h_3}\frac{\partial\psi}{\partial u^3}\right] \\ &= \frac{1}{h_1 h_2 h_3}\left[\frac{\partial}{\partial u^1}\left(\frac{h_2 h_3}{h_1}\frac{\partial\psi}{\partial u^1}\right) + \frac{\partial}{\partial u^2}\left(\frac{h_1 h_3}{h_2}\frac{\partial\psi}{\partial u^2}\right)\right. \\ &\qquad \left. + \frac{\partial}{\partial u^3}\left(\frac{h_1 h_2}{h_3}\frac{\partial\psi}{\partial u^3}\right)\right].\end{aligned} \tag{1.86}$$

From Eqs. (1.80), (1.82), (1.84), and (1.86), we see that the essential task in determining the explicit forms of $\nabla\psi$, $\nabla\cdot\mathbf{V}$, $\nabla\times\mathbf{V}$, and $\nabla^2\psi$ in orthogonal curvilinear coordinates is that of determining the scale factors which are related to the metric coefficients.

In Sections 1.6.6, 1.6.7, and 1.6.8, we give the details for three extremely useful coordinate transformations.

1.6.6 Plane Polar Coordinates (r, θ)

In **plane polar coordinates,** we have $x = r\cos\theta$ and $y = r\sin\theta$. See Fig. 1.22.

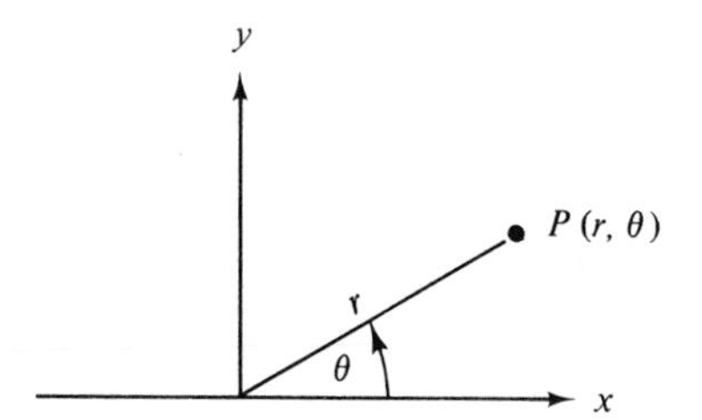

Figure 1.22

1.6.7 Right Circular Cylindrical Coordinates (ρ, ϕ, z)

For **right circular cylindrical coordinates** (or simply cylindrical), we have $x = \rho\cos\phi$, $y = \rho\sin\phi$, and $z = z$. See Fig. 1.23.

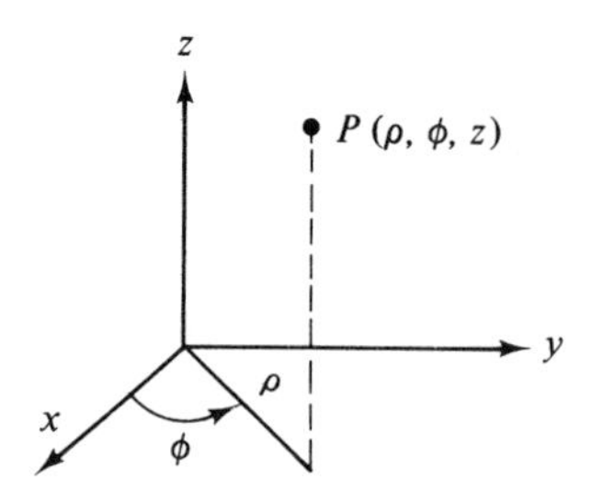

Figure 1.23

1.6.8 Spherical Polar Coordinates (r, θ, ϕ)

In **spherical polar coordinates** (or simply spherical), we have $\rho = r\sin\theta$, $x = \rho\cos\phi$, $y = \rho\sin\phi$, and $z = r\cos\theta$. See Fig. 1.24.

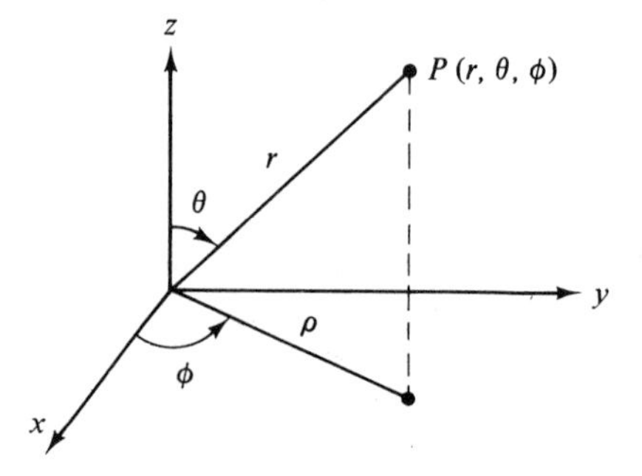

Figure 1.24

EXAMPLE 1.14 Find the expressions for $\nabla\psi$, $\nabla\cdot\mathbf{V}$, $\nabla\times\mathbf{V}$, and $\nabla^2\psi$ in cylindrical coordinates.

Solution The required transformation is $x, y, z \to \rho, \phi, z$. From Fig. 1.23, we have

$$u^1 = \rho, \qquad u^2 = \phi, \quad \text{and} \quad u^3 = z.$$

We also have

$$x = x^1 = \rho\cos\phi, \qquad y = x^2 = \rho\sin\phi, \quad \text{and} \quad x^3 = z.$$

In Cartesian coordinates $g_{11} = g_{22} = g_{33} = 1$ and

$$g_{ii} = \sum_{k=1}^{3}\left(\frac{\partial x^k}{\partial u^i}\right)^2 \qquad (i = 1, 2, 3).$$

The scale factors are determined as follows:

$$\begin{aligned}
g_{11} &= \sum_{k=1}^{3}\left(\frac{\partial x^k}{\partial u^1}\right)^2 \\
&= \left(\frac{\partial x^1}{\partial u^1}\right)^2 + \left(\frac{\partial x^2}{\partial u^1}\right)^2 + \left(\frac{\partial x^3}{\partial u^1}\right)^2 \\
&= \left[\frac{\partial}{\partial\rho}(\rho\cos\phi)\right]^2 + \left[\frac{\partial}{\partial\rho}(\rho\sin\phi)\right]^2 + \left[\frac{\partial}{\partial\rho}(z)\right]^2 \\
&= \cos^2\phi + \sin^2\phi \\
&= 1 = h_1^2 \\
g_{22} &= \left(\frac{\partial x^1}{\partial u^2}\right)^2 + \left(\frac{\partial x^2}{\partial u^2}\right)^2 + \left(\frac{\partial x^3}{\partial u^2}\right)^2 \\
&= \left[\frac{\partial}{\partial\phi}(\rho\cos\phi)\right]^2 + \left[\frac{\partial}{\partial\phi}(\rho\sin\phi)\right]^2 + \left[\frac{\partial}{\partial\phi}(z)\right]^2 \\
&= \rho^2\sin^2\phi + \rho^2\cos^2\phi \\
&= \rho^2 = h_2^2
\end{aligned}$$

and

$$\begin{aligned}
g_{33} &= \left(\frac{\partial x^1}{\partial u^3}\right)^2 + \left(\frac{\partial x^2}{\partial u^3}\right)^2 + \left(\frac{\partial x^3}{\partial u^3}\right)^2 \\
&= \left[\frac{\partial}{\partial z}(\rho\cos\phi)\right]^2 + \left[\frac{\partial}{\partial z}(\rho\sin\phi)\right]^2 + \left(\frac{\partial z}{\partial z}\right)^2 \\
&= 1 = h_3^2.
\end{aligned}$$

For circular cylindrical coordinates, we may therefore write

$$\nabla\psi = \hat{\mathbf{e}}_\rho\frac{\partial\psi}{\partial\rho} + \frac{\hat{\mathbf{e}}_\phi}{\rho}\frac{\partial\psi}{\partial\phi} + \hat{\mathbf{e}}_z\frac{\partial\psi}{\partial z} \tag{1.87}$$

$$\nabla \cdot \mathbf{V} = \frac{1}{\rho}\left[\frac{\partial}{\partial \rho}(\rho V_\rho) + \frac{\partial}{\partial \phi}(V_\phi) + \frac{\partial}{\partial z}(\rho V_z)\right] \tag{1.88}$$

$$\nabla \times \mathbf{V} = \frac{1}{\rho}\begin{vmatrix} \hat{\mathbf{e}}_\rho & \rho\hat{\mathbf{e}}_\phi & \hat{\mathbf{e}}_z \\ \dfrac{\partial}{\partial \rho} & \dfrac{\partial}{\partial \phi} & \dfrac{\partial}{\partial z} \\ V_\rho & \rho V_\phi & V_z \end{vmatrix} \tag{1.89}$$

and

$$\nabla^2 \psi = \frac{1}{\rho}\left[\frac{\partial}{\partial \rho}\left(\rho\frac{\partial \psi}{\partial \rho}\right) + \frac{\partial}{\partial \phi}\left(\frac{1}{\rho}\frac{\partial \psi}{\partial \phi}\right) + \frac{\partial}{\partial z}\left(\rho\frac{\partial \psi}{\partial z}\right)\right]. \tag{1.90}$$

1.7 PROBLEMS

1.1 Find $\mathbf{A} + \mathbf{B}$ and $\mathbf{A} - \mathbf{B}$ for $\mathbf{A} = 2\mathbf{i} - \mathbf{j} + \mathbf{k}$ and $\mathbf{B} = \mathbf{i} - 3\mathbf{j} - 5\mathbf{k}$.

1.2 (a) Calculate the distance from $P_1(-1, -4, 5)$ to $P_2(3, -2, 2)$.
(b) What are the direction cosines of $\overline{P_1P_2}$?
(c) What is the equation of the line connecting P_1 and P_2?

1.3 What are the direction cosines of the vector $2\mathbf{i} - 2\mathbf{j} + \mathbf{k}$?

1.4 Show that $\mathbf{A} = \mathbf{i} + 4\mathbf{j} + 3\mathbf{k}$ and $\mathbf{B} = 4\mathbf{i} + 2\mathbf{j} - 4\mathbf{k}$ are perpendicular.

1.5 Find the angle between $2\mathbf{i}$ and $3\mathbf{i} + 4\mathbf{j}$.

1.6 Show that $|\mathbf{A}||\mathbf{B}| \sin \theta$ is the area of the parallelogram formed by $\mathbf{A}$ and $\mathbf{B}$.

1.7 A force of 6 lb acts through the point $P(4, -1, 7)$ in the direction of the vector $(9, 6, -2)$. Find the moment of this force about the point $A(1, -3, 2)$.

1.8 By use of the scalar product of $\mathbf{R}$ and the appropriate $\mathbf{i}, \mathbf{j}, \mathbf{k}$ base vector, show that $x = R \cos \alpha$, $y = R \cos \beta$, and $z = R \cos \gamma$.

1.9 Find the area of the parallelogram formed by $\mathbf{A} = 3\mathbf{i} + 2\mathbf{j}$ and $\mathbf{B} = 2\mathbf{j} - 4\mathbf{k}$.

1.10 Find the component of $8\mathbf{i} + \mathbf{j}$ in the direction of $\mathbf{i} + 2\mathbf{j} - 2\mathbf{k}$.

1.11 Calculate the work done by the force $\mathbf{F} = 2\mathbf{i} - 2\mathbf{j} + 10\mathbf{k}$ lb in moving an object a displacement of $\mathbf{D} = 4\mathbf{i} + 5\mathbf{j} - \mathbf{k}$ ft.

1.12 By use of determinants, find $\mathbf{k} \times \mathbf{i}$.

1.13 Find a unit vector perpendicular to $\mathbf{i} - 2\mathbf{j} - 3\mathbf{k}$ and $\mathbf{i} + 2\mathbf{j} - \mathbf{k}$.

1.14 If $a\mathbf{i} - 2\mathbf{j} + \mathbf{k}$ is perpendicular to $\mathbf{i} - 2\mathbf{j} - 3\mathbf{k}$, find a.

1.15 Show that $\mathbf{A} = 2\mathbf{i} - \mathbf{j} + \mathbf{k}$, $\mathbf{B} = \mathbf{i} - 3\mathbf{j} - 5\mathbf{k}$, and $\mathbf{C} = 3\mathbf{i} - 4\mathbf{j} - 4\mathbf{k}$ form the sides of a right triangle.

1.16 Find $\mathbf{A}\cdot\mathbf{B} \times \mathbf{C}$ if $\mathbf{A} = 2\mathbf{i}$, $\mathbf{B} = 3\mathbf{j}$, and $\mathbf{C} = 4\mathbf{k}$.

1.17 Show that $\mathbf{A}\cdot(\mathbf{B} \times \mathbf{C})$ is the volume of the parallelepiped formed by $\mathbf{A}$, $\mathbf{B}$, and $\mathbf{C}$.

1.18 Compute $(\mathbf{A} \times \mathbf{B}) \times \mathbf{C}$ and $\mathbf{A} \times (\mathbf{B} \times \mathbf{C})$ directly for

$$\mathbf{A} = 2\mathbf{i} + 2\mathbf{j},\ \mathbf{B} = 3\mathbf{i} - \mathbf{j} + \mathbf{k},\ \text{and}\ \mathbf{C} = 8\mathbf{i}.$$

1.19 By use of diagrams, show that $\mathbf{A} \times (\mathbf{B} + \mathbf{C}) = \mathbf{A} \times \mathbf{B} + \mathbf{A} \times \mathbf{C}$.

1.20 By use of Cartesian components, show that:

(a) $\nabla \cdot \mathbf{r} = 3$

(b) $\nabla \times \mathbf{R} = 0$

(c) $\nabla r^n = nr^{n-2}\mathbf{r}$

(d) $(\mathbf{U} \cdot \nabla)\mathbf{R} = \mathbf{U}$

(e) $\nabla^2 r^n = n(n + 1)r^{n-2}$

(f) $\oint \mathbf{R} \cdot d\mathbf{R} = 0$.

1.21 By use of the form of the gradient given in Eq. (1.50), show that:

(a) $\nabla r^m = mr^{m-2}\mathbf{r}$

(b) $\nabla^2 r^m = m(m + 1)r^{m-2}$

(c) $\nabla^2(1/r) = 0$;

hence $u = 1/r$ is a solution of $\nabla^2 u = 0$ (**Laplace's equation**).

1.22 Show that $(\mathbf{p} - e\mathbf{A}) \times (\mathbf{p} - e\mathbf{A}) = ie\hbar\mathbf{B}$, where

$$\mathbf{p} = -i\hbar\nabla,\ \mathbf{B} = \nabla \times \mathbf{A},\ i = \sqrt{-1},\ e$$

and $\hbar$ are constants.

1.23 If $\mathbf{v}$ is the velocity of a particle in a rigid body whose angular velocity is $\boldsymbol{\omega}$, show that $\nabla \times \mathbf{v} = 2\boldsymbol{\omega}$. If

$$\mathbf{v} = \frac{d\mathbf{R}}{dt} \quad \text{and} \quad \mathbf{a} = \frac{d\mathbf{v}}{dt} = \frac{d^2\mathbf{R}}{dt^2}$$

show that

$$\mathbf{v}(t) = \mathbf{v}(0) + \int_0^t \mathbf{a}(t')\,dt'$$

and

$$\mathbf{R}(t) = \mathbf{R}(0) + \int_0^t \mathbf{v}(t')\,dt'.$$

1.24 If $d|\mathbf{A}|/dt = 0$, show that $\mathbf{A}$ is perpendicular to $d\mathbf{A}/dt$. Explain the physical meaning of this result.

1.25 At time t, the position vector of a moving particle is $\mathbf{R} = \mathbf{A}\cos\omega t + \mathbf{B}\sin\omega t$ where $\mathbf{A}$ and $\mathbf{B}$ are constant vectors and ω is a constant scalar.

(a) Find the velocity, $\mathbf{v}$, and show that $\mathbf{R} \times \mathbf{v}$ = a constant vector.

(b) Show that the acceleration is directed toward the origin and is proportional to the distance (**simple harmonic motion** $\equiv$ SHM).

1.26 Suppose the radius vector for a particle moving in a plane is given by $\mathbf{R} = r(t)\hat{\mathbf{r}}$ with polar coordinates (r, θ) where $|\hat{\mathbf{r}}| = 1$.

(a) Show that the velocity of the particle is given by

$$\mathbf{v} = \frac{dr}{dt}\hat{\mathbf{r}} + r\frac{d\theta}{dt}\hat{\mathbf{T}}$$

where $\hat{\mathbf{T}} = -\mathbf{i}\sin\theta + \mathbf{j}\cos\theta$.

(b) Identify the radial and tangential components of this velocity.

(c) Show that the acceleration of the particle is given by

$$\mathbf{a} = \left[\frac{d^2r}{dt^2} - r\left(\frac{d\theta}{dt}\right)^2\right]\hat{\mathbf{r}} + \left[r\frac{d^2\theta}{dt^2} + 2\frac{dr}{dt}\frac{d\theta}{dt}\right]\hat{\mathbf{T}}$$

(d) Identify the tangential, centripetal, and radial components of this acceleration. Note that $2(dr/dt)(d\theta/dt)$ is called the **Coriolis acceleration.**

1.27 Compute the line integral along the line segment joining (0, 0, 0) and (1, 2, 4) if $\mathbf{F} = x^2\mathbf{i} + y\mathbf{j} + (xz - y)\mathbf{k}$.

1.28 A force with x, y, and z components of 3, 4, and 12 lb, respectively, displaces an object from the point (1, 2, 3) to (2, −3, 1). How much work does this force do?

1.29 Show that the work done by a conservative force during the displacement of an object from A to B is independent of the path taken between these two points.

1.30 Show that the work done by a conservative force is equal to the loss in potential energy of the object on which the force acts.

1.31 Show that the sum of the kinetic and potential energies is a constant if a particle is subject to a conservative force only.

1.32 Show that $\mathbf{F} = xy^2\mathbf{i} + x^3y\mathbf{j}$ is not conservative.

1.33 Show that $\oint \mathbf{F} \times d\mathbf{R} = 0$, where $\mathbf{F}$ is a constant vector.

1.34 Given: $\mathbf{A} = x^2y\mathbf{i} + (x - y)\mathbf{k}$, $\mathbf{B} = x\mathbf{i}$, and $\phi = xy^2z^3$. Find

(a) curl $\mathbf{A}$

(b) div $\mathbf{B}$

(c) grad ϕ.

1.35 For $f(x, y) = \sin x + e^{xy} + z$, find grad f.

1.36 Find the directional derivative of $\phi(x, y, z) = 2x^3 - 3yz$ at the point (2, 1, 3) in the direction parallel to (2, 1, −2).

1.37 Find a vector normal to the surface $x^2 + yz$ at (2, 1, 1).

1.38 If $\mathbf{A}$ is irrotational, show that $\mathbf{A} \times \mathbf{R}$ is solenoidal.

1.39 If $\nabla^2\phi = 0$, show that $\nabla\phi$ is both solenoidal and irrotational.

1.40 By use of Maxwell's (1831–1879) equations (for a vacuum), show that

$$\nabla^2 E_x = \epsilon_0 \mu_0 \frac{\partial^2 E_x}{\partial t^2}$$

$$\nabla^2 E_y = \epsilon_0 \mu_0 \frac{\partial^2 E_y}{\partial t^2}$$

$$\nabla^2 E_z = \epsilon_0 \mu_0 \frac{\partial^2 E_z}{\partial t^2}.$$

1.41 Show that

$$\oint_\lambda f\, d\boldsymbol{\lambda} = \iint_\sigma \hat{\mathbf{n}} \times \nabla f\, d\sigma.$$

1.42 Show that

$$\oiint_\sigma f\hat{\mathbf{n}}\, d\sigma = \iiint_\tau \nabla f\, d\tau.$$

1.43 By use of **Ampère's** (1775–1836) **law**

$$\oint_\lambda \mathbf{H}\cdot d\boldsymbol{\lambda} = \iint_\sigma \mathbf{J}\cdot d\boldsymbol{\sigma}$$

where **H** is the **magnetic field strength** and **J** is the **electric current density vector**, show that $\nabla \times \mathbf{H} = \mathbf{J}$.

1.44 By use of Stokes' theorem, prove that $\nabla \times \nabla V = \mathbf{0}$.

1.45 By use of Green's theorem, establish the conditions for

$$\iiint_\tau \chi \nabla^2 \psi\, d\tau = \iiint_\tau \psi \nabla^2 \chi\, d\tau$$

where χ and ψ are two scalar functions of x, y, and z.

1.46 If σ is a closed surface which bounds a volume τ, prove that

(a) $\displaystyle\iiint_\tau \nabla\cdot\hat{\mathbf{n}}\, d\tau = \sigma$

(b) $\displaystyle\iiint_\tau \nabla \times \hat{\mathbf{n}}\, d\tau = 0$

(c) $\displaystyle\oiint_\sigma \nabla \times \mathbf{A}\cdot\hat{\mathbf{n}}\, d\sigma = 0.$

1.47 Prove the relations expressed by the following equations:

(a) Eq. (1.45) (c) Eq. (1.63)

(b) Eq. (1.46) (d) Eq. (1.64)

1.48 Develop Laplace's equation in spherical polar coordinates.

1.49 Find $\nabla^2\psi$ if $\psi(r, \theta, \phi) = 2r \sin\theta + r^2 \cos\phi$.

1.50 Verify Eqs. (1.83) and (1.85). *Hint:* First show that $\hat{e}_i = h_i \nabla u^i$ for $i = 1, 2, 3$.

2

Operator and Matrix Analysis

2.1 INTRODUCTION

In quantum mechanics and quantum statistics (quantum statistical mechanics) as well as in certain other areas of physics, physical quantities are represented by linear operators on a vector (Hilbert) space. While this approach may seem far removed from an experimental process, it makes possible the development of these subject areas by use of rigorous mathematical procedures. Moreover, agreement between calculated results obtained by use of this formal (or abstract) approach with experimentally measured values has given credence to this formulation.

Mathematical operations involving linear operators are often carried out by use of matrices. Hence a knowledge of vector spaces, linear operators, and matrix analysis is required in many areas of physics. We therefore begin this chapter with a brief discussion of vector spaces and linear operators; the remainder of the chapter is devoted to a development of matrix analysis.

2.2 RUDIMENTS OF VECTOR SPACES

2.2.1 Definition of a Vector Space

A **vector space** (linear space or linear manifold) V_n is a set of elements, $\psi_1, \psi_2, \ldots, \psi_n = \{\psi_i\}$, called vectors for which the operations of addition and multiplication by a scalar defined below are valid. The term "vector," in this context, is used in an abstract mathematical sense; it is a generalization of the definition given in Chapter 1 to cases of arbitrary dimensions. In a strict mathematical sense, a **space** is a set of elements (vectors, points, functions, or any abstract quantities) for which certain defined mathematical operations are valid. We will call the set of elements vectors and require that they satisfy the following relations.

A. Addition of Vectors

1. For any two vectors ψ_i and ψ_k in V_n, there exists the sum $\psi_i + \psi_k$ in V_n such that

$$\psi_i + \psi_k = \psi_k + \psi_i.$$

2. For vectors ψ_i, ψ_j, and ψ_k in V_n, there exists $\psi_i + (\psi_j + \psi_k)$ in V_n such that

$$\psi_i + (\psi_j + \psi_k) = (\psi_i + \psi_j) + \psi_k.$$

3. There exists a unique vector 0 (zero or null vector) in V_n such that

$$\psi + 0 = \psi$$

for any ψ in V_n.

4. For each vector ψ in V_n, there exists a unique vector $-\psi$ in V_n such that

$$\psi + (-\psi) = 0.$$

B. Multiplication of Vectors by Scalars

5. For vectors ψ_i and ψ_k in V_n, there exists vectors $\alpha(\psi_i + \psi_k)$, $(\alpha + \beta)\psi_i$, and $\alpha(\beta\psi_i)$ in V_n such that

$$\alpha(\psi_i + \psi_k) = \alpha\psi_i + \alpha\psi_k$$
$$(\alpha + \beta)\psi_i = \alpha\psi_i + \beta\psi_i$$
$$\alpha(\beta\psi_i) = (\alpha\beta)\psi_i$$

where α and β are two scalars (real or complex numbers).

6. For the zero and unit vectors in V_n, the following respective products exist:

$$0 \cdot \psi = 0 \quad \text{and} \quad 1 \cdot \psi = \psi.$$

2.2.2 Linear Dependence

A set of vectors $\{\psi_i\}$ is said to be **linearly dependent** if there exists a corresponding set of scalars $\{\alpha_i\}$, not all zero, such that

$$\sum_{i=1}^{n} \alpha_i \psi_i = 0. \tag{2.1}$$

If

$$\sum_{i=1}^{n} \beta_i \phi_i = 0$$

implies that $\beta_i = 0$ for all i, then the set of vectors $\{\phi_i\}$ is said to be **linearly independent.** Here $\{\beta_i\}$ is a set of scalars.

2.2.3 Dimensionality of a Vector Space

A vector space is said to be n-dimensional if it contains precisely n linearly independent vectors. A vector space is called infinite-dimensional if there exists an arbitrarily large (but countable) number of linearly independent vectors in the space.

If an arbitrary vector ϕ in V_n can be represented as a linear combination of the vectors $\{\psi_i\}$ in V_n and scalars $\{\alpha_i\}$

$$\phi = \sum_{i=1}^{n} \alpha_i \psi_i \tag{2.2}$$

then $\{\psi_i\}$ is said to **span** the vector space V_n.

A linearly independent set of vectors $\{\psi_i\}$ that spans a vector space V_n is called a **basis** for V_n. For example, the **i, j, k** unit vectors described in Chapter 1 are the basis for the three-dimensional Cartesian space (a three-dimensional vector space).

2.2.4 Inner Product

A **Euclidean (Euclid about 300 B.C.) space** E_n is a vector space on which an inner (scalar) product is defined. The **inner product** of two vectors ψ and ϕ is denoted by

$$(\psi, \phi) \tag{2.3}$$

The following properties are valid for the inner product:

$$(\psi, \phi + \xi) = (\psi, \phi) + (\psi, \xi)$$
$$(\psi + \phi, \xi) = (\psi, \xi) + (\phi, \xi)$$

and

$$(\psi, \psi) > 0 \qquad (\text{unless } \psi = 0).$$

The **norm** (length), $||\psi||$, of a vector ψ is defined as

$$||\psi|| = (\psi, \psi)^{1/2}. \tag{2.4}$$

If the inner product of two vectors equals zero

$$(\psi_i, \psi_k) = 0 \qquad \text{for} \quad \begin{cases} i \neq j; & i, j = 1, 2, \ldots, n \\ \psi_i \neq 0 & \text{and} \quad \psi_k \neq 0 \end{cases}$$

the vectors are said to form an **orthogonal set.** If the norm within an orthogonal set is unity

$$||\psi_i|| = 1$$

the set is called **orthonormal.**

2.2.5 Hilbert (1862–1943) Space

For a vector space V_n to be **complete**, it is required that each **Cauchy sequence** in V_n,

$$||\psi_i - \psi_k|| \longrightarrow 0 \qquad (\text{for } i \text{ and } k \longrightarrow \infty),$$

converges to a limit in V_n.

A complete and infinite-dimensional complex Euclidean space is called a **Hilbert space.**

The notations used thus far in this section are those preferred by mathematicians. In physics (quantum mechanics), a vector in Hilbert space is denoted by $|\psi\rangle$, and the inner product (ψ, ϕ) is written as $\langle\psi|\phi\rangle$ where $|\phi\rangle$ and $\langle\psi|$ are called **ket** and **bra** vectors, respectively. This latter notation is due to Dirac (1902–).

2.2.6 Linear Operators

A **linear operator** on a vector space V_n is a procedure for obtaining a unique vector, ϕ_i, in V_n for each ψ_i in V_n. For example,

$$\phi_i = A\psi_i \tag{2.5a}$$

where A is a linear operator. Using the Dirac notation, we write

$$|\phi\rangle = A|\psi\rangle \tag{2.5b}$$

For linear operators A and B, it is required that

$$A(|\psi\rangle + |\phi\rangle) = A|\psi\rangle + A|\phi\rangle$$
$$(A + B)|\psi\rangle = A|\psi\rangle + B|\psi\rangle$$
$$(AB)|\psi\rangle = A(B|\psi\rangle)$$

and

$$A\alpha|\psi\rangle = \alpha A|\psi\rangle$$

where α is a scalar. The vectors $|\psi\rangle$ and $|\phi\rangle$ are two arbitrary vectors in the vector space.

EXAMPLE 2.1 *A Quantum Mechanical Illustration of a Linear Operator* Linear operators, in contrast to ordinary numbers and functions, do not always commute, that is, AB is not always equal to BA. The difference $AB - BA$ which is symbolically written as $[A, B]$ is called the **commutator** of A and B.

In quantum mechanics, linear operators play a central role, and it is understood that simultaneous specification of the physical quantities represented by two noncommuting operators, $[A, B] \neq 0$, cannot be made. Consider, for example, $[x, p_x]$ in the x-representation. Here x and p_x, $p_x = -i\hbar\partial/\partial x$ for $i = \sqrt{-1}$ and $\hbar = h/2\pi$, represent the position and x-component of the momentum of a particle, respectively. The value of the commutator $[x, p_x]$ is obtained by operating on some function $\psi(x)$; we obtain

$$\begin{aligned}[x, p_x]\psi(x) &= (xp_x - p_x x)\psi(x) \\ &= -i\hbar\left\{x\frac{\partial\psi}{\partial x} - \frac{\partial(x\psi)}{\partial x}\right\} \\ &= i\hbar\psi(x)\end{aligned}$$

or

$$[x, p_x] = i\hbar.$$

The matrix representation of a linear operator is consistent with the fact that matrix multiplication is not in general commutative, as can be seen from the definition in Eq. (2.17).

2.3 MATRIX ANALYSIS AND NOTATIONS

The word "matrix" was introduced in 1850 by Sylvester, and matrix theory was developed by Hamilton (1805–1865) and Cayley (1821–1895) in the study of simultaneous linear equations.

Matrices were rarely used by physicists prior to 1925. Today, they are used in most areas of physics. A **matrix** is a rectangular array of quantities,

$$A = \begin{pmatrix} a_{11} & a_{12} & \cdots & a_{1n} \\ a_{21} & a_{22} & \cdots & a_{2n} \\ \cdot & & & \cdot \\ \cdot & & & \cdot \\ \cdot & & & \cdot \\ a_{m1} & a_{m2} & \cdots & a_{mn} \end{pmatrix} \tag{2.6}$$

where the a_{ij} are called **elements** (of the ith row and jth column); they may be real (or complex) numbers or functions. The matrix A has m rows and n columns and is called a matrix of **order** $m \times n$ (m by n). If $m = n$, the matrix is called a **square matrix.** The **main diagonal** of a square matrix consists of the elements $a_{11}, a_{22}, \ldots, a_{nn}$.

The row matrix

$$A = (a_{11} \quad a_{12} \quad \cdots \quad a_{1n}) \tag{2.7}$$

is called a **row vector**, and the column matrix,

$$A = \begin{pmatrix} a_{11} \\ a_{21} \\ \cdot \\ \cdot \\ \cdot \\ a_{n1} \end{pmatrix} \tag{2.8}$$

is called a **column vector.**

Two matrices of the same order (A and B) are said to be equal if and only if $a_{ij} = b_{ij}$ for all i and j. For example,

$$A = \begin{pmatrix} 2i \\ 1 \end{pmatrix} \quad \text{and} \quad B = \begin{pmatrix} 2i \\ 1 \end{pmatrix}. \tag{2.9}$$

If $a_{ij} = 0$ for all i and j, then A is called a **null matrix.** For example,

$$A = \begin{pmatrix} 0 & 0 & 0 \\ 0 & 0 & 0 \\ 0 & 0 & 0 \end{pmatrix}. \tag{2.10}$$

The multiplication of a matrix, A, by a scalar, k, is given by

$$kA = Ak \tag{2.11}$$

where the elements of kA are ka_{ij} for all i and j. For example,

$$2\begin{pmatrix} 1 & 2 \\ 3 & 1 \end{pmatrix} = \begin{pmatrix} 2 & 4 \\ 6 & 2 \end{pmatrix}. \tag{2.12}$$

2.4 MATRIX OPERATIONS

In this section, we discuss the important operations for matrices.

2.4.1 Addition (Subtraction)

The operation of **addition** (or **subtraction**) for two $n \times m$ matrices is defined as

$$C = A \pm B \tag{2.13}$$

where $c_{ij} = a_{ij} \pm b_{ij}$ for all i and j. For example,

$$\begin{pmatrix} 3 & 1 & 4 \\ 4 & 0 & 0 \end{pmatrix} = \begin{pmatrix} 1 & -1 & 2 \\ 3 & 0 & 1 \end{pmatrix} + \begin{pmatrix} 2 & 2 & 2 \\ 1 & 0 & -1 \end{pmatrix}. \tag{2.14}$$

The following laws are also valid for addition of matrices of the same order:

$$A + B = B + A \qquad \text{(commutative)} \tag{2.15}$$

$$(A + B) + C = A + (B + C) \qquad \text{(associative)} \tag{2.16}$$

2.4.2 Multiplication

For two matrices A and B, two kinds of products are defined; they are called the matrix product, AB, and direct product, $A \otimes B$.

The **matrix product** $C = AB$ is obtained by use of the following definition:

$$c_{ij} = \sum_{k=1}^{s} a_{ik} b_{kj} \tag{2.17}$$

where the orders of A, B, and C are $n \times s$, $s \times m$, and $n \times m$, respectively. Note that the matrix product is defined for conformable matrices only. This means that the number of columns of A must equal the number of rows of B. Consider

$$A = \begin{pmatrix} a_{11} & a_{12} \\ a_{21} & a_{22} \end{pmatrix} \quad \text{and} \quad B = \begin{pmatrix} b_{11} & b_{12} \\ b_{21} & b_{22} \end{pmatrix}.$$

Here AB becomes

$$AB = \begin{pmatrix} a_{11} & a_{12} \\ a_{21} & a_{22} \end{pmatrix} \begin{pmatrix} b_{11} & b_{12} \\ b_{21} & b_{22} \end{pmatrix}$$

$$= \begin{pmatrix} a_{11}b_{11} + a_{12}b_{21} & a_{11}b_{12} + a_{12}b_{22} \\ a_{21}b_{11} + a_{22}b_{21} & a_{21}b_{12} + a_{22}b_{22} \end{pmatrix}. \tag{2.18}$$

For example,

$$\begin{pmatrix} 0 & 0 \\ 0 & 1 \end{pmatrix} \begin{pmatrix} 0 & 0 \\ 2 & 3 \end{pmatrix} = \begin{pmatrix} 0 & 0 \\ 2 & 3 \end{pmatrix}. \tag{2.19}$$

Consider the following system of equations:

$$\begin{aligned} 3x + y + 2z &= 3 \\ 2x - 3y - z &= -3 \\ x + 2y + z &= 4. \end{aligned} \tag{2.20}$$

In matrix form, we write Eq. (2.20) as

$$\begin{pmatrix} 3 & 1 & 2 \\ 2 & -3 & -1 \\ 1 & 2 & 1 \end{pmatrix} \begin{pmatrix} x \\ y \\ z \end{pmatrix} = \begin{pmatrix} 3 \\ -3 \\ 4 \end{pmatrix}. \tag{2.21}$$

As can be shown by use of the definition of the matrix product, the commutative law of multiplication is not, in general, valid for the matrix product, $AB \neq BA$. However, the associative law of multiplication is valid for the matrix product, $A(BC) = (AB)C$.

The direct product (tensor product) is defined for general matrices. If A is an $n \times n$ matrix and B is an $m \times m$ matrix, then the **direct product** $A \otimes B$ is an $nm \times nm$ matrix and is defined by

$$C = A \otimes B \tag{2.22}$$

where

$$C_{ik,jl} = a_{ij}b_{kl}. \tag{2.23}$$

If

$$A = \begin{pmatrix} a_{11} & a_{12} \\ a_{21} & a_{22} \end{pmatrix} \quad \text{and} \quad B = \begin{pmatrix} b_{11} & b_{12} \\ b_{21} & b_{22} \end{pmatrix}$$

then

$$A \otimes B = \begin{pmatrix} a_{11}B & a_{12}B \\ a_{21}B & a_{22}B \end{pmatrix}$$

$$= \begin{pmatrix} a_{11}b_{11} & a_{11}b_{12} & a_{12}b_{11} & a_{12}b_{12} \\ a_{11}b_{21} & a_{11}b_{22} & a_{12}b_{21} & a_{12}b_{22} \\ a_{21}b_{11} & a_{21}b_{12} & a_{22}b_{11} & a_{22}b_{12} \\ a_{21}b_{21} & a_{21}b_{22} & a_{22}b_{21} & a_{22}b_{22} \end{pmatrix}. \tag{2.24}$$

For

$$\sigma_1 = \begin{pmatrix} 0 & 1 \\ 1 & 0 \end{pmatrix} \quad \text{and} \quad \sigma_3 = \begin{pmatrix} 1 & 0 \\ 0 & -1 \end{pmatrix}$$

the direct product $\sigma_1 \otimes \sigma_3$ is given by

$$\sigma_1 \otimes \sigma_3 = \begin{pmatrix} 0 \cdot \sigma_3 & 1 \cdot \sigma_3 \\ 1 \cdot \sigma_3 & 0 \cdot \sigma_3 \end{pmatrix}$$

$$= \begin{pmatrix} 0 & 0 & 1 & 0 \\ 0 & 0 & 0 & -1 \\ 1 & 0 & 0 & 0 \\ 0 & -1 & 0 & 0 \end{pmatrix}. \tag{2.25}$$

2.4.3 Division

The division by a general matrix is not uniquely defined.

2.4.4 The Derivative of a Matrix

The derivative of a matrix with respect to a variable x is equal to the derivative of each element with respect to x separately. For example,

$$\frac{d}{dx}\begin{pmatrix} x & x^3 & 2 \\ e^{-x} & 0 & 3x^2 \end{pmatrix} = \begin{pmatrix} 1 & 3x^2 & 0 \\ -e^{-x} & 0 & 6x \end{pmatrix}. \tag{2.26}$$

2.4.5 The Integral of a Matrix

The integral of a matrix with respect to a variable x is equal to the integral of each element with respect to x separately. For example,

$$\int \begin{pmatrix} x & x^3 & 2 \\ e^{-x} & 0 & 3x^2 \end{pmatrix} dx = \begin{pmatrix} \frac{x^2}{2} & \frac{x^4}{4} & 2x \\ -e^{-x} & c & x^3 \end{pmatrix} + \begin{pmatrix} c_1 & c_2 & c_3 \\ c_4 & 0 & c_5 \end{pmatrix}. \tag{2.27}$$

2.4.6 Partitioned Matrices

Thus far, we have assumed that the elements of matrices are numbers or functions. However, the elements may themselves be matrices. That is to say, a matrix may be partitioned. The following is one way of partitioning a 3×3 matrix:

$$A = \left(\begin{array}{cc|c} a_{11} & a_{12} & a_{13} \\ a_{21} & a_{22} & a_{23} \\ \hline a_{31} & a_{32} & a_{33} \end{array}\right) = \begin{pmatrix} b_{11} & b_{12} \\ b_{21} & b_{22} \end{pmatrix} \tag{2.28}$$

where

$$b_{11} = \begin{pmatrix} a_{11} & a_{12} \\ a_{21} & a_{22} \end{pmatrix}, \qquad b_{12} = \begin{pmatrix} a_{13} \\ a_{23} \end{pmatrix}$$

$$b_{21} = (a_{31} \quad a_{32}) \quad \text{and} \quad b_{22} = a_{33}.$$

The b's are called **submatrices.** For partitioned matrices, the usual operations are valid.

2.5 PROPERTIES OF ARBITRARY MATRICES

2.5.1 Transpose Matrix: A^T

The **transpose** of an arbitrary matrix A is written as A^T and is obtained by interchanging corresponding rows and columns of A, that is, $A = a_{ij}$ and $A^T = a_{ji}$. For example,

$$A = \begin{pmatrix} 1 & -1 & 2 \\ 3 & 0 & 1 \end{pmatrix} \quad \text{and} \quad A^T = \begin{pmatrix} 1 & 3 \\ -1 & 0 \\ 2 & 1 \end{pmatrix}. \tag{2.29}$$

2.5.2 Complex-Conjugate Matrix: A^*

The **complex conjugate** of an arbitrary matrix A is formed by taking the complex conjugate of each element. Hence we have

$$A^* = a_{ij}^* \qquad \text{(for all } i \text{ and } j\text{)}. \tag{2.30}$$

For example,

$$A = \begin{pmatrix} 2+3i & 4-5i \\ 3 & 4i \end{pmatrix} \quad \text{and} \quad A^* = \begin{pmatrix} 2-3i & 4+5i \\ 3 & -4i \end{pmatrix}. \tag{2.31}$$

If $A^* = A$, then A is a real matrix.

2.5.3 Hermitian Conjugate: $A^\dagger$

The **Hermitian conjugate** of an arbitrary matrix A is obtained by taking the complex conjugate of the matrix and then the transpose of the complex conjugate matrix. For example,

$$A = \begin{pmatrix} 2+3i & 4-5i \\ 3 & 4i \end{pmatrix} \tag{2.32}$$

$$\begin{aligned} A^\dagger &= \begin{pmatrix} 2-3i & 4+5i \\ 3 & -4i \end{pmatrix}^T \\ &= \begin{pmatrix} 2-3i & 3 \\ 4+5i & -4i \end{pmatrix}. \end{aligned} \tag{2.33}$$

2.6 SPECIAL SQUARE MATRICES

Here we define certain important square matrices.

2.6.1 Unit Matrix: I

The **unit matrix** is given by

$$I = \delta_{ij} \tag{2.34}$$

where

$$IA = AI = A. \tag{2.35}$$

The **Kronecker (1823–1891) delta,** δ_{ij}, has the following property:

$$\delta_{ij} = \begin{cases} 1; & i = j \\ 0; & i \neq j \end{cases}. \tag{2.36}$$

The 3×3 unit matrix is given by

$$I = \begin{pmatrix} 1 & 0 & 0 \\ 0 & 1 & 0 \\ 0 & 0 & 1 \end{pmatrix}. \tag{2.37}$$

2.6.2 Diagonal Matrix

Here we write

$$D = D_{ij}\,\delta_{ij}. \tag{2.38}$$

The following is an example of a 3 × 3 **diagonal matrix:**

$$D = \begin{pmatrix} 2 & 0 & 0 \\ 0 & -1 & 0 \\ 0 & 0 & 4 \end{pmatrix}. \tag{2.39}$$

2.6.3 Singular Matrix†

If $|A| = 0$, then A is said to be a **singular matrix.** For example,

$$A = \begin{pmatrix} 1 & 0 \\ 0 & 0 \end{pmatrix} \quad \text{or} \quad |A| = 0. \tag{2.40}$$

2.6.4 Cofactor Matrix†

The **cofactor matrix** is written as A^c and is defined by

$$A^c = A^{ij}. \tag{2.41}$$

For example,

$$A = \begin{pmatrix} a_{11} & a_{12} & a_{13} \\ a_{21} & a_{22} & a_{23} \\ a_{31} & a_{32} & a_{33} \end{pmatrix} \qquad A^c = \begin{pmatrix} A^{11} & A^{12} & A^{13} \\ A^{21} & A^{22} & A^{23} \\ A^{31} & A^{32} & A^{33} \end{pmatrix} \tag{2.42}$$

where

$$A^{11} = (-1)^{1+1}\begin{vmatrix} a_{22} & a_{23} \\ a_{32} & a_{33} \end{vmatrix}, \qquad A^{12} = (-1)^{1+2}\begin{vmatrix} a_{21} & a_{23} \\ a_{31} & a_{33} \end{vmatrix}$$

$$A^{13} = (-1)^{1+3}\begin{vmatrix} a_{21} & a_{22} \\ a_{31} & a_{32} \end{vmatrix}, \qquad A^{21} = (-1)^{2+1}\begin{vmatrix} a_{12} & a_{13} \\ a_{32} & a_{33} \end{vmatrix}$$

$$A^{22} = (-1)^{2+2}\begin{vmatrix} a_{11} & a_{13} \\ a_{31} & a_{33} \end{vmatrix}, \qquad A^{23} = (-1)^{2+3}\begin{vmatrix} a_{11} & a_{12} \\ a_{31} & a_{32} \end{vmatrix}$$

$$A^{31} = (-1)^{3+1}\begin{vmatrix} a_{12} & a_{13} \\ a_{22} & a_{23} \end{vmatrix}, \qquad A^{32} = (-1)^{3+2}\begin{vmatrix} a_{11} & a_{13} \\ a_{21} & a_{23} \end{vmatrix}$$

and

$$A^{33} = (-1)^{3+3}\begin{vmatrix} a_{11} & a_{12} \\ a_{21} & a_{22} \end{vmatrix}.$$

†See the appendix at the end of this chapter for a discussion of determinants.

2.6.5 Adjoint of a Matrix

The **adjoint of a matrix** is written as adj A; it is defined as the **cofactor transpose**, that is,

$$\text{adj } A = A^{cT}. \tag{2.43}$$

For example,

$$A = \begin{pmatrix} 1 & 3 \\ 2 & 1 \end{pmatrix}, \quad A^c = \begin{pmatrix} 1 & -2 \\ -3 & 1 \end{pmatrix}, \quad A^{cT} = \begin{pmatrix} 1 & -3 \\ -2 & 1 \end{pmatrix}. \tag{2.44}$$

2.6.6 Self-Adjoint Matrix

If adj $A = A$, A is said to be **self-adjoint.** For example,

$$A = \begin{pmatrix} -1 & 0 \\ 0 & -1 \end{pmatrix}, \quad A^c = \begin{pmatrix} -1 & 0 \\ 0 & -1 \end{pmatrix}$$

$$A^{cT} = \text{adj } A = \begin{pmatrix} -1 & 0 \\ 0 & -1 \end{pmatrix} = A. \tag{2.45}$$

2.6.7 Symmetric Matrix

If $A^T = A$, A is said to be a **symmetric matrix.** For example,

$$\sigma_1 = \begin{pmatrix} 0 & 1 \\ 1 & 0 \end{pmatrix}$$

hence

$$\sigma_1^T = \begin{pmatrix} 0 & 1 \\ 1 & 0 \end{pmatrix} = \sigma_1. \tag{2.46}$$

2.6.8 Antisymmetric (Skew) Matrix

If $A^T = -A$, A is said to be an **antisymmetric** (skew) **matrix.** For example,

$$\sigma_2 = \begin{pmatrix} 0 & -i \\ i & 0 \end{pmatrix}$$

hence

$$\sigma_2^T = \begin{pmatrix} 0 & i \\ -i & 0 \end{pmatrix} = -\sigma_2. \tag{2.47}$$

2.6.9 Hermitian Matrix

If $A^\dagger = A$, A is said to be a **Hermitian matrix.** For example,

$$\sigma_2 = \begin{pmatrix} 0 & -i \\ i & 0 \end{pmatrix} \quad \text{and} \quad \sigma_2^* = \begin{pmatrix} 0 & i \\ -i & 0 \end{pmatrix}$$

hence

$$(\sigma_2^*)^T = \sigma_2^\dagger = \begin{pmatrix} 0 & -i \\ i & 0 \end{pmatrix} = \sigma_2. \tag{2.48}$$

In quantum mechanics, all physical observables are represented by Hermitian operators (matrices).

2.6.10 Unitary Matrix

If $AA^\dagger = I$, A is said to be a **unitary matrix.** For example,

$$\sigma_1 = \begin{pmatrix} 0 & 1 \\ 1 & 0 \end{pmatrix}, \qquad \sigma_1^* = \begin{pmatrix} 0 & 1 \\ 1 & 0 \end{pmatrix}, \qquad (\sigma_1^*)^T = \begin{pmatrix} 0 & 1 \\ 1 & 0 \end{pmatrix} = \sigma_1^\dagger,$$

and

$$\sigma_1\sigma_1^\dagger = \begin{pmatrix} 0 & 1 \\ 1 & 0 \end{pmatrix}\begin{pmatrix} 0 & 1 \\ 1 & 0 \end{pmatrix} = \begin{pmatrix} 1 & 0 \\ 0 & 1 \end{pmatrix} = I. \tag{2.49}$$

2.6.11 Orthogonal Matrix

If $AA^T = I$, A is said to be an **orthogonal matrix.** For example,

$$\sigma_1 = \begin{pmatrix} 0 & 1 \\ 1 & 0 \end{pmatrix}, \qquad \sigma_1^T = \begin{pmatrix} 0 & 1 \\ 1 & 0 \end{pmatrix}$$

and

$$\sigma_1\sigma_1^T = \begin{pmatrix} 0 & 1 \\ 1 & 0 \end{pmatrix}\begin{pmatrix} 0 & 1 \\ 1 & 0 \end{pmatrix} = \begin{pmatrix} 1 & 0 \\ 0 & 1 \end{pmatrix} = I. \tag{2.50}$$

2.6.12 The Trace of a Matrix

The **trace** of a matrix is given by

$$\operatorname{Tr} A = \sum_k a_{kk} \tag{2.51}$$

For example,

$$A = \begin{pmatrix} 2 & 4 \\ 3 & 7 \end{pmatrix} \quad \text{and} \quad \text{Tr } A = 2 + 7 = 9. \tag{2.52}$$

2.6.13 The Inverse Matrix

For the **inverse matrix,** A^{-1}, we require that

$$AA^{-1} = I. \tag{2.53}$$

We now develop the explicit expression for A^{-1}. In the Laplace development (see the appendix) for the value of a determinant, we have

$$|A| = \sum_{j=1}^{n} a_{ij}A^{ij}$$

or

$$|A|\,\delta_{ik} = \sum_{j=1}^{n} a_{ij}A^{kj}. \tag{2.54}$$

Let $b_{jk} = A^{kj}$, that is, $B = A^{cT}$. Equation (2.54) now becomes

$$|A|\,\delta_{ik} = \sum_{j=1}^{n} a_{ij}b_{jk}$$

or

$$I|A| = AB = AA^{cT} \tag{2.55}$$

where $\delta_{ik} = I$. On dividing Eq. (2.55) by $|A|$, we obtain

$$I = A\left[\frac{A^{cT}}{|A|}\right]. \tag{2.56}$$

The quantity in brackets must be A^{-1} because of Eq. (2.53).

EXAMPLE 2.2 Find the inverse of

$$A = \begin{pmatrix} 1 & 3 \\ 2 & 1 \end{pmatrix}.$$

Solution For A, we have $|A| = 1 - 6 = -5$,

$$A^c = \begin{pmatrix} A^{11} & A^{12} \\ A^{21} & A^{22} \end{pmatrix} = \begin{pmatrix} 1 & -2 \\ -3 & 1 \end{pmatrix}$$

and

$$A^{cT} = \begin{pmatrix} 1 & -3 \\ -2 & 1 \end{pmatrix}.$$

Hence the inverse of A is

$$A^{-1} = \frac{A^{cT}}{|A|} = -\frac{1}{5}\begin{pmatrix} 1 & -3 \\ -2 & 1 \end{pmatrix}.$$

Check:

$$\begin{aligned} AA^{-1} &= -\frac{1}{5}\begin{pmatrix} 1 & 3 \\ 2 & 1 \end{pmatrix}\begin{pmatrix} 1 & -3 \\ -2 & 1 \end{pmatrix} \\ &= -\frac{1}{5}\begin{pmatrix} -5 & 0 \\ 0 & -5 \end{pmatrix} \\ &= I. \end{aligned}$$

2.7 SOLUTION OF A SYSTEM OF LINEAR EQUATIONS

The matrix method may be used to solve a system of linear equations. Consider the following system of equations to illustrate the method:

$$\begin{aligned} x + 3y &= 2 \\ 2x + y &= 3. \end{aligned} \tag{2.57}$$

In matrix form, we write

$$AX = C \tag{2.58}$$

where

$$A = \begin{pmatrix} 1 & 3 \\ 2 & 1 \end{pmatrix}, \quad X = \begin{pmatrix} x \\ y \end{pmatrix}, \quad \text{and} \quad C = \begin{pmatrix} 2 \\ 3 \end{pmatrix}.$$

The solution of Eq. (2.58) is

$$X = A^{-1}C = \frac{A^{cT}}{|A|}C$$

where

$$|A| = -5, \qquad A^{cT} = \begin{pmatrix} 1 & -3 \\ -2 & 1 \end{pmatrix}.$$

On substituting the value of $|A|$ into the expression for X, we obtain

$$\begin{aligned} X &= \begin{pmatrix} x \\ y \end{pmatrix} \\ &= -\frac{1}{5}\begin{pmatrix} 1 & -3 \\ -2 & 1 \end{pmatrix}\begin{pmatrix} 2 \\ 3 \end{pmatrix} \\ &= \begin{pmatrix} \frac{7}{5} \\ \frac{1}{5} \end{pmatrix} \end{aligned}$$

or $x = \frac{7}{5}$ and $y = \frac{1}{5}$.

2.8 THE EIGENVALUE PROBLEM

The importance of eigenvalue problems in mathematical physics cannot be overemphasized. In general, it is assumed that associated with each linear operator is a set of functions and corresponding numbers such that

$$Au_i = \lambda_i u_i \tag{2.59}$$

where A is a linear operator, u_i are called **eigenfunctions,** and the λ_i are known as **eigenvalues.** The matrix form of an eigenvalue problem is

$$AX = \lambda IX$$

or

$$(A - \lambda I)X = 0 \tag{2.60}$$

where A is represented by a square matrix.

In solving the general matrix equation $BX = C$ for X, we obtain

$$X = \frac{B^{cT}}{|B|}C.$$

If C is a null (zero) matrix and $|B| \neq 0$, then the solution, X, is a trivial one. For C equal to a null matrix, a necessary condition for a nontrivial solution of $BX = 0$ is that $|B| = 0$. Hence the condition for a nontrivial solution of Eq. (2.60) is that

$$|A - \lambda I| = 0. \tag{2.61}$$

Equation (2.61) is called the **secular** (or **characteristic**) equation of A. The eigenvalues are just the roots of the equation obtained by expanding the determinant in Eq. (2.61). That is,

$$|A - \lambda I| = \begin{vmatrix} a_{11} - \lambda & a_{12} & \cdots & a_{1n} \\ a_{21} & a_{22} - \lambda & \cdots & a_{2n} \\ \vdots & \vdots & & \vdots \\ a_{n1} & a_{n2} & \cdots & a_{nn} - \lambda \end{vmatrix}. \tag{2.62}$$

EXAMPLE 2.3 Find the eigenvalues of

$$A = \begin{pmatrix} 3 & 1 \\ 2 & 2 \end{pmatrix}.$$

Solution In this case, the secular equation reduces to

$$\begin{vmatrix} 3 - \lambda & 1 \\ 2 & 2 - \lambda \end{vmatrix} = (3 - \lambda)(2 - \lambda) - 2$$
$$= (\lambda - 1)(\lambda - 4)$$
$$= 0$$

Therefore the eigenvalues are $\lambda = 1$ and $\lambda = 4$.

Equation (2.61) may be written as

$$|A - \lambda I| = 0$$
$$= \phi(\lambda) \tag{2.63}$$

where $\phi(\lambda)$ is an nth degree polynomial (matrix) which can be represented as

$$\phi(\lambda) = \phi_0 I + \phi_1 \lambda + \cdots + \phi_{n-1}\lambda^{n-1} + \phi_n \lambda^n. \tag{2.64}$$

The **Cayley-Hamilton theorem** states that $\lambda = A$ satisfies the above nth degree polynomial. Loosely stated, we say that a matrix satisfies its own characteristic equation, $\phi(A) = 0$.

EXAMPLE 2.4 Illustrate the Cayley-Hamilton theorem for the matrix A where

$$A = \begin{pmatrix} 1 & 2 & 0 \\ 2 & -1 & 0 \\ 0 & 0 & 1 \end{pmatrix}.$$

Solution For this matrix, we have

$$\phi(\lambda) = \begin{vmatrix} 1-\lambda & 2 & 0 \\ 2 & -1-\lambda & 0 \\ 0 & 0 & 1-\lambda \end{vmatrix}$$
$$= -5 + 5\lambda + \lambda^2 - \lambda^3.$$

The corresponding third-degree polynomial (matrix) is

$$\phi(\lambda) = \phi_0 I + \phi_1\lambda + \phi_2\lambda^2 + \phi_3\lambda^3$$

where $\phi_0 = -5$, $\phi_1 = 5$, $\phi_2 = 1$, and $\phi_3 = -1$. By use of the Cayley-Hamilton theorem, we may write

$$\phi(A) = \phi_0 I + \phi_1 A + \phi_2 A^2 + \phi_3 A^3$$
$$= \begin{pmatrix} -5 & 0 & 0 \\ 0 & -5 & 0 \\ 0 & 0 & -5 \end{pmatrix} + \begin{pmatrix} 5 & 10 & 0 \\ 10 & -5 & 0 \\ 0 & 0 & 5 \end{pmatrix} + \begin{pmatrix} 5 & 0 & 0 \\ 0 & 5 & 0 \\ 0 & 0 & 1 \end{pmatrix}$$
$$+ \begin{pmatrix} -5 & -10 & 0 \\ -10 & 5 & 0 \\ 0 & 0 & -1 \end{pmatrix}$$
$$= 0.$$

2.9 COORDINATE TRANSFORMATIONS

2.9.1 Rotation in Two Dimensions

Consider the rotation of the two-dimensional Cartesian system illustrated in Fig. 2.1. In the figure, $\beta = \theta$ since $\theta + \gamma = \pi/2$ and $\gamma + \beta = \pi/2$. The relations between the prime and unprime axes are

$$\begin{aligned} x' &= x\cos\theta + l_1\sin\theta + l_2\sin\theta \\ &= x\cos\theta + (l_1 + l_2)\sin\theta \\ &= x\cos\theta + y\sin\theta \end{aligned} \tag{2.65}$$

and

$$\begin{aligned} y' &= l_1\cos\theta \\ &= (y - l_2)\cos\theta \\ &= y\cos\theta - l_2\cos\theta \\ &= y\cos\theta - x\sin\theta \end{aligned} \tag{2.66}$$

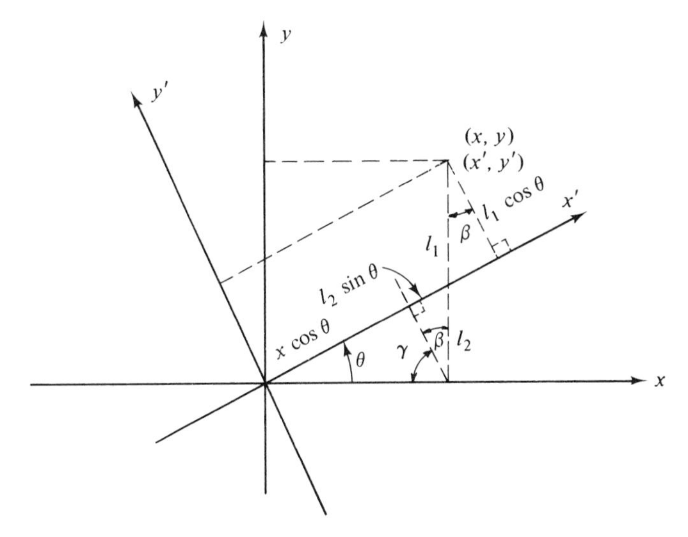

Figure 2.1

In matrix form, Eqs. (2.65) and (2.66) become

$$\begin{pmatrix} x' \\ y' \end{pmatrix} = \begin{pmatrix} \cos\theta & \sin\theta \\ -\sin\theta & \cos\theta \end{pmatrix} \begin{pmatrix} x \\ y \end{pmatrix}$$

or

$$X' = R_2 X \tag{2.67}$$

where

$$X' = \begin{pmatrix} x' \\ y' \end{pmatrix}, \qquad R_2 = \begin{pmatrix} \cos\theta & \sin\theta \\ -\sin\theta & \cos\theta \end{pmatrix}, \quad \text{and} \quad X = \begin{pmatrix} x \\ y \end{pmatrix}.$$

The matrix R_2 is called the 2×2 **rotation matrix.**

Equations (2.65) and (2.66) may be put in the following form:

$$x_1' = x_1 \cos\theta + x_2 \sin\theta$$
$$x_2' = -x_1 \sin\theta + x_2 \cos\theta$$

or

$$x_1' = x_1 \lambda_{11} + x_2 \lambda_{12}$$
$$x_2' = -x_1 \lambda_{21} + x_2 \lambda_{22}$$

or

$$x_i' = \sum_{j=1}^{2} \lambda_{ij} x_j \qquad (i = 1, 2). \tag{2.68}$$

where $x \rightarrow x_1$, $y \rightarrow x_2$, $x' \rightarrow x_1'$, $y' \rightarrow x_2'$, and $\lambda_{ij} = \cos(x_i', x_j)$.

2.9.2 Rotation in Three Dimensions

In three dimensions, we have

$$x_i' = \sum_{j=1}^{3} \lambda_{ij} x_j \qquad (i = 1, 2, 3). \tag{2.69}$$

where

$$\lambda_{ij} = \cos(x_i', x_j) \qquad (\text{for } i, j = 1, 2, 3).$$

In general, a **rotation** is defined as a transformation whose matrix R is determined by a set of parameters, $\alpha, \beta, \ldots, \gamma$, such that the following conditions are satisfied:

1. $R(\alpha, \beta, \ldots, \gamma)$ is a continuous function of $\alpha, \beta, \ldots, \gamma$.
2. $R(0, 0, \ldots, 0) = I$ (the unit matrix).
3. $\det(R) = 1$.

EXAMPLE 2.5 Find the transformation equations for a 90-degree rotation about the x_3-axis (see Fig. 2.2).

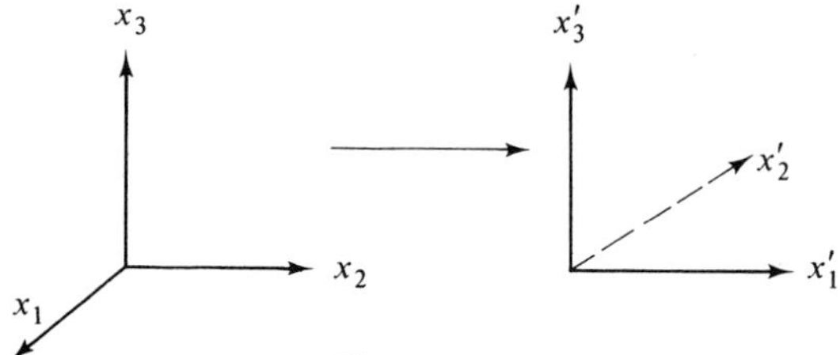

Figure 2.2

Solution Here the elements, λ_{ij}, of the rotation matrix are given by

$$\begin{pmatrix} \lambda_{11} & \lambda_{12} & \lambda_{13} \\ \lambda_{21} & \lambda_{22} & \lambda_{23} \\ \lambda_{31} & \lambda_{32} & \lambda_{33} \end{pmatrix} = \begin{pmatrix} 0 & 1 & 0 \\ -1 & 0 & 0 \\ 0 & 0 & 1 \end{pmatrix}$$

since

$$\lambda_{11} = \cos(x_1', x_1) = \cos\frac{\pi}{2} = 0 \qquad \lambda_{23} = \cos(x_2', x_3) = \cos\frac{\pi}{2} = 0$$

$$\lambda_{12} = \cos(x_1', x_2) = \cos 0° = 1 \qquad \lambda_{31} = \cos(x_3', x_1) = \cos\frac{\pi}{2} = 0$$

$$\lambda_{13} = \cos(x_1', x_3) = \cos\frac{\pi}{2} = 0 \qquad \lambda_{32} = \cos(x_3', x_2) = \cos\frac{\pi}{2} = 0$$

$$\lambda_{21} = \cos(x_2', x_1) = \cos\pi = -1 \qquad \lambda_{33} = \cos(x_3', x_3) = \cos 0° = 1$$

$$\lambda_{22} = \cos(x_2', x_2) = \cos\frac{\pi}{2} = 0$$

By use of Eq. (2.69), we obtain

$$\begin{pmatrix} x_1' \\ x_2' \\ x_3' \end{pmatrix} = \begin{pmatrix} 0 & 1 & 0 \\ -1 & 0 & 0 \\ 0 & 0 & 1 \end{pmatrix} \begin{pmatrix} x_1 \\ x_2 \\ x_3 \end{pmatrix} = \begin{pmatrix} x_2 \\ -x_1 \\ x_3 \end{pmatrix}$$

Hence

$$x_1' = x_2, \qquad x_2' = -x_1, \quad \text{and} \quad x_3' = x_3.$$

A more extensive treatment of coordinate transformation properties is given in Chapter 9.

2.10 PROBLEMS

2.1 Given

$$A = \begin{pmatrix} 3 & 2 \\ 1 & 0 \end{pmatrix} \quad \text{and} \quad B = \begin{pmatrix} 2 & 4 \\ 1 & 2 \end{pmatrix}.$$

Find:
(a) $A + B$
(b) $B - A$
(c) AB
(d) BA.

2.2 Does $AB = AC$ imply that $B = C$? Illustrate by means of an example.

2.3 Show that $AB \neq BA$ for general A and B.

2.4 For the **Pauli** (1900–1958) **spin matrices**

$$\sigma_1 = \begin{pmatrix} 0 & 1 \\ 1 & 0 \end{pmatrix}, \quad \sigma_2 = \begin{pmatrix} 0 & -i \\ i & 0 \end{pmatrix} \quad \text{and} \quad \sigma_3 = \begin{pmatrix} 1 & 0 \\ 0 & -1 \end{pmatrix}$$

show that
(a) $\sigma_1^2 = \sigma_2^2 = \sigma_3^2 = I$
(b) $[\sigma_1, \sigma_2] = 2i\sigma_3$ (cyclic).

2.5 Show that

$$\frac{d}{dx}[A(x)B(x)] = \frac{dA(x)}{dx}B(x) + A(x)\frac{dB(x)}{dx}$$

where $A(x)$ and $B(x)$ are two matrices.

2.6 Write the two coupled (dependent) differential equations (equations involving derivatives) indicated by

$$i\hbar\frac{d}{dt}\begin{pmatrix} a(t) \\ b(t) \end{pmatrix} = \begin{pmatrix} \omega_0 & \omega_1 \cos \omega t \\ \omega_1 \cos \omega t & -\omega_0 \end{pmatrix} \begin{pmatrix} a(t) \\ b(t) \end{pmatrix}.$$

2.7 The **rank** of a matrix is equal to the order of the largest nonvanishing determinant contained within the matrix. Find the rank of the following matrices:

(a) $\begin{pmatrix} 1 & 2 & -1 \\ 4 & 1 & 5 \\ 3 & -1 & 6 \end{pmatrix}$ (b) $\begin{pmatrix} 0 & 0 & 0 \\ 0 & 0 & 0 \\ 0 & 0 & 0 \end{pmatrix}$

(c) $\begin{pmatrix} 0 & 0 & 2 \\ 0 & 0 & 0 \end{pmatrix}$ (d) $\begin{pmatrix} 2 & 4 \\ -4 & -8 \\ 1 & 2 \end{pmatrix}$.

2.8 Calculate the trace (where it exists) of the matrices in Problem 2.7.

2.9 Find the transpose of:

(a) $\begin{pmatrix} 2 \\ 4 \\ 6 \end{pmatrix}$

(b) $(3 \quad 4 \quad -2)$

(c) $\begin{pmatrix} 3 & 2 & 1 \\ 2 & 0 & -6 \\ 1 & -6 & 1 \end{pmatrix}$.

2.10 Determine which of the following matrices are symmetric and which are antisymmetric:

(a) $\begin{pmatrix} 1 & 2 & 5 \\ 2 & 2 & -1 \\ 5 & -1 & 3 \end{pmatrix}$ (b) $\begin{pmatrix} 2 & 3 \\ 3 & 4 \\ 0 & 0 \end{pmatrix}$

(c) $\begin{pmatrix} 2 & -1 & -2 \\ 1 & 2 & -1 \\ 2 & 1 & 2 \end{pmatrix}$ (d) $\begin{pmatrix} 2 & 1 \\ 2 & 4 \end{pmatrix}$.

2.11 Show that

$$\begin{pmatrix} i & 0 \\ 0 & 1 \end{pmatrix}$$

is a unitary matrix.

2.12 Find A such that

$$A^{-1} = \begin{pmatrix} 3 & 2 \\ 1 & 6 \end{pmatrix}.$$

2.13 Given

$$A = \begin{pmatrix} i & 2 & 3+i \\ -2 & 2i & 0 \\ -3+i & 0 & -i \end{pmatrix}.$$

Find $A^{\dagger}$.

2.14 Given

$$\begin{Bmatrix} a_{11}x_1 + a_{12}x_2 = k_1 \\ a_{21}x_1 + a_{22}x_2 = k_2 \end{Bmatrix}.$$

Using the determinant method, solve for x_1 and x_2.

2.15 Show that

$$\begin{vmatrix} 2\cos\theta & 1 & 0 \\ 1 & 2\cos\theta & 1 \\ 0 & 1 & 2\cos\theta \end{vmatrix} = \frac{\sin 4\theta}{\sin\theta}.$$

2.16 If A is antisymmetric, prove that:

(a) $AA^T = A^TA$

(b) A^2 is symmetric.

2.17 Prove that:

(a) $\mathrm{Tr}(AB) = \mathrm{Tr}(BA)$

(b) $\mathrm{Tr}(A + B) = \mathrm{Tr}A + \mathrm{Tr}\,B$.

2.18 If $C = AA^T$, prove that C is symmetric.

2.19 Prove the following:

(a) $(A^T)^T = A$

(b) $(AB)^T = B^TA^T$

(c) $(A + B)^T = A^T + B^T$

(d) $(cA)^T = cA^T$ where c is a scalar.

2.20 Prove that (a) $A + A^\dagger$, (b) $i(A - A^\dagger)$, and (c) $AA^\dagger$ are all Hermitian for any A.

2.21 If A and B are two Hermitian matrices, prove that AB is Hermitian only if A and B commute.

2.22 If A is a Hermitian matrix, show that e^{iA} is unitary.

2.23 Show that the eigenvalues of a Hermitian matrix are all real.

2.24 By use of the matrix method, solve the following system of equations:

(a) $\begin{Bmatrix} a_{11}x_1 + a_{12}x_2 = k_1 \\ a_{21}x_1 + a_{22}x_2 = k_2 \end{Bmatrix}$

(b) $\begin{Bmatrix} x + 3y = 4 \\ 2x - 2y = 6 \end{Bmatrix}$

(c) $\begin{Bmatrix} x - 4y = 2 \\ 2x - y = 1 \end{Bmatrix}$

(d) $\begin{Bmatrix} x + 5y + 3z = 1 \\ 3x + y + 2z = 1 \\ x + 2y + z = 0 \end{Bmatrix}$.

2.25 Find the eigenvalues of the following matrices:

(a) $\begin{pmatrix} 4 & -2 \\ 1 & 1 \end{pmatrix}$

(b) $\begin{pmatrix} 1 & 2 \\ -8 & 11 \end{pmatrix}$

(c) $\begin{pmatrix} 1 & 0 \\ 2 & -1 \end{pmatrix}$

(d) $\begin{pmatrix} 13 & -3 & 5 \\ 0 & 4 & 0 \\ -15 & 9 & -7 \end{pmatrix}$.

2.26 Verify the Cayley-Hamilton theorem for

$$\begin{pmatrix} 0 & 1 \\ 1 & 0 \end{pmatrix}.$$

2.27 By use of the Cayley-Hamilton theorem, compute the inverse of

$$\begin{pmatrix} 0 & 1 \\ 1 & 0 \end{pmatrix}.$$

2.28 Solve $X' = R_2 X$ for x and y where R_2 is the 2×2 rotation matrix.

2.29 By use of the matrix method, find

(a) R_3

(b) $|R_3|$

(c) the required transformation equations for the indicated inversion transformation in Fig. 2.3.

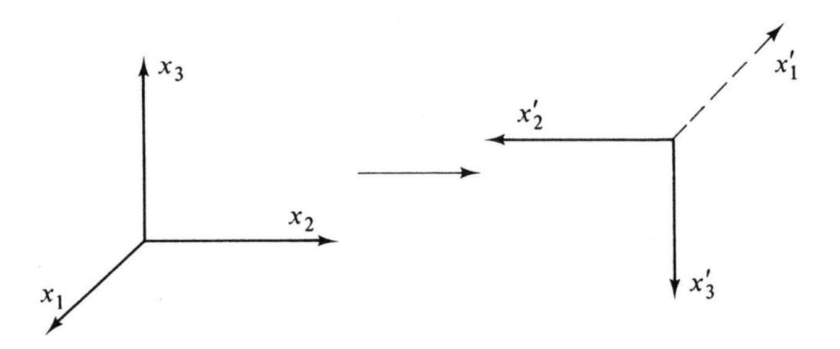

Figure 2.3

2.30 By use of the matrix method, find

(a) R_3

(b) $|R_3|$

(c) the required transformation equations for the indicated reflection transformation in Fig. 2.4.

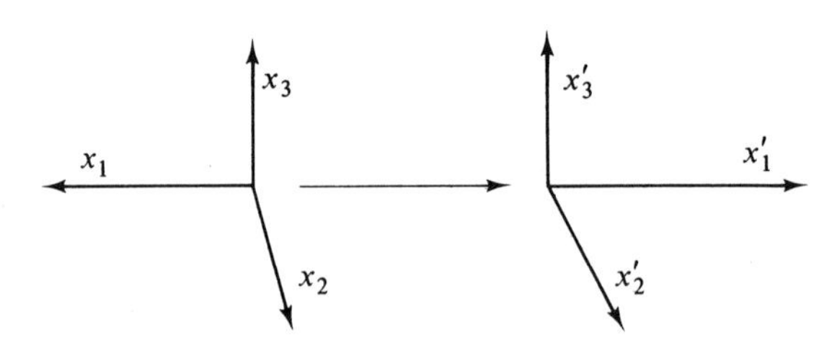

Figure 2.4

APPENDIX: RUDIMENTS OF DETERMINANTS

This appendix contains a summary of the essential properties of determinants.

A 2.1 Introduction

A **determinant** is a square array of quantities called **elements** which may be combined according to the rules given below. In symbolic form, we write

$$\Delta = \begin{vmatrix} a_1 & b_1 & c_1 & \cdots & r_1 \\ a_2 & b_2 & c_2 & \cdots & r_2 \\ \cdot & \cdot & \cdot & & \cdot \\ \cdot & \cdot & \cdot & & \cdot \\ \cdot & \cdot & \cdot & & \cdot \\ a_n & b_n & c_n & \cdots & r_n \end{vmatrix}. \tag{A2.1}$$

Here n is called the **order** of the determinant. The value of the determinant in terms of the elements $a_i, b_j, \ldots, r_l$ is defined as

$$\Delta = \sum_{i,j\cdots l}^{n} \epsilon_{ijk\cdots l} a_i b_j \cdots r_l \tag{A2.2}$$

where the Levi-Civita symbol, $\epsilon_{ijk\cdots l}$, has the following property:

$$\epsilon_{ij\ldots l} = \begin{cases} +1 & \text{for an even permutation of } (i, j, \ldots, l) \\ -1 & \text{for an odd permutation of } (i, j, \ldots, l) \\ 0 & \text{if an index is repeated.} \end{cases} \tag{A2.3}$$

For example, $\epsilon_{ijk} = \begin{cases} 1 & (\text{if } ijk = 123, 231, 312) \\ -1 & (\text{if } ijk = 321, 213, 132) \\ 0 & (\text{otherwise}) \end{cases}$

On applying Eq. (A2.2) to the third-order determinant,

$$\Delta = \begin{vmatrix} a_1 & b_1 & c_1 \\ a_2 & b_2 & c_2 \\ a_3 & b_3 & c_3 \end{vmatrix}$$

we obtain

$$\begin{aligned} \Delta &= \sum_{ijk}^{3} \epsilon_{ijk} a_i b_j c_k \\ &= \sum_{jk}^{3} [\epsilon_{1jk} a_1 b_j c_k + \epsilon_{2jk} a_2 b_j c_k + \epsilon_{3jk} a_3 b_j c_k]. \end{aligned} \tag{A2.4}$$

Equation (A2.4) reduces to

$$\begin{aligned}\Delta = \sum_{k}^{3} [&\epsilon_{11k}a_1b_1c_k + \epsilon_{12k}a_1b_2c_k + \epsilon_{13k}a_1b_3c_k \\ &+ \epsilon_{21k}a_2b_1c_k + \epsilon_{22k}a_2b_2c_k + \epsilon_{23k}a_2b_3c_k \\ &+ \epsilon_{31k}a_3b_1c_k + \epsilon_{32k}a_3b_2c_k + \epsilon_{33k}a_3b_3c_k]. \end{aligned} \tag{A2.5}$$

Since $\epsilon_{11k} = \epsilon_{22k} = \epsilon_{33k} = 0$, Eq. (A2-5) becomes

$$\begin{aligned}\Delta = \quad &\epsilon_{121}a_1b_2c_1 + \epsilon_{122}a_1b_2c_2 + \epsilon_{123}a_1b_2c_3 \\ &+ \epsilon_{131}a_1b_3c_1 + \epsilon_{132}a_1b_3c_2 + \epsilon_{133}a_1b_3c_3 \\ &+ \epsilon_{211}a_2b_1c_1 + \epsilon_{212}a_2b_1c_2 + \epsilon_{213}a_2b_1c_3 \\ &+ \epsilon_{231}a_2b_3c_1 + \epsilon_{232}a_2b_3c_2 + \epsilon_{233}a_2b_3c_3 \\ &+ \epsilon_{311}a_3b_1c_1 + \epsilon_{312}a_3b_1c_2 + \epsilon_{313}a_3b_1c_3 \\ &+ \epsilon_{321}a_3b_2c_1 + \epsilon_{322}a_3b_2c_2 + \epsilon_{323}a_3b_2c_3. \end{aligned} \tag{A2.6}$$

With the aid of Eq. (A2.3), Eq. (A2.6) reduces to

$$\begin{aligned}\Delta = &a_1b_2c_3 - a_1b_3c_2 - a_2b_1c_3 \\ &+ a_2b_3c_1 + a_3b_1c_2 - a_3b_2c_1. \end{aligned} \tag{A2.7}$$

The following scheme may also be used to obtain the result in Eq. (A2.7),

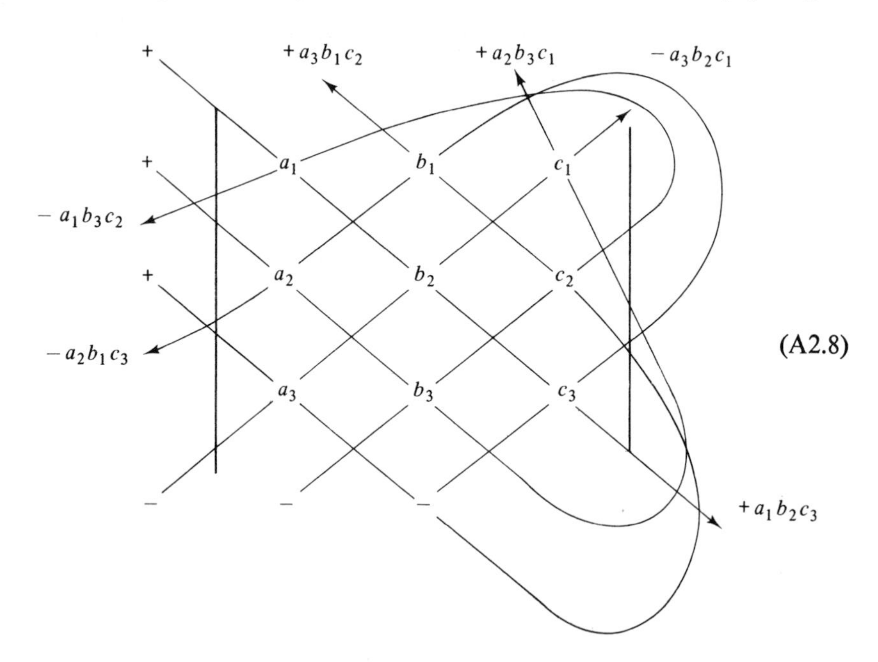

(A2.8)

that is,

$$\Delta = a_1b_2c_3 + a_2b_3c_1 + a_3b_1c_2 - (a_3b_2c_1 + a_1b_3c_2 + a_2b_1c_3). \tag{A2.9}$$

A2.2 Laplace Development by Minors

The result in Eqs. (A2.7) and (A2.9) may be written in the form

$$\begin{aligned}\Delta &= a_1(b_2c_3 - b_3c_2) - a_2(b_1c_3 - b_3c_1) + a_3(b_1c_2 - b_2c_1) \\ &= a_1\begin{vmatrix} b_2 & c_2 \\ b_3 & c_3 \end{vmatrix} - a_2\begin{vmatrix} b_1 & c_1 \\ b_3 & c_3 \end{vmatrix} + a_3\begin{vmatrix} b_1 & c_1 \\ b_2 & c_2 \end{vmatrix}\end{aligned} \tag{A2.10}$$

where

$$\begin{vmatrix} b_2 & c_2 \\ b_3 & c_3 \end{vmatrix} = b_2c_3 - b_3c_2.$$

The procedure of expressing Δ in the form given in Eq. (A2.10) may be generalized to obtain the value of an nth-order determinant. In Eq. (A2.10), we see that the expansion of a third-order determinant is expressed as a linear combination of the product of an element and a second-order determinant. Careful examination of Eq. (A2.10) reveals that the second-order determinant is the determinant obtained by omitting the elements in the row and column in which the multiplying element (the element in front of the second-order determinant) appears in the original determinant. The resulting second-order determinant is called a **minor.** Thus the minor of a_1 is obtained in the following way:

$$\begin{vmatrix} a_1 & b_1 & c_1 \\ a_2 & b_2 & c_2 \\ a_3 & b_3 & c_3 \end{vmatrix}$$

(first row struck through → out; first column struck through ↓ out)

The minor of a_1 is therefore

$$\begin{vmatrix} b_2 & c_2 \\ b_3 & c_3 \end{vmatrix}$$

In the general nth-order determinant, the sign $(-1)^{i+j}$ is associated with the minor of the element in the ith row and the jth column. The minor with its sign $(-1)^{i+j}$ is called the **cofactor.** For the general determinant

$$|A| = \det A = \begin{vmatrix} a_{11} & a_{12} & \cdots & a_{1n} \\ a_{21} & & & \\ \cdot & & & \cdot \\ \cdot & & & \cdot \\ \cdot & & & \cdot \\ a_{n1} & a_{n2} & \cdots & a_{nn} \end{vmatrix}$$

the value in terms of cofactors is given by

$$|A| = \det A = \sum_{j=1}^{n} a_{ij} A^{ij} \text{ for any } i. \tag{A2.11}$$

The relation in Eq. (A2.11) is called the **Laplace development.** Expanding it along the first row gives $i = 1$. For example,

$$|A| = \begin{vmatrix} a_{11} & a_{12} \\ a_{21} & a_{22} \end{vmatrix}$$

becomes

$$|A| = \sum_{j=1}^{2} a_{1j} A^{1j} = a_{11}A^{11} + a_{12}A^{12}$$

where

$$A^{11} = (-1)^{1+1}|a_{22}| = a_{22} \quad \text{and} \quad A^{12} = (-1)^{1+2}|a_{21}| = -a_{21}. \tag{A2.12}$$

On substituting Eq. (A2.12) into the expression for $|A|$, we obtain

$$|A| = a_{11}a_{22} - a_{12}a_{21}.$$

Unlike matrices, determinants may be evaluated to yield a number.

A2.3 Summary of the Properties of Determinants

The following are properties of determinants which may be readily proved:

1. The value of a determinant is not changed if corresponding rows and columns are interchanged.

$$\Delta = \begin{vmatrix} a_1 & b_1 & c_1 \\ a_2 & b_2 & c_2 \\ a_3 & b_3 & c_3 \end{vmatrix} = \begin{vmatrix} a_1 & a_2 & a_3 \\ b_1 & b_2 & b_3 \\ c_1 & c_2 & c_3 \end{vmatrix} \tag{A2.13}$$

2. If a multiple of one column is added (row by row) to another column or if a multiple of one row is added (column by column) to another row, the value of the determinant is unchanged.

$$\begin{vmatrix} a_1 & b_1 & c_1 \\ a_2 & b_2 & c_2 \\ a_3 & b_3 & c_3 \end{vmatrix} = \begin{vmatrix} a_1 + kb_1 & b_1 & c_1 \\ a_2 + kb_2 & b_2 & c_2 \\ a_3 + kb_3 & b_3 & c_3 \end{vmatrix} \tag{A2.14}$$

3. If each element of a column or row is zero, the value of the determinant is zero.

$$\begin{vmatrix} a_1 & b_1 & c_1 \\ 0 & 0 & 0 \\ a_3 & b_3 & c_3 \end{vmatrix} = 0 \tag{A2.15}$$

4. If two columns or rows are identical, the value of the determinant is zero.

$$\begin{vmatrix} a_1 & b_1 & c_1 \\ a_2 & b_2 & c_2 \\ a_1 & b_1 & c_1 \end{vmatrix} = 0 \tag{A2.16}$$

5. If two columns or rows are proportional, the value of the determinant is zero.

$$\begin{vmatrix} 2 & 4 & 5 \\ 3 & 6 & -2 \\ 1 & 2 & 7 \end{vmatrix} = 0 \tag{A2.17}$$

6. If two columns or rows are interchanged, the sign of the determinant is changed.

$$\begin{vmatrix} a_1 & b_1 & c_1 \\ a_2 & b_2 & c_2 \\ a_3 & b_3 & c_3 \end{vmatrix} = -\begin{vmatrix} c_1 & b_1 & a_1 \\ c_2 & b_2 & a_2 \\ c_3 & b_3 & a_3 \end{vmatrix} \tag{A2.18}$$

7. If each element of a column or row is multiplied by the same number, the resulting determinant is multiplied by that same number.

$$\begin{vmatrix} a_1 & b_1 & kc_1 \\ a_2 & b_2 & kc_2 \\ a_3 & b_3 & kc_3 \end{vmatrix} = k\begin{vmatrix} a_1 & b_1 & c_1 \\ a_2 & b_2 & c_2 \\ a_3 & b_3 & c_3 \end{vmatrix} \tag{A2.19}$$

A2.4 Problems

A2.1 Prove property 4, Eq. (A2.16), for determinants by use of the Levi-Civita symbol, ϵ_{ijk}.

3

Functions of a Complex Variable

3.1 INTRODUCTION

Concepts from complex variable theory are used in analyzing many kinds of physical problems. For example, they are used in (1) potential theory, (2) response theory, (3) power series solutions of certain types of differential equations (Frobenius-Fuchs method), (4) evaluating certain definite integrals, (5) quantum theory, and (6) investigations of stability conditions (dispersion relations).

This elementary presentation of complex variable analysis assumes no prior exposure to the subject. It is assumed that students who plan to do advanced work in physics or mathematics will pursue further study in complex variable analysis from a more rigorous mathematical point of view. We have therefore elected to omit the following important topics: (1) evaluation of indefinite integrals, (2) the method of steepest descents, (3) multivalued functions and Riemann surfaces, (4) the Schwarz-Christoffel transformation, and (5) functions of several complex variables. These are special-purpose topics and, with the exception of (5), can be found in most treatments of complex variable analysis.

3.2 COMPLEX VARIABLES AND REPRESENTATIONS

One's first exposure to complex numbers probably came in connection with a study of the general solution,

$$\frac{-B \pm (B^2 - 4AC)^{1/2}}{2A}$$

of the quadratic equation,

$$AX^2 + BX + C = 0.$$

When $4AC > B^2$, we write

$$\frac{-B \pm i(4AC - B^2)^{1/2}}{2A} \qquad (i = \sqrt{-1}).$$

The latter development is made necessary by the requirement that every quadratic equation must have two roots (solutions).

A **complex variable,** $Z = x + iy$, is an ordered pair of real variables x and y which satisfies certain laws of operation. In general, $x + iy \neq y + ix$; hence the term "ordered pair" is important. The **real** and **imaginary** parts of Z are $\operatorname{Re} Z = x$ and $\operatorname{Im} Z = y$, respectively. Note also that $Z = 0$ implies that $x = y = 0$, and that $Z_1 = Z_2$ means that $x_1 = x_2$ and $y_1 = y_2$.

3.2.1 Algebraic Operations

The algebraic operations for two complex variables are as follows.

1. Addition (subtraction):

$$\begin{aligned} Z_1 + Z_2 &= (x_1 + iy_1) + (x_2 + iy_2) \\ &= (x_1 + x_2) + i(y_1 + y_2). \end{aligned} \tag{3.1}$$

2. Multiplication:

$$\begin{aligned} (Z_1 Z_2) &= (x_1 + iy_1)(x_2 + iy_2) \\ &= (x_1 x_2 - y_1 y_2) + i(x_1 y_2 + x_2 y_1). \end{aligned} \tag{3.2}$$

3. Division ($Z_2 \neq 0$):

$$\begin{aligned} \frac{Z_1}{Z_2} &= \frac{x_1 + iy_1}{x_2 + iy_2} \\ &= \frac{(x_1 + iy_1)}{(x_2 + iy_2)} \cdot \frac{(x_2 - iy_2)}{(x_2 - iy_2)} \\ &= \frac{x_1 x_2 + y_1 y_2}{x_2^2 + y_2^2} + i\left\{\frac{x_2 y_1 - x_1 y_2}{x_2^2 + y_2^2}\right\}. \end{aligned} \tag{3.3}$$

We can also show that the following laws hold.

1. Associative law of addition:

$$Z_1 + (Z_2 + Z_3) = (Z_1 + Z_2) + Z_3. \tag{3.4}$$

2. Commutative law of multiplication:

$$Z_1Z_2 = Z_2Z_1. \tag{3.5}$$

3. Associative law of multiplication:

$$Z_1(Z_2Z_3) = (Z_1Z_2)Z_3. \tag{3.6}$$

4. Distributive law:

$$Z_1(Z_2 + Z_3) = Z_1Z_2 + Z_1Z_3. \tag{3.7}$$

In fact, complex variables are subject to the same algebraic laws of addition, subtraction, multiplication, and division as are real variables.

3.2.2 Argand Diagram: Vector Representation

A convenient graphical representation for complex variables was devised by Argand (1768–1822); it is illustrated in Fig. 3.1.

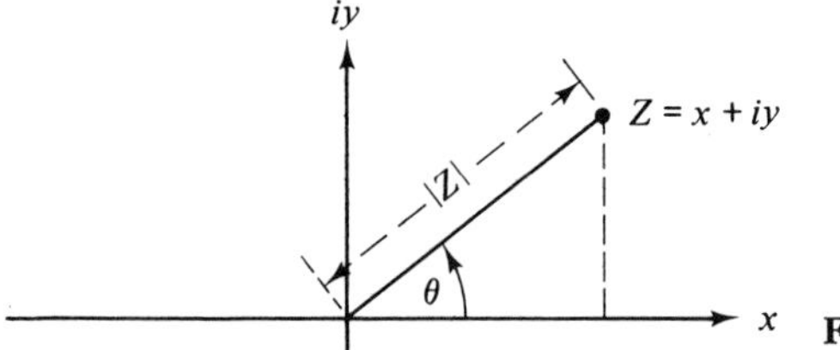

Figure 3.1

A student of physics should recognize Fig. 3.1 **(Argand diagram)** as simply a vector representation for Z where the y-axis is an imaginary axis. That is to say, Z is just the vector sum of x and iy.

In plane polar coordinates $(x, y) \to (r, \theta)$, we have

$$x = r\cos\theta \qquad \text{and} \qquad y = r\sin\theta.$$

In terms of r and θ, Z becomes

$$Z = x + iy = r(\cos\theta + i\sin\theta). \tag{3.8}$$

Also note that

$$r = \sqrt{x^2 + y^2} \equiv |Z| \tag{3.9}$$

and

$$|Z|^2 = (\text{Re } Z)^2 + (\text{Im } Z)^2 \tag{3.10}$$

where r, $|Z|$, is called the **modulus (absolute value)** of Z and θ denotes the **argument (phase)** of Z. The respective notations are $r = |Z| = \text{Mod } Z$ and $\theta = \arg Z$. We note that the argument of Z is not unique (determined up to a multiple of 2π). Hence we may write

$$\theta = \theta_p + 2\pi k \qquad (k = 0, \pm 1, \pm 2, \ldots) \tag{3.11}$$

where

$$0 \leq \theta_p < 2\pi.$$

The **principal argument** of Z is denoted by θ_p.

3.2.3 Complex Conjugate

The **complex conjugate** of Z, Z^*, is $Z^* = x - iy$, and its graphical representation is given in Fig. 3.2. To find the complex conjugate of a complex quantity, change the sign of the imaginary part (or imaginary terms).

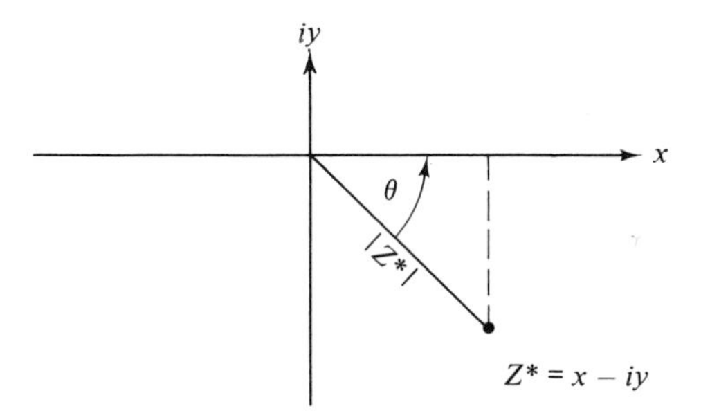

Figure 3.2

In connection with the complex conjugate, we have

$$ZZ^* = (x + iy)(x - iy) = |Z|^2 \tag{3.12}$$

$$(Z_1 + Z_2)^* = Z_1^* + Z_2^* \tag{3.13}$$

$$(Z_1 Z_2)^* = Z_1^* Z_2^* \tag{3.14}$$

and

$$\left(\frac{Z_1}{Z_2}\right)^* = \frac{Z_1^*}{Z_2^*}. \tag{3.15}$$

Comparing Eq. (3.12) with Eq. (3.10), we observe that ZZ^* is always real.

EXAMPLE 3.1 Given $Z = (1 + i)/(2 - 3i)$. (a) Put Z in the standard form, $x + iy$. Find (b) $\text{Re } Z$, (c) $\text{Im } Z$, (d) $\text{Mod } Z$, and (e) $\arg Z$. (f) Give the graphical representation of Z.

Solution On multiplying the numerator and denominator of Z by the complex conjugate of the denominator of Z, we obtain

(a)
$$\begin{aligned} Z &= \frac{1+i}{2-3i}\cdot\frac{2+3i}{2+3i} \\ &= \frac{2+3i+2i-3}{4+6i-6i+9} \\ &= -\frac{1}{13}+i\frac{5}{13} \qquad \text{(standard form)} \end{aligned}$$

(b)
$$\operatorname{Re} Z = -\frac{1}{13}$$

(c)
$$\operatorname{Im} Z = \frac{5}{13}.$$

(d)
$$\begin{aligned} \operatorname{Mod} Z &= \sqrt{(\operatorname{Re} Z)^2 + (\operatorname{Im} Z)^2} = r \\ &= \sqrt{\left(-\frac{1}{13}\right)^2 + \left(\frac{5}{13}\right)^2} \\ &= \frac{1}{13}\sqrt{26} \\ &= 0.392 \end{aligned}$$

$$\left.\begin{aligned} x &= r\cos\theta = -\frac{1}{13} \\ y &= r\sin\theta = \frac{5}{13} \end{aligned}\right\} \quad \text{See Fig. 3.3.}$$

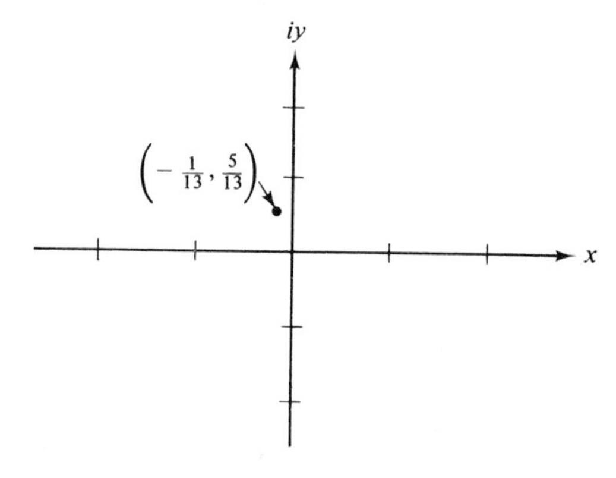

Figure 3.3

(e)
$$\begin{aligned} \arg Z = \theta &= \tan^{-1}\frac{y}{x} \\ &= \tan^{-1}(-5) \end{aligned}$$

EXAMPLE 3.2 By use of an algebraic analysis, show that $|Z_1 + Z_2| \leq |Z_1| + |Z_2|$.

Solution The square of the left-hand side of this inequality is

$$|Z_1 + Z_2|^2 = (Z_1 + Z_2)(Z_1^* + Z_2^*)$$
$$= Z_1Z_1^* + Z_1Z_2^* + Z_1^*Z_2 + Z_2Z_2^*.$$

It can be shown (see problem 3.5) that

$$Z_1Z_2^* + Z_1^*Z_2 \leq 2|Z_1Z_2^*| = 2|Z_1||Z_2|.$$

Hence the expression $|Z_1 + Z_2|^2$ reduces to

$$|Z_1 + Z_2|^2 \leq |Z_1|^2 + 2|Z_1||Z_2| + |Z_2|^2$$
$$= (|Z_1| + |Z_2|)^2.$$

We therefore obtain (see Fig. 3.4)

$$|Z_1 + Z_2| \leq |Z_1| + |Z_2|.$$

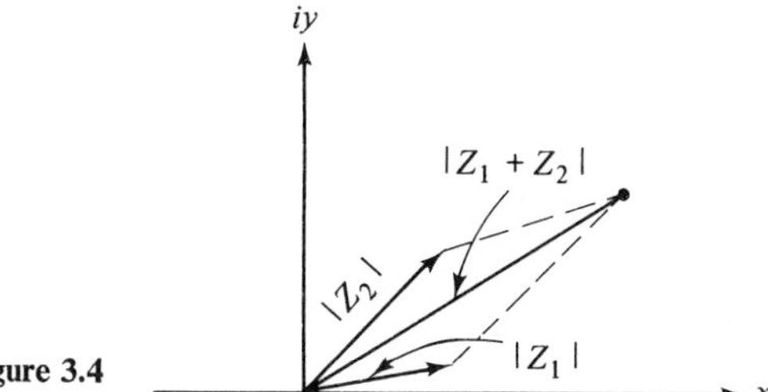

Figure 3.4

3.2.4 Euler's (1707–1783) Formula

We now develop an extremely useful relation, Euler's formula. For real θ, we may write

$$\sin\theta = \theta - \frac{\theta^3}{3!} + \frac{\theta^5}{5!} - \frac{\theta^7}{7!} + \cdots$$

and

$$\cos\theta = 1 - \frac{\theta^2}{2!} + \frac{\theta^4}{4!} - \frac{\theta^6}{6!} + \cdots.$$

For complex variables, we assume that the power-series expansion is similar to that for real variables,

$$e^x = \sum_{n=0}^{\infty} \frac{x^n}{n!}.$$

Hence we write

$$e^Z = \sum_{n=0}^{\infty} \frac{Z^n}{n!}. \tag{3.16}$$

When $Z = i\theta$ in the above expansion, it becomes

$$\begin{aligned} e^{i\theta} &= \sum_{n=0}^{\infty} \frac{(i\theta)^n}{n!} \\ &= 1 + i\theta - \frac{\theta^2}{2!} - \frac{i\theta^3}{3!} + \frac{\theta^4}{4!} + \frac{i\theta^5}{5!} + \cdots \\ &= \left(1 - \frac{\theta^2}{2!} + \frac{\theta^4}{4!} - \cdots\right) + i\left(\theta - \frac{\theta^3}{3!} + \frac{\theta^5}{5!} - \cdots\right) \\ &= \cos\theta + i\sin\theta. \end{aligned} \tag{3.17}$$

The rearrangement of the terms in the series is valid since the series is uniformly convergent (this property of series is discussed in the appendix at the end of this chapter). Equation (3.17) is the well-known **Euler formula.** On substituting Eq. (3.17) into Eq. (3.8), we obtain the polar coordinate representation of Z in the form

$$Z = r(\cos\theta + i\sin\theta) = re^{i\theta}. \tag{3.18}$$

Equation (3.18) is an extremely useful relation.

3.2.5 De Moivre's (1667–1754) Theorem

The product of two complex variables, Z_1 and Z_2, may be written as

$$\begin{aligned} Z_1 Z_2 &= r_1(\cos\theta_1 + i\sin\theta_1) r_2(\cos\theta_2 + i\sin\theta_2) \\ &= r_1 r_2[\cos(\theta_1 + \theta_2) + i\sin(\theta_1 + \theta_2)] \\ &= r_1 r_2 e^{i(\theta_1+\theta_2)}. \end{aligned}$$

For the case of n terms, the product is

$$Z_1 Z_2 \cdots Z_n = r_1 r_2 \cdots r_n[\cos(\theta_1 + \theta_2 + \cdots + \theta_n) + i\sin(\theta_1 + \theta_2 + \cdots + \theta_n)].$$

When $Z_1 = Z_2 = \cdots = Z_n$, the product becomes

$$Z^n = r^n(\cos n\theta + i\sin n\theta) = r^n e^{in\theta}. \tag{3.19}$$

Equation (3.19) is known as **De Moivre's theorem.** We can easily show that it is true for negative and fractional n as well as positive n.

It is clear from Eqs. (3.9) and (3.18) that

$$|Z_1Z_2| = |Z_1||Z_2| \tag{3.20}$$

$$\arg(Z_1Z_2) = \arg Z_1 + \arg Z_2 \tag{3.21}$$

and

$$\arg\left(\frac{Z_1}{Z_2}\right) = \arg Z_1 - \arg Z_2. \tag{3.22}$$

3.2.6 The *n*th Root or Power of a Complex Number

De Moivre's theorem may be used to develop a general relation for extracting the nth root of a complex number or the raising of a complex quantity to the nth power. The nth root of Z may be written as

$$\begin{aligned} Z^{1/n} &= r^{1/n}\left(\cos\frac{\theta}{n} + i\sin\frac{\theta}{n}\right) \\ &= r^{1/n}\left[\cos\left(\frac{\theta_p + 2\pi k}{n}\right) + i\sin\left(\frac{\theta_p + 2\pi k}{n}\right)\right] \end{aligned} \tag{3.23}$$

where $0 \leq \theta_p < 2\pi$ and $k = 0, 1, 2, \ldots, (n-1)$ since we require only n distinct roots of Z.

EXAMPLE 3.3 Find the square root of i.

Solution The specific equation for this problem is

$$(i)^{1/2} = r^{1/2}\left[\cos\left(\frac{\theta_p + 2\pi k}{2}\right) + i\sin\left(\frac{\theta_p + 2\pi k}{2}\right)\right].$$

The principal argument is $\theta_p = \pi/2$ since $Z = i$, $x = 0$, $y = 1$, and $r = 1$.

For $k = 0$, we obtain

$$\begin{aligned} (i)^{1/2}_{k=0} &= \cos\frac{\pi}{4} + i\sin\frac{\pi}{4} \\ &= \frac{1}{\sqrt{2}} + \frac{i}{\sqrt{2}} \\ &= \frac{1}{\sqrt{2}}(1 + i). \end{aligned}$$

For $k = 1$, we obtain

$$\begin{aligned} (i)^{1/2}_{k=1} &= \cos\left(\frac{\pi}{4} + \pi\right) + i\sin\left(\frac{\pi}{4} + \pi\right) \\ &= \cos\frac{\pi}{4}\cos\pi + i\sin\frac{\pi}{4}\cos\pi \\ &= -\frac{1}{\sqrt{2}}(1 + i). \end{aligned}$$

The result is

$$(i)^{1/2} = \pm\frac{1}{\sqrt{2}}(1 + i).$$

The graphical representation of these roots is given in Fig. 3.5.

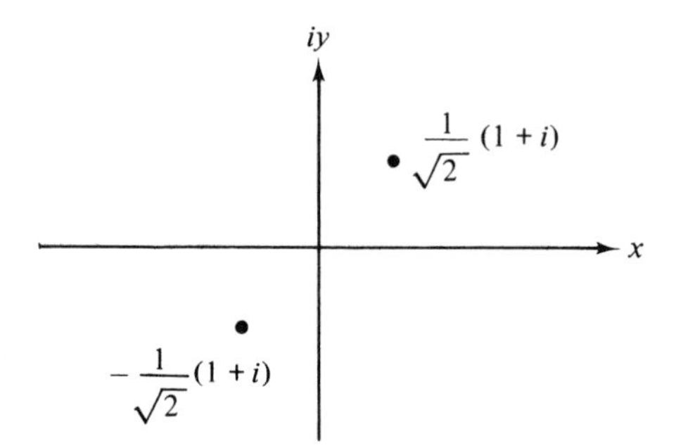

Figure 3.5

3.3 ANALYTIC FUNCTIONS OF A COMPLEX VARIABLE

3.3.1 The Derivative of $f(Z)$ and Analyticity

A single-valued function (only single-valued functions will be considered in this chapter) $f(Z)$ is said to be **analytic (regular** or **holomorphic)** in a certain region of the complex plane if it has a unique derivative at every point in the region. The derivative of $f(Z)$, like the function of a real variable, is defined by

$$\begin{aligned} f'(Z) = \frac{df(Z)}{dZ} &= \lim_{\Delta Z \to 0}\left(\frac{\Delta f}{\Delta Z}\right) \\ &= \lim_{\Delta Z \to 0}\left\{\frac{f(Z + \Delta Z) - f(Z)}{\Delta Z}\right\} \\ &= \lim_{\Delta Z \to 0}\left\{\frac{\Delta u + i\,\Delta v}{\Delta x + i\,\Delta y}\right\} \end{aligned} \tag{3.24}$$

where

$$f(Z) = u(x, y) + iv(x, y)$$
$$\Delta f = \Delta u + i\,\Delta v$$

and

$$Z = x + iy$$
$$\Delta Z = \Delta x + i\,\Delta y.$$

There is an infinite number of ways of obtaining $\Delta Z \to 0$ in the Z-plane. For convenience, we select the simple scheme illustrated in Fig. 3.6.

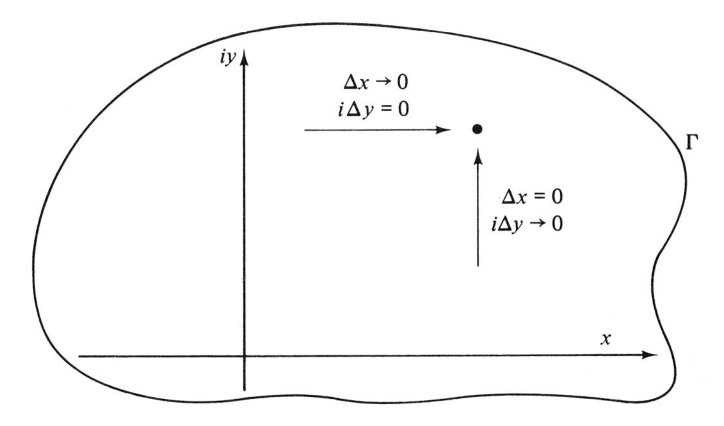

Figure 3.6

Case 1 $\Delta x = 0$ and $i\,\Delta y \rightarrow 0$.

$$f_{\mathrm{I}}'(Z) = \lim_{i\Delta y \to 0} \left\{\frac{\Delta u + i\,\Delta v}{i\,\Delta y}\right\}$$

$$= -i\frac{\partial u}{\partial y} + \frac{\partial v}{\partial y} \tag{3.25}$$

The partial derivative is indicated in Eq. (3.25) since u and v are functions of both x and y.

Case 2 $i\,\Delta y = 0$ and $\Delta x \rightarrow 0$.

$$f_{\mathrm{II}}'(Z) = \lim_{\Delta x \to 0} \left\{\frac{\Delta u + i\,\Delta v}{\Delta x}\right\}$$

$$= \frac{\partial u}{\partial x} + i\frac{\partial v}{\partial x} \tag{3.26}$$

If $f(Z)$ is analytic in the region $\mathbf{\Gamma}$ (see Fig. 3.6), we require that $f_{\mathrm{I}}'(Z) = f_{\mathrm{II}}'(Z)$. As a result of this requirement, we obtain the following relations:

$$\frac{\partial u}{\partial x} = \frac{\partial v}{\partial y} \tag{3.27}$$

and

$$\frac{\partial u}{\partial y} = -\frac{\partial v}{\partial x}. \tag{3.28}$$

The relations in Eqs. (3.27) and (3.28) are the well-known **Cauchy** (1789–1857)-**Riemann** (1826–1866) **conditions**; they constitute the necessary conditions for a unique derivative of $f(Z)$ in $\mathbf{\Gamma}$.

3.3.2 Harmonic Functions

On differentiating Eqs. (3.27) and (3.28) with respect to x and y, respectively, we obtain

$$\frac{\partial^2 u}{\partial x^2} = \frac{\partial^2 v}{\partial x\, \partial y} \quad \text{and} \quad \frac{\partial^2 u}{\partial y^2} = -\frac{\partial^2 v}{\partial y\, \partial x}. \tag{3.29}$$

If u and v possess continuous partial derivatives up to second order, Eq. (3.29) leads to

$$\frac{\partial^2 u}{\partial x^2} + \frac{\partial^2 u}{\partial y^2} = 0. \tag{3.30}$$

Similarly, it can be shown that

$$\frac{\partial^2 v}{\partial x^2} + \frac{\partial^2 v}{\partial y^2} = 0. \tag{3.31}$$

Therefore u and v are solutions of Laplace's equation, Eqs. (3.30) and (3.31), in two dimensions; they are called **harmonic** (or **conjugate**) **functions.**

EXAMPLE 3.4 (a) Show that $v(x, y) = 3x^2y - y^3$ is harmonic. (b) Find the conjugate function, $u(x, y)$. (c) Find the analytic function $f(Z) = u(x, y) + iv(x, y)$.

Solution

(a) If

$$\frac{\partial^2 v}{\partial x^2} + \frac{\partial^2 v}{\partial y^2} = 0$$

then v is said to be harmonic.

$$\frac{\partial v}{\partial x} = 6xy \qquad \frac{\partial^2 v}{\partial x^2} = 6y$$

$$\frac{\partial v}{\partial y} = 3x^2 - 3y^2 \qquad \frac{\partial^2 v}{\partial y^2} = -6y$$

By adding the right-hand sides of $\frac{\partial^2 v}{\partial x^2}$ and $\frac{\partial^2 v}{\partial y^2}$, we see that v is harmonic.

(b)

$$\frac{\partial v}{\partial y} = 3x^2 - 3y^2 = \frac{\partial u}{\partial x}$$

and

$$\frac{\partial v}{\partial x} = 6xy = -\frac{\partial u}{\partial y}.$$

Integrating the two above equations with respect to x and y, respectively, we obtain

$$u(x, y) = x^3 - 3y^2x + f(y)$$

and

$$u(x, y) = -3xy^2 + g(x)$$

or

$$x^3 - 3y^2x + f(y) = -3xy^2 + g(x).$$

In the above equation, we must have $f(y) = 0$ and $g(x) = x^3$. The required conjugate function is therefore given by

$$u(x, y) = x^3 - 3y^2x.$$

(c) The corresponding analytic function is given by

$$f(Z) = x^3 - 3y^2x + i(3x^2y - y^3) = Z^3.$$

3.3.3 Contour Integrals

The integral (in the Riemannian sense) of the function $f(Z)$ is defined, as in the case of real variables, by

$$\int_C f(Z)\,dZ = \int_{Z_0}^{Z'} f(Z)\,dZ$$

$$\equiv \lim_{\substack{n\to\infty \\ |Z_j - Z_{j-1}|\to 0}} \sum_{j=1}^{n} f(\xi_j)(Z_j - Z_{j-1}). \tag{3.32}$$

where the path C has been divided into n segments and ξ_j is some point on the curve between Z_{j-1} and Z_j (see Fig. 3.7).

In complex variable theory, $\int_C f(Z)\,dZ$ is called a **contour integral** of $f(Z)$ along the contour C from Z_0 to Z'.

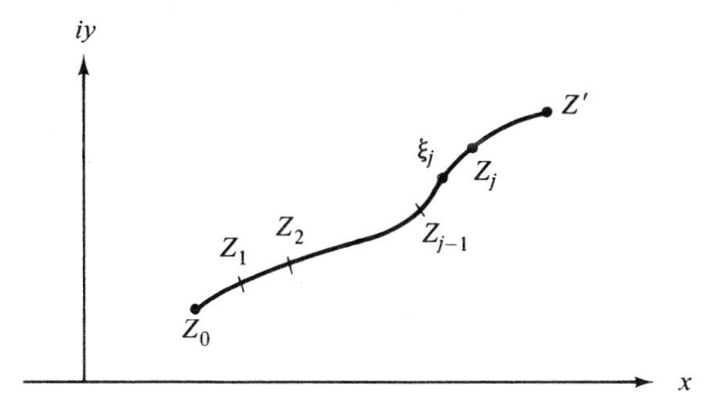

Figure 3.7

3.3.4 Cauchy's Integral Theorem

THEOREM If $f(Z)$ is analytic and its partial derivatives are continuous throughout some simply connected region, then for every closed path C within this region

$$\oint_C f(Z)\, dZ = 0. \tag{3.33}$$

The symbol $\oint$ means the integral is around a closed path, and the symbol $\ointctrclockwise$ indicates that the path is traversed in a positive (counterclockwise) manner. Our convention will be to traverse the path in such a way that the region of interest lies to the left. Equation (3.33) is called **Cauchy's integral theorem.**

The following are three equivalent definitions of a **simply connected** region: A region in the complex plane is called simply connected if (1) all closed paths within the region contain only points that belong to the region, (2) all closed paths within the region can be shrunk to a point, and (3) it has the property that every scissor cut starting at an arbitrary point, Z_1, on the boundary and finishing at another point, Z_2, on the boundary separates the region into unconnected (two) pieces. All other regions are said to be **multiply connected.** This classification is illustrated in Fig. 3.8. A closed curve drawn within C, Fig. 3.8(a), contains only points that belong to C. However, all points within C_3, Fig. 3.8(b), do not belong to the region between C_1 and C_2. Therefore the region within C is simply connected, and that between C_1 and C_2 is multiply connected.

The curl theorem will be used in developing a proof of Cauchy's integral theorem. On expanding the left-hand side of Eq. (3.33), we get

$$\begin{aligned}\oint_C f(Z)\, dZ &= \oint_C (u + iv)(dx + i\, dy) \\ &= \oint_C (u\, dx - v\, dy) + i \oint_C (v\, dx + u\, dy).\end{aligned} \tag{3.34}$$

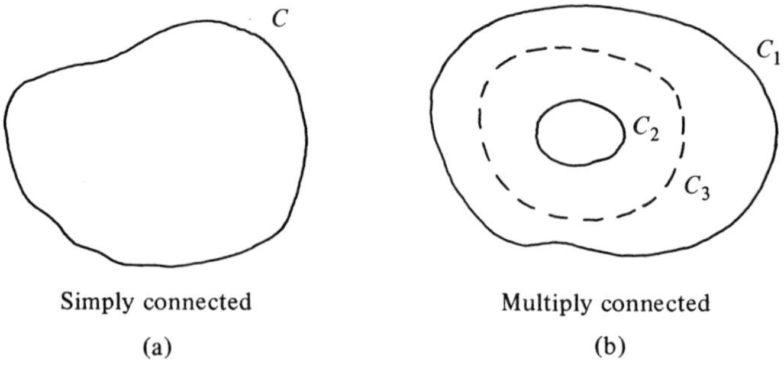

C — Simply connected (a)

C_1, C_2, C_3 — Multiply connected (b)

Figure 3.8

The curl theorem due to Stokes (Section 1.4.4)

$$\iint_\sigma \nabla \times \mathbf{F} \cdot d\boldsymbol{\sigma} = \oint_\lambda \mathbf{F} \cdot d\boldsymbol{\lambda}$$

in two dimensions becomes

$$\iint_\sigma \left(\frac{\partial F_y}{\partial x} - \frac{\partial F_x}{\partial y} \right) dx\, dy = \oint_\lambda (F_x\, dx + F_y\, dy) \tag{3.35}$$

where F_x and F_y are any two functions with continuous partial derivatives. Substituting Eq. (3.35) into the first and second terms on the right-hand side of Eq. (3.34) separately and applying the Cauchy-Riemann conditions, we obtain

$$\oint_C f(Z)\, dZ = -\iint_\sigma \left(\frac{\partial v}{\partial x} + \frac{\partial u}{\partial y} \right) dx\, dy + i \iint_\sigma \left(\frac{\partial u}{\partial x} - \frac{\partial v}{\partial y} \right) dx\, dy = 0. \tag{3.36}$$

It is clear that Cauchy's integral theorem tells us the integral of $f(Z)$ around a closed path in a simply connected region is independent of the path.

A less restrictive theorem due to Goursat (1858–1936) which eliminates the continuous partial derivative requirement can also be proved. The inverse of Cauchy's theorem is known as **Morera's** (1856–1909) **theorem.**

3.3.5 Cauchy's Integral Formula

Another important and extremely useful relation concerning the integral of a function of a complex variable is **Cauchy's integral formula.** It can be written as

$$\oint_C \frac{f(Z)\, dZ}{Z - Z_0} = 2\pi i f(Z_0) \tag{3.37}$$

where Z_0 is within C. The function $f(Z)$ is assumed to be analytic within C; however, $f(Z)/(Z - Z_0)$ is clearly not analytic at $Z = Z_0$. For $r \to 0$, Fig. 3.9(a) is equivalent to Fig. 3.9(b).

From Fig. 3.9(b), it can be seen that $f(Z)/(Z - Z_0)$ is analytic in the region between C and C'. Hence we may apply Cauchy's integral theorem and obtain

$$\oint_C \frac{f(Z)\, dZ}{Z - Z_0} + \oint_{C'} \frac{f(Z)\, dZ}{Z - Z_0} = 0$$

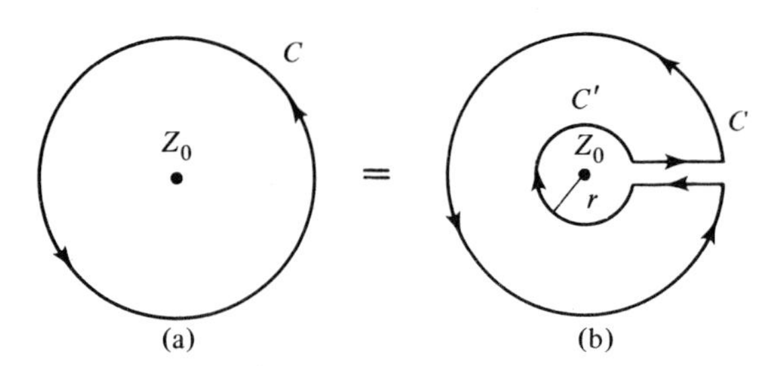

Figure 3.9

or

$$\oint_C \frac{f(Z)\,dZ}{Z - Z_0} = \oint_{C'} \frac{f(Z)\,dZ}{Z - Z_0}. \tag{3.38}$$

Around the path C', we set

$$Z - Z_0 = re^{i\theta}; \qquad dZ = ire^{i\theta}\,d\theta. \tag{3.39}$$

Substituting Eq. (3.39) into Eq. (3.38), we obtain

$$\begin{aligned} \oint_C \frac{f(Z)\,dZ}{Z - Z_0} &= \int_0^{2\pi} \frac{f(Z_0 + re^{i\theta})ire^{i\theta}\,d\theta}{re^{i\theta}} \\ &= i\int_0^{2\pi} f(Z_0 + re^{i\theta})\,d\theta \\ &= 2\pi i f(Z_0) \end{aligned}$$

for $r \to 0$. Hence the formula in Eq. (3.37) is established. In summary, we may write

$$\frac{1}{2\pi i}\oint_C \frac{f(Z)\,dZ}{Z - Z_0} = \begin{cases} f(Z_0) & (\text{for } Z_0 \text{ inside } C) \\ 0 & (\text{for } Z_0 \text{ outside } C). \end{cases} \tag{3.40}$$

3.3.6 Differentiation Inside the Sign of Integration

As indicated in Section 3.3.1, the derivative of $f(Z)$ with respect to Z is defined by

$$f'(Z) = \lim_{\Delta Z \to 0} \left\{\frac{f(Z + \Delta Z) - f(Z)}{\Delta Z}\right\}. \tag{3.41}$$

On applying Cauchy's integral formula to Eq. (3.41), we obtain

$$f'(Z) = \lim_{\Delta Z \to 0} \left[\frac{1}{\Delta Z} \{ f(Z + \Delta Z) - f(Z) \} \right]$$

$$= \lim_{\Delta Z \to 0} \left[\frac{1}{2\pi i \, \Delta Z} \left\{ \oint_C \frac{f(\xi)\, d\xi}{\xi - Z - \Delta Z} - \oint_C \frac{f(\xi)\, d\xi}{\xi - Z} \right\} \right]$$

$$= \frac{1}{2\pi i} \oint_C \frac{f(\xi)\, d\xi}{(\xi - Z)^2}.$$

For the nth derivative of $f(Z)$ with respect to Z, we have

$$f^{(n)}(Z_0) \equiv f^{(n)}(Z)|_{Z=Z_0} = \frac{n!}{2\pi i} \oint_C \frac{f(\xi)\, d\xi}{(\xi - Z_0)^{n+1}} \tag{3.42}$$

when $f(\xi)$ is analytic within C.

3.4 SERIES EXPANSIONS

The Taylor and Laurent expansions will be developed in this section.

3.4.1 Taylor's Series

If $f(Z)$ is analytic in some region Γ and C is a circle within Γ with center at Z_0, then $f(Z)$ can be expanded in a **Taylor's** (1685–1731) **series.** Here Z is any point interior to C (see Fig. 3.10), that is,

$$f(Z) = f(Z_0) + f'(Z_0)(Z - Z_0) + \cdots + \frac{f^{(n)}(Z_0)(Z - Z_0)^n}{n!}$$

$$= \sum_{n=0}^{\infty} \frac{(Z - Z_0)^n f^{(n)}(Z_0)}{n!}. \tag{3.43}$$

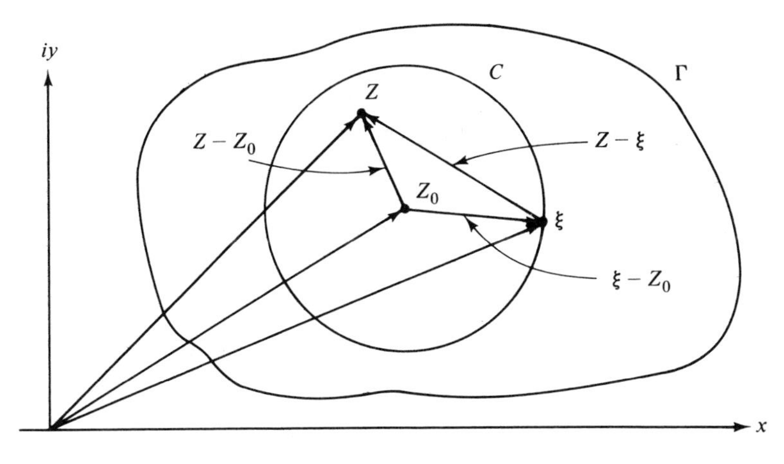

Figure 3.10

The series in Eq. (3.43) converges absolutely and uniformly for $|Z - Z_0| < R$, where R is the radius of convergence.

Proof From the Cauchy integral formula, we note that (see Fig. 3.10)

$$f(Z) = \frac{1}{2\pi i} \oint_C \frac{f(\xi)\, d\xi}{\xi - Z} \tag{3.44}$$

where $|Z - Z_0| < R$ and $R = |\xi - Z_0|$. To prove that the expansion holds, a series expansion for $1/(\xi - Z)$ will be developed. For convenience, we set

$$K_T = \frac{Z - Z_0}{\xi - Z_0}. \tag{3.45}$$

On subtracting unity from both sides of Eq. (3.45), we obtain

$$\begin{aligned} 1 - K_T &= 1 - \frac{(Z - Z_0)}{\xi - Z_0} \\ &= \frac{\xi - Z}{\xi - Z_0}. \end{aligned} \tag{3.46}$$

Inverting both sides of Eq. (3.46), we have

$$\begin{aligned} \frac{1}{1 - K_T} &= \frac{\xi - Z_0}{\xi - Z} \\ &= \sum_{n=0}^{\infty} K_T^n \qquad (|K_T| < 1). \end{aligned} \tag{3.47}$$

The required expansion for $1/(\xi - Z)$ is obtained by dividing Eq. (3.47) by $\xi - Z_0$; the result is

$$\begin{aligned} \frac{1}{\xi - Z} &= \frac{1}{\xi - Z_0} \sum_{n=0}^{\infty} \left(\frac{Z - Z_0}{\xi - Z_0}\right)^n \\ &= \sum_{n=0}^{\infty} \frac{(Z - Z_0)^n}{(\xi - Z_0)^{n+1}}. \end{aligned} \tag{3.48}$$

Substituting Eq. (3.48) into Eq. (3.44), we obtain

$$\begin{aligned} \frac{1}{2\pi i} \oint_C \frac{f(\xi)\, d\xi}{\xi - Z} &= \sum_{n=0}^{\infty} \left\{\frac{1}{2\pi i} \oint_C \frac{f(\xi)(Z - Z_0)^n\, d\xi}{(\xi - Z_0)^{n+1}}\right\} \\ &= \sum_{n=0}^{\infty} \left\{(Z - Z_0)^n \left(\frac{1}{2\pi i} \oint_C \frac{f(\xi)\, d\xi}{(\xi - Z_0)^{n+1}}\right)\right\} \\ &= \sum_{n=0}^{\infty} \frac{(Z - Z_0)^n f^{(n)}(Z_0)}{n!}. \end{aligned} \tag{3.49}$$

which is the required expansion in Eq. (3.43). When $Z_0 = 0$, the Taylor series becomes the **Maclaurin** (1698–1746) **series.**

EXAMPLE 3.5 Expand $1/(1 - Z)$ in a Taylor's series about $Z_0 = i$ and find the radius of convergence.

Solution We have

$$\begin{aligned} f(Z) &= (1 - Z)^{-1} \\ f'(Z) &= (1 - Z)^{-2} \\ f''(Z) &= 2(1 - Z)^{-3} \\ f'''(Z) &= 3 \cdot 2(1 - Z)^{-4} \end{aligned}$$

The general term is

$$f^{(n)}(Z) = n!(1 - Z)^{-(n+1)}$$

The required expansion is therefore

$$\begin{aligned} f(Z) &= \sum_{n=0}^{\infty} \frac{(Z - Z_0)^n f^{(n)}(Z_0)}{n!} \\ &= \sum_{n=0}^{\infty} \frac{(Z - i)^n n!(1 - i)^{-(n+1)}}{n!} \\ &= \sum_{n=0}^{\infty} \frac{(Z - i)^n}{(1 - i)^{n+1}} \end{aligned}$$

The **radius** of **convergence** of the above series is

$$\begin{aligned} R &= \lim_{n \to \infty} \left| \frac{a_n}{a_{n+1}} \right| \\ &= |1 - i| = \sqrt{2} \end{aligned}$$

by use of Eq. (3.12), where

$$a_n = \frac{1}{(1 - i)^{n+1}}.$$

3.4.2 Laurent's Expansion

If $f(Z)$ is analytic in the interior and on the boundary of the circular ring between C_1 and C_2 (see Fig. 3.11), it can be represented as a series expansion about Z_0 in the form

$$f(Z) = \sum_{n=0}^{\infty} a_n (Z - Z_0)^n + \sum_{n'=1}^{\infty} \frac{b_{n'}}{(Z - Z_0)^{n'}} \tag{3.50}$$

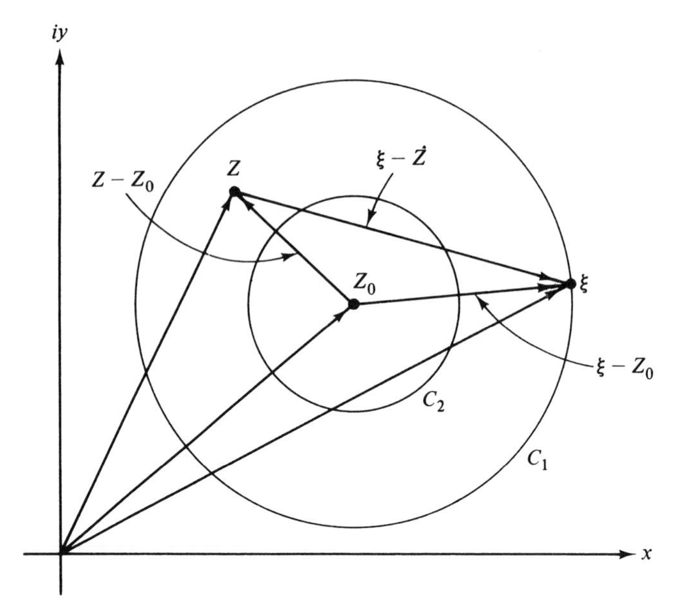

Figure 3.11

where

$$a_n = \frac{1}{2\pi i}\oint_{C_1} \frac{f(\xi)\,d\xi}{(\xi - Z_0)^{n+1}} \qquad (n = 0, 1. \ldots) \tag{3.51}$$

and

$$b_{n'} = \frac{1}{2\pi i}\oint_{C_2} \frac{f(\xi)\,d\xi}{(\xi - Z_0)^{-n'+1}} \qquad (n' = 1, 2, \ldots). \tag{3.52}$$

The expansion in Eq. (3.50) is called the **Laurent** (1841–1908) **expansion.** In Fig. 3.11, we make a scissor cut from C_1 to C_2 as shown in Fig. 3.12.

Proof On applying the Cauchy integral formula in Fig. 3.12, we find that

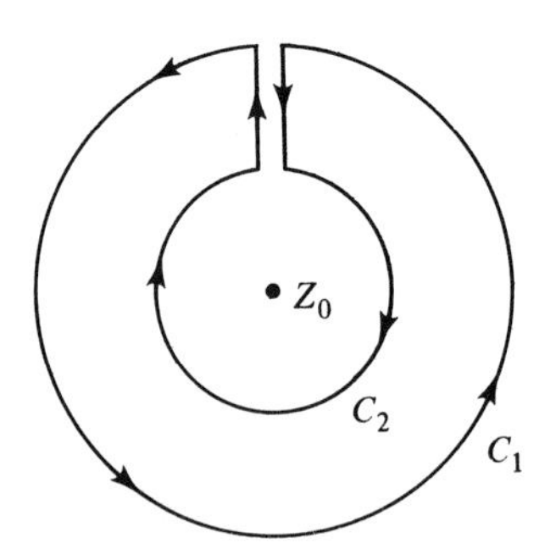

Figure 3.12

$$f(Z) = \frac{1}{2\pi i}\oint_{C_1}\frac{f(\xi)\,d\xi}{\xi - Z} - \frac{1}{2\pi i}\oint_{C_2}\frac{f(\xi)\,d\xi}{\xi - Z}$$
$$= f_A(Z) + f_P(Z) \tag{3.53}$$

where

$$f_A(Z) = \frac{1}{2\pi i}\oint_{C_1}\frac{f(\xi)\,d\xi}{\xi - Z}$$

and

$$f_P(Z) = \frac{1}{2\pi i}\oint_{C_2}\frac{f(\xi)\,d\xi}{Z - \xi}.$$

Consider $f_A(Z)$ where ξ is on C_1 and Z is inside the region between C_1 and C_2. Using the expansion for $1/(\xi - Z)$ developed in Section 3.4.1, we may write

$$f_A(Z) = \frac{1}{2\pi i}\oint_{C_1}\frac{f(\xi)\,d\xi}{\xi - Z} = \sum_{n=0}^{\infty} a_n(Z - Z_0)^n \tag{3.54}$$

where

$$|Z - Z_0| < |\xi - Z_0|$$

$$a_n = \frac{1}{2\pi i}\oint_{C_1}\frac{f(\xi)\,d\xi}{(\xi - Z_0)^{n+1}}. \tag{3.51$'$}$$

Note that a_n cannot be represented by $f^{(n)}(Z_0)/n!$ since $f_A(Z)$ is not analytic at $Z = Z_0$ as in the case of the Taylor expansion.

Now consider the integral over C_2, $f_P(Z)$, where it is assumed that ξ is now on C_2 and Z is inside the ring (see Fig. 3.13). We need the appropriate expansion for $1/(Z - \xi)$; it is obtained by the same method used in the Taylor series case. Here we set

$$K_L = \frac{\xi - Z_0}{Z - Z_0}. \tag{3.55}$$

Subtracting unity from both sides of Eq. (3.55) and inverting the resulting equation, we obtain

$$\frac{1}{1 - K_L} = \frac{Z - Z_0}{Z - \xi} = \sum_{n=0}^{\infty} K_L^n$$

or

$$\frac{1}{Z - \xi} = \sum_{n=0}^{\infty}\frac{(\xi - Z_0)^n}{(Z - Z_0)^{n+1}} \tag{3.56}$$

where

$$|K_L| < 1 \quad \text{or} \quad |\xi - Z_0| < |Z - Z_0|.$$

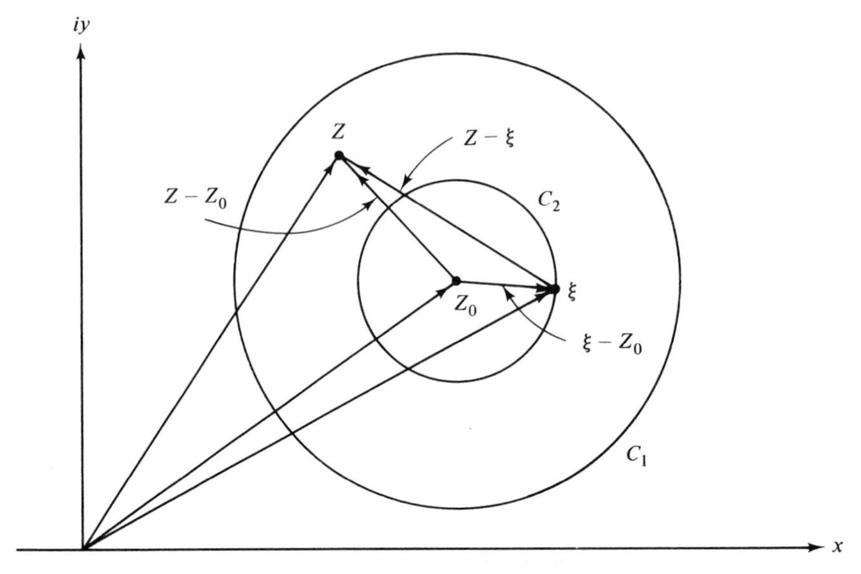

Figure 3.13

On substituting Eq. (3.56) into $f_P(Z)$, we obtain

$$f_P(Z) = \frac{1}{2\pi i}\oint_{C_2}\frac{f(\xi)\,d\xi}{Z-\xi}$$

$$= \sum_{n=0}^{\infty}\left[\frac{1}{(Z-Z_0)^{n+1}}\frac{1}{2\pi i}\oint_{C_2} f(\xi)(\xi - Z_0)^n\,d\xi\right]$$

or

$$f_P(Z) = \sum_{n'=1}^{\infty}\left[\frac{1}{(Z-Z_0)^{n'}}\frac{1}{2\pi i}\oint_{C_2}\frac{f(\xi)\,d\xi}{(\xi-Z_0)^{-n'+1}}\right]$$

$$= \sum_{n'=1}^{\infty}\frac{b_{n'}}{(Z-Z_0)^{n'}} \tag{3.57}$$

where

$$b_{n'} = \frac{1}{2\pi i}\oint_{C_2}\frac{f(\xi)\,d\xi}{(\xi-Z_0)^{-n'+1}}. \tag{3.52$'$}$$

Hence the Laurent expansion in Eq. (3.50) is established.

The Laurent expansion consists of two series. The first series in Eq. (3.58) is called the **analytic part** of the expansion, and it converges everywhere inside C_1. The second series is referred to as the **principal part**, and it converges everywhere outside C_2. If $f(Z)$ is analytic within C_2, the principal part equals zero, and the Laurent expansion reduces to the Taylor expansion since $a_n = f^{(n)}(Z_0)/n!$ in this case. It is important to note that a real variable analog for the full Laurent expansion does not exist.

The expansion in Eq. (3.50), Laurent's expansion, can also be written in the following useful forms:

$$f(Z) = \sum_{n=0}^{\infty} a_n(Z - Z_0)^n + \sum_{n=1}^{\infty} \frac{a_{-n}}{(Z - Z_0)^n} \tag{3.58}$$

where

$$\begin{aligned} a_n &= \frac{1}{2\pi i} \oint_{C_1} \frac{f(\xi)\, d\xi}{(\xi - Z_0)^{n+1}} \qquad (n = 0, 1, 2, \ldots) \\ a_{-n} &= \frac{1}{2\pi i} \oint_{C_2} \frac{f(\xi)\, d\xi}{(\xi - Z_0)^{-n+1}} \qquad (n = 1, 2, 3, \ldots) \end{aligned} \tag{3.59}$$

and

$$f(Z) = \sum_{n=-\infty}^{\infty} A_n(Z - Z_0)^n \tag{3.60}$$

where

$$A_n = \frac{1}{2\pi i} \oint_C \frac{f(\xi)\, d\xi}{(\xi - Z_0)^{n+1}} \qquad (n = 0, \pm 1, \pm 2, \ldots). \tag{3.61}$$

In Eq. (3.61), C is any circle between C_1 and C_2.

EXAMPLE 3.6 By direct evaluation of A_n, find the Laurent expansion for

$$f(Z) = \frac{1}{Z(Z - 1)}$$

about $Z_0 = 0$.

Solution Here we have

$$\begin{aligned} f(Z) &= \sum_{n=-\infty}^{\infty} A_n(Z - Z_0)^n \\ &= \sum_{n=-\infty}^{\infty} A_n Z^n \end{aligned}$$

where

$$\begin{aligned} A_n &= \frac{1}{2\pi i} \oint_C \frac{f(\xi)\, d\xi}{(\xi - Z_0)^{n+1}} \\ &= \frac{1}{2\pi i} \oint_C \frac{f(\xi)\, d\xi}{\xi^{n+1}} \\ &= \frac{1}{2\pi i} \oint_C \frac{d\xi}{\xi^{n+1}\, \xi(\xi - 1)} \\ &= -\frac{1}{2\pi i} \oint_C \frac{d\xi}{\xi^{n+2}(1 - \xi)} \\ &= -\frac{1}{2\pi i} \oint_C \frac{d\xi}{\xi^{n+2}} \sum_{k=0}^{\infty} \xi^k \qquad (|\xi| < 1) \end{aligned}$$

or

$$A_n = -\frac{1}{2\pi i}\sum_{k=0}^{\infty}\oint_C \frac{d\xi}{\xi^{n-k+2}}$$

$$= -\frac{1}{2\pi i}\sum_{k=0}^{\infty} 2\pi i\delta_{n-k+2,\,1}$$

$$= \begin{cases} -1 & (\text{for } n \geq -1) \\ 0 & (\text{for } n < -1). \end{cases}$$

The required expansion becomes

$$f(Z) = \sum_{n=-\infty}^{\infty} A_n Z^n$$

$$= \sum_{n=-1}^{\infty} A_n Z^n$$

$$= -\sum_{n=-1}^{\infty} Z^n.$$

The problem of developing the Laurent expansion for a function by evaluating the A_n coefficient (or a_n and a_{-n}) is, in general (except for simple functions), tedious. It is sometimes advantageous to use a procedure similar to that illustrated in Examples 3.7 and 3.8 to obtain the required Laurent expansion for a function.

EXAMPLE 3.7 Find the Laurent expansion for $f(Z) = 1/Z(Z-1)$ by use of a geometric series.

Solution

$$f(Z) = \frac{1}{Z(Z-1)}$$

$$= -\frac{1}{Z}\left(\frac{1}{1-Z}\right)$$

or

$$f(Z) = -\frac{1}{Z}\sum_{n=0}^{\infty} Z^n \qquad (|Z| < 1)$$

$$= -\frac{1}{Z}(1 + Z + Z^2 + \cdots)$$

$$= -\frac{1}{Z} - 1 - Z - Z^2 - \cdots$$

$$= -\sum_{n=-1}^{\infty} Z^n.$$

Note that this result is the same as that obtained in Example 3.6.

EXAMPLE 3.8 Find the Laurent expansion for $f(Z) = 1/Z(Z - 2)$ in the regions (a) $0 < |Z| < 2$ and (b) $2 < |Z| < \infty$.

Solution

(a)
$$f(Z) = \frac{1}{Z(Z-2)}$$
$$= -\frac{1}{2}\frac{1}{Z}\left(\frac{1}{1 - Z/2}\right)$$
$$= -\frac{1}{2}\frac{1}{Z}\sum_{n=0}^{\infty}\left(\frac{Z}{2}\right)^n \qquad \left(\left|\frac{Z}{2}\right| < 1;\ |Z| < 2\right)$$

or

$$f(Z) = -\frac{1}{2Z}\left(1 + \frac{Z}{2} + \frac{Z^2}{2^2} + \frac{Z^3}{2^3} + \cdots\right)$$
$$= -\frac{1}{2Z} - \frac{1}{2^2} - \frac{Z}{2^3} - \frac{Z^2}{2^4} - \cdots.$$

(b)
$$f(Z) = \frac{1}{Z(Z-2)}$$
$$= \frac{1}{Z^2(1 - 2/Z)}$$
$$= \frac{1}{Z^2}\sum_{n=0}^{\infty}\left(\frac{2}{Z}\right)^2 \qquad \left(\left|\frac{2}{Z}\right| < 1;\ |Z| > 2\right)$$
$$= \frac{1}{Z^2}\left(1 + \frac{2}{Z} + \frac{2^2}{Z^2} + \cdots\right)$$
$$= \frac{1}{Z^2} + \frac{2}{Z^3} + \frac{2^2}{Z^4} + \cdots.$$

3.5 PROBLEMS

3.1 Show that both ZZ^* and $Z + Z^*$ are real quantities, whereas $Z - Z^*$ is imaginary.

3.2 Find the real and imaginary parts of:

(a) Z^2 (b) $\dfrac{1}{Z}$

(c) $\dfrac{Z-1}{Z+1}$ (d) $\dfrac{1}{Z^2}$.

3.3 Show that $|\cos\theta + i\sin\theta| = 1$.

3.4 Show that $|Z| = |Z^*|$.

3.5 Show that:

(a) $|Z_1Z_2^*| = |Z_1||Z_2|$

(b) $Z_1Z_2^* + Z_1^*Z_2 \leq 2|Z_1Z_2^*|$.

3.6 Find the real part, modulus, and argument of:

(a) $\dfrac{1+Z}{1-Z}$ (b) $1+i\sqrt{3}$

(c) $\dfrac{1+i}{1-i}$ (d) $\left(\dfrac{3}{i-\sqrt{3}}\right)^2$.

3.7 If $Z + 1/Z$ is real, prove that either $\operatorname{Im} Z = 0$ or $|Z| = 1$.

3.8 Write in polar form:

(a) $Z = 2i$

(b) $Z = -2 + 2i$.

3.9 Prove that:

(a) $\arg\left(\dfrac{Z_1}{Z_2}\right) = \arg Z_1 - \arg Z_2$

(b) $\arg(Z_1 Z_2) = \arg Z_1 + \arg Z_2$.

3.10 If $Z = x + iy$, show that $|x| \leq |Z|$ and $|y| \leq |Z|$.

3.11 Show that DeMoivre's theorem is valid for (a) $n < 0$ and (b) n fractional.

3.12 Find and graph all the roots of:

(a) $Z^3 - 1 = 0$

(b) $\sqrt[3]{i}$

(c) $\sqrt{-1}$

(d) $\sqrt{3+4i}$.

3.13 Prove that $e^{Z_1} \cdot e^{Z_2} = e^{Z_1+Z_2}$.

3.14 Show that for $Z \neq 0$:

(a) $\dfrac{d}{dZ} e^Z = e^Z$

(b) $\dfrac{d}{dZ} \ln Z = \dfrac{1}{Z}$.

3.15 In polar coordinates, show that the Cauchy-Riemann conditions become

$$\frac{\partial u}{\partial r} = \frac{1}{r}\frac{\partial v}{\partial \theta} \quad \text{and} \quad \frac{1}{r}\frac{\partial u}{\partial \theta} = -\frac{\partial v}{\partial r}.$$

3.16 Show that e^Z, e^{iZ}, and e^{-iZ} are analytic functions of Z.

3.17 If $f(Z) = u(x, y) + iv(x, y) = Z^2 + 3Z + 4$, find u and v.

3.18 Show that the following functions are harmonic, and find the corresponding conjugate function:

(a) $u = x^3 - 3xy^2 + 3x - 3y$

(b) $v = 2xy + 3y$

(c) $u = x^3 - 3y^2x$

(d) $u = e^x \cos y$.

3.19 Find A and B so that $x^2 + Axy + By^2$ is harmonic.

3.20 Suppose the position of a particle (moving in a plane) is represented by $Z = re^{i\theta}$.

(a) From the expression of the complex velocity of the particle, identify the radial and tangential components of this complex velocity.

(b) From the expression for the complex acceleration, identify the tangential, centripetal, radial, and Coriolis components of the complex acceleration.

3.21 Show that the sum, s, of n terms of a geometric progression is given by

$$s = \frac{a(1 - r^n)}{1 - r}$$

where a is the first term and r is the common ratio.

3.22 Find the radius of convergence of the power series

$$\sum_{r=0}^{\infty} \frac{Z^r}{r!}.$$

3.23 Find the radius of convergence of

$$\sum_{n=0}^{\infty} Z^n.$$

3.24 Find the radius of convergence of

$$\sum_{n=0}^{\infty} n!Z^n.$$

3.25 Show that

$$\sum_{n=0}^{\infty} c_n Z^n$$

and its derivative have the same radius of convergence.

3.26 Develop the Taylor expansion and find the radius of convergence for:

(a) $\ln Z$ about $Z_0 = 1$

(b) $\ln (1 + Z)$ about $Z_0 = 0$

(c) $1/(1 - Z)$ about $Z_0 = i$.

3.27 Show that the principal part of the Laurent expansion equals zero when $f(Z)$ is analytic within C_2 (see Fig. 3.13).

3.28 Develop the Laurent expansion for:

(a) $\dfrac{1}{Z(1 - Z)^2}$ about $Z_0 = 0$ and $Z_0 = 1$

(b) $\dfrac{e^Z}{(Z - 2)^3}$ about $Z_0 = 3$

(c) $\dfrac{1}{(Z - 1)(Z - 2)}$ in the region $|Z| < 1$.

Hint for part (c)*:* Use the method of partial fractions.

APPENDIX: RUDIMENTS OF SERIES

A3.1 Introduction

In this appendix, we summarize the essentials of infinite series without detailed proofs. The mathematical proofs may be found in standard treatments of infinite series.

Infinite sequences and series are important in mathematical physics. Let $w_1, w_2, \ldots, w_n, \ldots$ be a sequence of numbers (real or complex). If

$$\begin{aligned} s &= \lim_{k\to\infty} s_k \\ &= \lim_{k\to\infty} \sum_{n=1}^{k} w_n \end{aligned}$$

then the series $\sum_{n=1}^{\infty} w_n$ is said to be **convergent.** The number s is called the **sum** (or **value**) of the series and is given by

$$s = a + ib$$

where

$$a = \sum_{n=1}^{\infty} u_n$$

and

$$b = \sum_{n=1}^{\infty} v_n$$

for $w_n = u_n + iv_n$.

If the series

$$\sum_{n=1}^{\infty} |w_n| = |w_1| + |w_2| + \cdots$$

is convergent, then $\sum_{n=1}^{\infty} w_n$ is said to be **absolutely convergent.** If $\sum_{n=1}^{\infty} w_n$ is convergent but $\sum_{n=1}^{\infty} |w_n|$ diverges, then $\sum_{n=1}^{\infty} w_n$ is said to be **conditionally convergent.**

A3.2 Simple Convergence Tests

A. Comparison Test If $\sum_n a_n$ is a convergent series and $u_n \le a_n$ for all n, then $\sum_n u_n$ is a convergent series.

If $\sum_n b_n$ is a divergent series and $v_n \ge b_n$ for all n, then $\sum_n v_n$ is also a divergent series.

B. Ratio Test Consider the series $\sum_n a_n$. If

$$\lim_{n\to\infty}\left(\frac{a_{n+1}}{a_n}\right) = \begin{cases} < 1, & \text{then the series is convergent.} \\ > 1, & \text{then the series is divergent.} \\ = 1, & \text{then the test is indeterminate.} \end{cases}$$

A3.3 Some Important Series in Mathematical Physics

A. Geometric Series The sequence

$$s_n = 1 + Z + Z^2 + \cdots + Z^{n-1}$$

is called a **geometric sequence.** On multiplying this sequence by Z and subtracting the resulting sequence Zs_n from s_n, we obtain

$$(1 - Z)s_n = 1 - Z^n$$

or

$$s_n = \frac{1 - Z^n}{1 - Z}; \qquad (Z \neq 1).$$

The corresponding infinite geometric series converges, for $|Z| < 1$, to

$$\begin{aligned} \underset{n\to\infty}{s_n} &= \sum_{n=0}^{\infty} Z^n \\ &= \frac{1}{1 - Z} \qquad (|Z| < 1). \end{aligned}$$

The region $|Z| < 1$ is called the **circle of convergence** of the infinite geometric series.

B. Series of Functions Consider

$$s_n(Z) = U_1(Z) + U_2(Z) + \cdots + U_n(Z)$$

and

$$f(Z) = \lim_{n\to\infty} s_n(Z) = \sum_{n=1}^{\infty} U_n(Z).$$

If

$$|f(Z) - s_n(Z)| < \epsilon \qquad (\text{for all } n \geq N)$$

where N is independent of Z in the region $a \leq |Z| \leq b$ and ϵ is an arbitrarily small quantity greater than zero, then the series $s_n(Z)$ is said to be **uniformly convergent** in the closed region $a \leq |Z| \leq b$.

If the individual terms, $U_n(Z)$, of a uniformly convergent series are continuous, the series may be integrated term by term, and the resultant series will always be convergent. Thus

$$\int_a^b f(Z)\,dZ = \int_a^b \sum_{n=1}^{\infty} U_n(Z)\,dZ = \sum_{n=1}^{\infty} \int_a^b U_n(Z)\,dZ.$$

The derivative of $f(Z)$,

$$\frac{df(Z)}{dZ} = \frac{d}{dZ} \sum_{n=1}^{\infty} U_n(Z)$$

is equal to

$$\sum_{n=1}^{\infty} \frac{d}{dZ} U_n(Z)$$

only if

$$U_n(Z) \quad \text{and} \quad \frac{dU_n(Z)}{dZ}$$

are continuous in the region and

$$\sum_{n=1}^{\infty} \frac{dU_n(Z)}{dZ}$$

is uniformly convergent in the region.

4

Calculus of Residues

This chapter contains an introductory treatment of (1) the singularities of functions of a complex variable, (2) the evaluation of residues and techniques for the evaluation of certain kinds of definite integrals which occur in various branches of physics, (3) the concept of the principal value of an integral and dispersion relations, and (4) conformal mapping.

4.1 ZEROS

Consider the Taylor expansion for the function $f(Z)$

$$f(Z) = \sum_{n=0}^{\infty} a_n(Z - Z_0)^n$$

where

$$a_n = \frac{f^{(n)}(Z_0)}{n!}.$$

Here the function $f(Z)$ is analytic within some region Γ. If $f(Z)$ vanishes at

$Z = Z_0$, the point Z_0 is said to be a **zero** of $f(Z)$. If $a_0 = a_1 = \cdots = a_{m-1} = 0$ but $a_m \neq 0$, then the Taylor expansion becomes

$$f(Z) = a_m(Z - Z_0)^m + a_{m+1}(Z - Z_0)^{m+1} + \cdots = \sum_{n=m}^{\infty} a_n(Z - Z_0)^n.$$

In this case, $f(Z)$ is said to have a **zero of order *m*** at $Z = Z_0$. Throughout this chapter, m is considered to be a positive integer. A zero of order 1 ($m = 1$) is called a **simple zero.** In this connection ($m = 1$), note that

$$\begin{aligned} f(Z) &= \sum_{n=0}^{\infty} a_n(Z - Z_0)^n \\ &= a_0 + a_1(Z - Z_0) + a_2(Z - Z_0)^2 + \cdots \\ &= f(Z_0) + f'(Z_0)(Z - Z_0) + \frac{f''(Z_0)}{2!}(Z - Z_0)^2 + \cdots. \end{aligned}$$

It is clear that the conditions

$$f(Z_0) = 0 \quad \text{and} \quad f'(Z_0) \neq 0 \tag{4.1}$$

indicate the existence of a simple zero for $f(Z)$ at $Z = Z_0$.

4.2 ISOLATED SINGULAR POINTS

Points at which the function $f(Z)$ is not analytic are called **singular points (singularities).** For example, $Z = 0$ is a singular point of the function $f(Z) = 1/Z$.

If $f(Z)$ is analytic throughout the neighborhood of a point Z_0, $|Z - Z_0| < \epsilon$, but is not analytic at Z_0, then the point Z_0 is called an **isolated singularity** of $f(Z)$. That is to say, the singularity at Z_0 is isolated if a circle, containing no other singularities, can be drawn with Z_0 as its center. By use of this definition, we see that $Z = 0$ for $f(Z) = 1/Z$ is an isolated singular point. The function $f(Z) = 1/[\sin(1/Z)]$ has an isolated singularity when $Z = 1/(n\pi)$ for $n = 1, 2, \ldots$. However, the origin $Z = 0$ is not an isolated singular point since every neighborhood contains other singular points (an infinite number) as a result of $Z = 1/(n\pi)$. In other words, the singularities from $Z = 1/(n\pi)$ for $n \to \infty$ are arbitrarily close to the origin ($Z = 0$).

The Laurent expansion of $f(Z)$ with an isolated singularity at Z_0 is given by

$$f(Z) = \sum_{n=0}^{\infty} a_n(Z - Z_0)^n + \sum_{n'=1}^{\infty} \frac{b_{n'}}{(Z - Z_0)^{n'}}.$$

This expansion converges for $r_2 < |Z - Z_0| < r_1$, where r_1 is the radius of

C_1 and r_2 is the radius of C_2 (see Fig. 3.11). A singular point, Z_0, of the analytic part of the Laurent expansion is called a **removable singularity** since $f_A(Z)$ may be redefined at

$$Z = Z_0, \qquad a_0 = \lim_{Z \to Z_0} f_A(Z)$$

so that $f_A(Z)$ becomes analytic at $Z = Z_0$. The physical importance of singularities of this type is not great. However, singularities of $f_P(Z)$ are of immense importance in the mathematical analysis of physical problems.

If $b_m \neq 0$ but $b_{m+1} = b_{m+2} = \cdots = 0$ in the principal part of the Laurent expansion of $f(Z)$, then $f_P(Z)$ becomes

$$\begin{aligned} f_P(Z) &= \frac{b_1}{Z - Z_0} + \cdots + \frac{b_m}{(Z - Z_0)^m} \\ &= \sum_{n'=1}^{m} \frac{b_{n'}}{(Z - Z_0)^{n'}}. \end{aligned}$$

Here $f(Z)$ is said to have a **pole** of order m at $Z = Z_0$. When $m = 1$, the singularity at $Z = Z_0$ is called a **simple pole.**

EXAMPLE 4.1 Find the pole and its order for

$$f(Z) = \frac{\sin Z}{Z^4}.$$

Solution

$$\begin{aligned} f(Z) &= \frac{\sin Z}{Z^4} \\ &= \frac{Z - \frac{Z^3}{3!} + \frac{Z^5}{5!} - \frac{Z^7}{7!} + \cdots}{Z^4} \\ &= \frac{1}{Z^3} - \frac{1}{3!}\frac{1}{Z} + \frac{Z}{5!} - \frac{Z^3}{7!} + \cdots. \end{aligned}$$

In this example, we see that $f(Z)$ has a pole of order 3 at $Z = 0$.

The function $f(Z)$ is said to have an **essential singularity** (pole of infinite order) at $Z = Z_0$ if there exists no m such that $b_s = 0$ for all $s > m$. That is, $f_P(Z)$ contains an infinite number of terms when $Z = Z_0$ is an essential singularity. The peculiar behavior of the function $f(Z)$ in the neighborhood of an essential singularity is revealed by a theorem due to Picard (1856–1941) (often called the Weierstrass-Casorati theorem). This theorem states that the function $f(Z)$ oscillates so rapidly in the neighborhood of an essential singularity that it comes arbitrarily close to any possible complex number. In symbolic form, we write

$$|f(Z) - g| < \epsilon \qquad (\text{for } |Z - Z_0| < \delta)$$

where ϵ and δ are arbitrary positive numbers and g is an arbitrary complex number. We restrict our discussion of analytic functions to this limited information concerning the behavior of functions in the vicinity of an essential singularity.

EXAMPLE 4.2 Classify the singularity of the function

$$f(Z) = Z \exp \frac{1}{Z}\cdot$$

Solution

$$\begin{aligned} f(Z) &= Ze^{1/Z} \\ &= Z \sum_{n=0}^{\infty} \frac{(1/Z)^n}{n!} \\ &= Z\left(1 + \frac{1}{Z} + \frac{1}{2!Z^2} + \frac{1}{3!Z^3} + \cdots\right) \\ &= Z + 1 + \frac{1}{2!Z} + \frac{1}{3!Z^2} + \cdots. \end{aligned}$$

There is an essential singularity at $Z = 0$.

A fourth kind of singularity, **branch point**, which results from the consideration of multivalued functions (fractional powers in the expansion) will not be treated. The geometrical devices of **Riemann surfaces** are used to represent these multivalued functions in terms of single-valued functions.

4.3 EVALUATION OF RESIDUES

4.3.1 *m*th-Order Pole

The Laurent expansion of a function for the case of a pole of order m is

$$f(Z) = \sum_{n=0}^{\infty} a_n(Z - Z_0)^n + \frac{a_{-1}}{Z - Z_0} + \cdots + \frac{a_{-m}}{(Z - Z_0)^m}\cdot \tag{4.2}$$

The coefficient a_{-1} is called the **residue** of $f(Z)$ at $Z = Z_0$; it is given by

$$a_{-n} = \frac{1}{2\pi i} \oint_{C_2} \frac{f(\xi)\, d\xi}{(\xi - Z_0)^{-n+1}}$$

or

$$a_{-1} = \frac{1}{2\pi i} \oint_C f(Z)\, dZ \tag{4.3}$$

Here C encloses Z_0. Except for the numerical factor $2\pi i$, a_{-1} is the value of the integral of $f(Z)$ over C. In those cases where a_{-1} can be determined directly (without carrying out the integration), we have a method of evaluating the definite integral of $f(Z)$ over C indirectly. We now develop an important expression for evaluating a_{-1} directly.

On multiplying Eq. (4.2) by $(Z - Z_0)^m$, we obtain

$$\begin{aligned} \phi(Z) &= \sum_{n=0}^{\infty} a_n(Z - Z_0)^{n+m} + \sum_{n=1}^{m} a_{-n}(Z - Z_0)^{m-n} \\ &= a_0(Z - Z_0)^m + a_1(Z - Z_0)^{m+1} + \cdots \\ &\quad + a_{-1}(Z - Z_0)^{m-1} + a_{-2}(Z - Z_0)^{m-2} + \cdots + a_{-m} \end{aligned} \tag{4.4}$$

where

$$\phi(Z) = (Z - Z_0)^m f(Z)$$

$$\begin{pmatrix}\text{For minimum } m, \\ \text{we must have}\end{pmatrix} \quad \phi(Z)\,|_{Z \to Z_0} \quad \begin{cases} \neq 0 \\ \text{analytic.} \end{cases} \tag{4.5}$$

Since $\phi(Z)$ is analytic at $Z = Z_0$, Eq. (4.4) may be thought of as the Taylor expansion of $\phi(Z)$ about $Z = Z_0$. The coefficient a_{-1} in this expansion must be the coefficient of the $(Z - Z_0)^{m-1}$ term. Thus, the residue is given by

$$a_{-1} = \frac{1}{(m-1)!} \frac{d^{m-1}}{dZ^{m-1}} [\phi(Z)]_{Z \to Z_0}. \tag{4.6}$$

4.3.2 Simple Pole

For a simple pole ($m = 1$) at $Z = Z_0$, the residue, Eq. (4.6), of $f(Z)$ reduces to

$$a_{-1} = \lim_{Z \to Z_0} (Z - Z_0) f(Z). \tag{4.7}$$

In this connection, suppose the function $f(Z)$ can be represented by

$$f(Z) = \frac{g(Z)}{h(Z)} \tag{4.8}$$

where $g(Z)$ and $h(Z)$ are analytic functions. If $g(Z_0) \neq 0$, $h(Z_0) = 0$, but $h'(Z_0) \neq 0$, the function $h(Z)$ has a simple zero at $Z = Z_0$ [See Eq. (4.1)], and $f(Z)$ has a simple pole at $Z = Z_0$. The residue of $f(Z)$ at $Z = Z_0$ is given by

$$
\begin{aligned}
a_{-1} &= \lim_{Z \to Z_0} \left[\frac{g(Z)(Z - Z_0)}{h(Z)} \right] \\
&= \lim_{Z \to Z_0} g(Z) \lim_{Z \to Z_0} \left[\frac{Z - Z_0}{h(Z) - h(Z_0)} \right] \\
&= g(Z_0) \left[\frac{1}{h'(Z_0)} \right] \\
&= \frac{g(Z_0)}{h'(Z_0)} \qquad (4.9)
\end{aligned}
$$

This expression for the residue ($m = 1$) is extremely useful.

The procedure for evaluating the residue of a function $f(Z)$ at $Z = Z_0$ may be summarized as follows.

1. The general expression for a_{-1} is

$$a_{-1} = \frac{1}{(m-1)!} \frac{d^{m-1}}{dZ^{m-1}} [\phi(Z)]_{Z \to Z_0}$$

where

$$\phi(Z) = (Z - Z_0)^m f(Z)$$

and

$$\phi(Z)|_{Z \to Z_0} \quad \begin{cases} \neq 0 \\ \text{analytic.} \end{cases}$$

The value of m is required before the general expression for a_{-1} can be used to evaluate the residue. The above conditions on $\phi(Z)|_{Z \to Z_0}$ may be used to determine the minimum m for a certain class of functions.

2. For a simple pole ($m = 1$), we have

$$a_{-1} = \lim_{Z \to Z_0} (Z - Z_0) f(Z).$$

3. If $f(Z) = g(Z)/h(Z)$ where $g(Z_0) \neq 0$, $h(Z_0) = 0$, but $h'(Z_0) \neq 0$, the residue is

$$a_{-1} = \frac{g(Z_0)}{h'(Z_0)}.$$

4. If the function $f(Z)$ is such that its residue at $Z = Z_0$ cannot be obtained by use of the above three procedures, we must rewrite the function by (1) expanding the transcendental functions that may appear in $f(Z)$, (2) developing a geometric series for the function, or (3) developing the full Laurent expansion for the function. The classification and location of the singularities and the value of the residue can always be obtained from the full Laurent expansion for the function under investigation.

EXAMPLE 4.3 Classify the singularities and calculate the residue for

$$f(Z) = \frac{1}{Z^2 - 1}.$$

Solution We have

$$f(Z) = \frac{1}{(Z^2 - 1)} = \frac{1}{(Z - 1)(Z + 1)}.$$

There are poles at $Z = +1$ and $Z = -1$.

$$\phi(Z)|_{Z_0 \to +1} = (Z - 1)^m f(Z)$$
$$= (Z - 1)^m \cdot \frac{1}{(Z - 1)(Z + 1)} = \frac{1}{Z + 1}.$$

Hence $Z = +1$ is a simple pole since $m = 1$, $\phi(Z)$ is analytic at $Z = +1$, and $\phi(+1) \neq 0$. Similarly, we find that $Z = -1$ is a simple pole.

$$a_{-1}|_{+1} = \lim_{Z \to +1} (Z - 1) \cdot \frac{1}{(Z - 1)(Z + 1)} = \frac{1}{2}$$

and

$$a_{-1}|_{-1} = \lim_{Z \to -1} (Z + 1) \cdot \frac{1}{(Z - 1)(Z + 1)} = -\frac{1}{2}.$$

EXAMPLE 4.4 Classify the singularities and calculate the residue for

$$f(Z) = \frac{1}{(Z^2 + a^2)^2} \qquad \text{(where } a > 0\text{)}.$$

Solution Here we have

$$f(Z) = \frac{1}{(Z^2 + a^2)^2}$$
$$= \frac{1}{(Z + ia)^2(Z - ia)^2}.$$

There are poles at $Z = ia$ and $Z = -ia$.

$$\phi(Z)|_{Z_0 \to ia} = (Z - ia)^m \cdot \frac{1}{(Z + ia)^2(Z - ia)^2} = \frac{1}{(Z + ia)^2}.$$

Hence $Z = ia$ is a second-order pole since $m = 2$, $\phi(Z)$ is analytic at $Z = ia$, and $\phi(ia) \neq 0$. Similarly, we find that $Z = -ia$ is a second-order pole.

$$a_{-1}|_{Z \to ia} = \frac{d}{dZ}\phi(Z)|_{Z \to ia}$$

$$= -\frac{2}{(Z + ia)^3}\bigg|_{Z \to ia} = \frac{1}{4ia^3}$$

and

$$a_{-1}|_{Z \to -ia} = -\frac{1}{4ia^3}.$$

EXAMPLE 4.5 Classify the singularities and calculate the residue for

$$f(Z) = \frac{Z^{-k}}{Z + 1} \qquad \text{(where } 0 < k < 1\text{)}.$$

Solution We have

$$f(Z) = \frac{Z^{-k}}{Z + 1}$$

$$= \frac{1}{Z^k(Z + 1)}.$$

There is a pole at $Z = -1$ and a branch point at $Z = 0$.

$$\phi(Z)|_{Z_0 \to -1} = (Z + 1)^m \cdot \frac{1}{Z^k(Z + 1)} = \frac{1}{Z^k}.$$

Hence $Z = -1$ is a simple pole since $m = 1$, $\phi(Z)$ is analytic at $Z = -1$, and $\phi(-1) \neq 0$.

$$a_{-1} = \lim_{Z \to -1}\left[(Z + 1) \cdot \frac{1}{Z^k(Z + 1)}\right]$$

$$= \frac{1}{(-1)^k} = e^{-ik\pi}.$$

EXAMPLE 4.6 Classify the singularities and calculate the residue for

$$f(Z) = \frac{A(Z)}{\sin Z}$$

where $A(Z)$ is analytic and contains no zeros.

Solution Let $h(Z) = \sin Z$. Since $h(\pm l\pi) = 0$ but $h'(\pm l\pi) \neq 0$, the zeros of $h(Z)$ at $Z = \pm l\pi$ are simple zeros; hence, the poles of the function $f(Z)$ at $Z = \pm l\pi$ are simple poles. The residue is

$$a_{-1} = \frac{A(Z)}{h'(Z)}\bigg|_{Z\to\pm l\pi} \qquad (l = 0, 1, 2\dots)$$

$$= \frac{A(Z)}{\cos Z}\bigg|_{Z\to\pm l\pi}$$

$$= \frac{A(\pm\, l\pi)}{\cos l\pi} = \frac{A(\pm\, l\pi)}{(-1)^l}.$$

EXAMPLE 4.7 Classify the singularities and calculate the residue for

$$f(Z) = \frac{\sin Z}{Z^4}.$$

Solution Since there is a zero in the numerator of $f(Z)$ for $Z = 0$, it is not clear from the use of the general expression for a_{-1} where (and to what order) the poles are located. To resolve the problem, we expand $\sin Z$ and obtain

$$f(Z) = \frac{Z - \dfrac{Z^3}{3!} + \dfrac{Z^5}{5!} - \dfrac{Z^7}{7!} + \cdots}{Z^4}$$

$$= \frac{1}{Z^3} - \frac{1}{3!Z} + \frac{Z}{5!} - \frac{Z^3}{7!} + \cdots.$$

This equation is the Laurent expansion for $f(Z)$ about $Z_0 = 0$. From this expansion, we observe that $Z = 0$ is a third-order pole with residue given by

$$a_{-1} = -\frac{1}{3!}.$$

EXAMPLE 4.8 Classify the singularities and calculate the residue for

$$f(Z) = Z \exp\frac{1}{Z}.$$

Solution Here we have

$$f(Z) = Ze^{1/Z}$$

$$= Z\sum_{n=0}^{\infty}\frac{(1/Z)^n}{n!}$$

$$= Z\left(1 + \frac{1}{Z} + \frac{1}{2!}\frac{1}{Z^2} + \frac{1}{3!}\frac{1}{Z^3} + \cdots\right).$$

There is an isolated essential singularity at $Z = 0$. The residue is just the coefficient of $1/Z$, that is, $a_{-1} = 1/2! = \frac{1}{2}$.

4.4 THE CAUCHY RESIDUE THEOREM

THE CAUCHY RESIDUE THEOREM If the function $f(Z)$ is analytic inside and on a closed region Γ (except at a finite number of isolated singular points within Γ), then

$$\oint_\Gamma f(Z)\,dZ = 2\pi i \sum (\text{enclosed residues})$$
$$= 2\pi i(a_{-1Z_1} + a_{-1Z_2} + \cdots + a_{-1Z_n})$$

or

$$\oint_\Gamma f(Z)\,dZ = 2\pi i \sum_{j=1}^{n} a_{-1Z_j} \tag{4.10}$$

where Z_j $(j = 1, 2, \ldots, n)$ are the enclosed singular points and a_{-1Z_j} are the corresponding residues.

Applying Cauchy's integral theorem in Fig. 4.1, we obtain

$$\oint_\Gamma f(Z)\,dZ - \oint_{C_1} f(Z)\,dZ - \cdots - \oint_{C_j} f(Z)\,dZ = 0. \tag{4.11}$$

The value of the circular integral around an isolated singular point [Eq. (4.3)] is $2\pi i a_{-1}$. Hence Eq. (4.11) becomes

$$\oint_\Gamma f(Z)\,dZ = 2\pi i(a_{-1Z_1} + a_{-1Z_2} + \cdots + a_{-1Z_n})$$
$$= 2\pi i \sum_{j=1}^{n} a_{-1Z_j}.$$

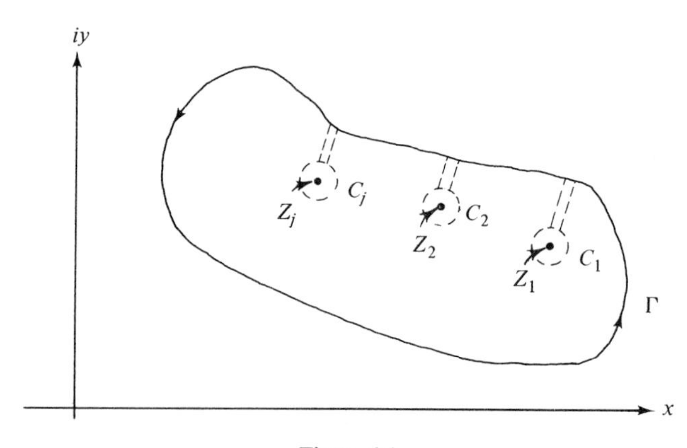

Figure 4.1

Thus, Cauchy's residue theorem is established. We will use this theorem in Section 4.6 to evaluate certain definite integrals.

4.5 THE CAUCHY PRINCIPAL VALUE

Thus far, we have considered contours that enclose and/or exclude singularities. We now consider the case where the path of integration passes through a singularity of the integrand. In a strict sense, this integral does not exist, and we must choose a path that circumvents the singularity. Since many physical problems involve the evaluation of integrals in which an isolated simple pole is on the contour of integration, it is useful to consider the details of dealing with this situation. We deform the contour to include or exclude the simple pole with a semicircular detour of infinitesimal radius as shown in Fig. 4.2.

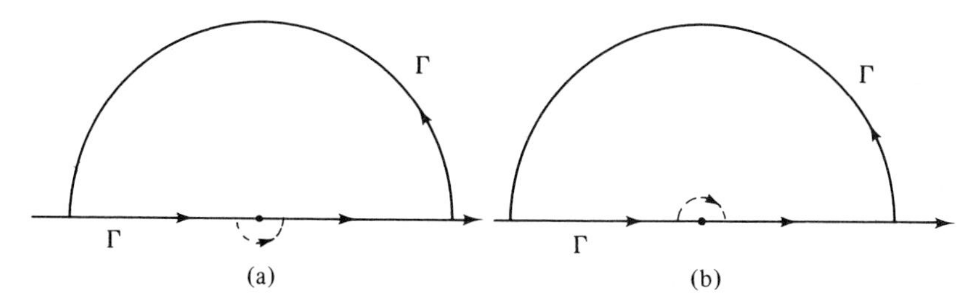

Figure 4.2

Counterclockwise and clockwise integrations over the dashed semicircles yield $\pi i a_{-1}$ and $-\pi i a_{-1}$, respectively. On using the residue theorem in Fig. 4.2(a), we obtain

$$\oint_{\Gamma} f(Z)\, dZ + \pi i a_{-1} = 2\pi i a_{-1}$$

or

$$\oint_{\Gamma} f(Z)\, dZ = \pi i a_{-1} \tag{4.12}$$

since the simple pole is now enclosed. Using the residue theorem in Fig. 4.2(b), we obtain

$$\oint_{\Gamma} f(Z)\, dZ - \pi i a_{-1} = 0$$

or

$$\oint_{\Gamma} f(Z)\, dZ = \pi i a_{-1} \tag{4.13}$$

since the simple pole is not enclosed. The net result for Fig. 4.2(a) or Fig. 4.2(b) is that a simple pole on the contour counts one-half of what it would be if it were within the contour (the Cauchy principal value).

The integral of a function $f(x)$ along the x-axis may be written as†

$$\lim_{\epsilon \to 0} \left[\int_a^{x_0-\epsilon} f(x)\,dx + \int_{x_0+\epsilon}^{b} f(x)\,dx \right] \equiv P \int_a^b f(x)\,dx \tag{4.14}$$

where P indicates the **Cauchy principal value**. Note that

$$P \int_{-\infty}^{\infty} f(x)\,dx$$

may exist even if

$$\lim_{a\to\infty} \int_{-a}^{0} f(x)\,dx + \lim_{a\to\infty} \int_0^a f(x)\,dx$$

does not exist. For example,

$$\lim_{a\to\infty} \int_{-a}^{a} x\,dx = \lim_{a\to\infty} \left(\frac{a^2}{2} - \frac{a^2}{2} \right) = 0$$

but

$$\lim_{a\to\infty} \int_0^a x\,dx \longrightarrow \infty.$$

4.6 EVALUATION OF DEFINITE INTEGRALS

4.6.1 Integrals of the Form: $\int_0^{2\pi} f(\sin\theta, \cos\theta)\,d\theta$.

Consider the real integral

$$I_1 = \int_0^{2\pi} f(\sin\theta, \cos\theta)\,d\theta \tag{4.15}$$

where $f(\sin\theta, \cos\theta)$ is a rational function (contains no isolated singularities other than poles) of $\sin\theta$ and/or $\cos\theta$. Let $Z = e^{i\theta}$ (unit circle). In this case,

$$\sin\theta = \frac{e^{i\theta} - e^{-i\theta}}{2i} = \frac{Z^2 - 1}{2iZ}$$

and

$$\cos\theta = \frac{e^{i\theta} + e^{-i\theta}}{2} = \frac{Z^2 + 1}{2Z}.$$

†Here $\int_a^b f(x)\,dx$ does not exist at $x = x_0$.

In terms of Z, Eq. (4.15) becomes

$$I_1 = -i \oint_{\substack{C \\ \text{unit} \\ \text{circle}}} f\left(\frac{Z^2 - 1}{2iZ}, \frac{Z^2 + 1}{2Z}\right) \frac{dZ}{Z} \tag{4.16}$$

since $d\theta = -idZ/Z$. Applying the residue theorem to Eq. (4.16), we obtain

$$\begin{aligned} I_1 &= (-i)(2\pi i) \sum_{j=1}^{n} a_{-1Z_j} \\ &= 2\pi \sum_{j=1}^{n} a_{-1Z_j} \end{aligned} \tag{4.17}$$

Here the contour is a unit circle.

EXAMPLE 4.9 Using the residue theorem, evaluate

$$I = \int_0^{2\pi} \frac{d\theta}{5 + 4\cos\theta}.$$

Solution Using Eq. (4.16), we obtain

$$\begin{aligned} I &= -i \oint_{\substack{\text{unit} \\ \text{circle}}} \frac{dZ}{Z\left[5 + 4\left(\dfrac{Z^2 + 1}{2Z}\right)\right]} \\ &= -i \oint_{\substack{\text{unit} \\ \text{circle}}} \frac{dZ}{5Z + 2Z^2 + 2} = -i \oint_{\substack{\text{unit} \\ \text{circle}}} \frac{dZ}{(2Z + 1)(Z + 2)}. \end{aligned}$$

We have singularities at $Z = -\frac{1}{2}$ and $Z = -2$. The second singularity, -2, is outside the unit circle. At $Z = -\frac{1}{2}$, we have a simple pole inside the unit circle since

$$\phi(Z) = \left(Z + \frac{1}{2}\right) \cdot \frac{1}{(2Z + 1)(Z + 2)} = \frac{1}{2(Z + 2)}$$

and

$$\phi\left(-\frac{1}{2}\right) \begin{cases} \text{analytic} \\ \neq 0 \end{cases}.$$

The residue is

$$\begin{aligned} a_{-1}\big|_{Z \to -1/2} &= \lim_{Z \to -1/2} \phi(Z) \\ &= \frac{1}{3}. \end{aligned}$$

By use of the residue theorem, we find that

$$I = (-i)(2\pi i)\left(\frac{1}{3}\right) = \frac{2\pi}{3}.$$

4.6.2 Integrals of the Form: $\int_{-\infty}^{\infty} f(x)\,dx.$

Consider definite integrals of the form

$$I_2 = \int_{-\infty}^{\infty} f(x)\,dx. \tag{4.18}$$

If $f(Z)$ (1) is analytic in the upper half-plane except for a finite number of poles and (2) has only simple poles on the real axis, and if (3) $Zf(Z) \to 0$ for all values of Z as $|Z| \to \infty$ for $0 \leq \arg Z \leq \pi$, then

$$\begin{aligned} I &= \lim_{R\to\infty} \int_{-R}^{R} f(x)\,dx + \lim_{R\to\infty} \int_{0}^{\pi} f(Re^{i\theta})iRe^{i\theta}\,d\theta \\ &= 2\pi i \sum_{j=1}^{n} a_{-1Z_j} + \pi i \sum_{k=1}^{m} a_{-1x_k} \end{aligned} \tag{4.19}$$

where $Z = Re^{i\theta}$. Here the contour of integration is the real axis and the semicircle in the upper half-plane (see Fig. 4.3). The second term on the right-hand side of Eq. (4.19) results from simple poles on the real axis.

Note that $|Z|^2 = Z^*Z = R^2$; $f(Re^{i\theta})iRe^{i\theta} = if(Z)Z \to 0$ for $R = |Z| \to \infty$ by condition (3). That is,

$$\lim_{R\to\infty} i \int_{0}^{\pi} f(Re^{i\theta})Re^{i\theta}\,d\theta \longrightarrow 0.$$

Therefore Eq. (4.19) becomes

$$I_2 = \int_{-\infty}^{\infty} f(x)\,dx = 2\pi i \sum_{j=1}^{n} a_{-1Z_j} + \pi i \sum_{k=1}^{m} a_{-1x_k}. \tag{4.20}$$

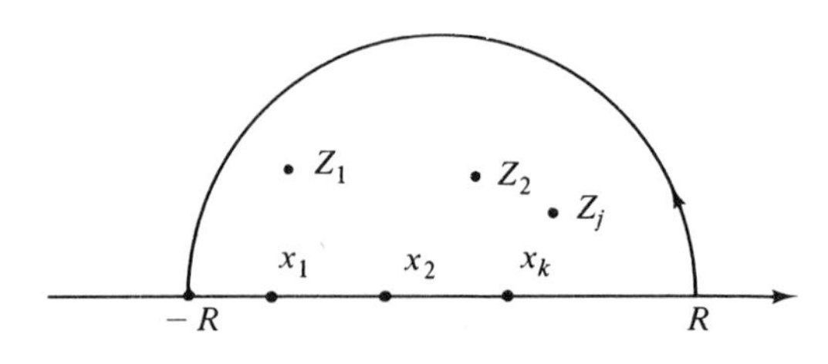

Figure 4.3

EXAMPLE 4.10 Using the residue theorem, evaluate

$$I = \int_{-\infty}^{\infty} \frac{dx}{1 + x^2}.$$

Solution By use of Eq. (4.20), we may write

$$I = \lim_{R\to\infty} \int_{-R}^{R} \frac{dx}{1 + x^2}$$

$$= \oint_{\substack{\text{semi-}\\\text{circle}}} \frac{dZ}{1 + Z^2} = \oint_{\substack{\text{semi-}\\\text{circle}}} \frac{dZ}{(Z - i)(Z + i)}.$$

Here there are singularities at $Z = -i$ and $Z = i$. Note that $Z = -i$ is not within the semicircle. We have a simple pole at $Z = i$ since

$$\phi(Z) = \frac{(Z - i)}{(Z - i)(Z + i)} \quad \text{and} \quad \phi(i) \begin{cases} \text{analytic} \\ \neq 0. \end{cases}$$

The corresponding residue is

$$a_{-1} = \frac{1}{Z + i}\bigg|_{Z=i} = \frac{1}{2i}.$$

Hence

$$I = \int_{-\infty}^{\infty} \frac{dx}{1 + x^2} = (2\pi i)\left(\frac{1}{2i}\right) = \pi.$$

4.6.3 A Digression on Jordan's (1838-1922) Lemma

Let Γ be a semicircle of radius R in the upper half-plane, and let $f(Z)$ be a function satisfying the following conditions: (1) $f(Z)$ is analytic in the upper half-plane except for a finite number of isolated singularities (and/or simple poles on the real axis) and (2) $f(Z) \to 0$ uniformly as $|Z| \to \infty$ such that $|f(Z)| \to 0$ for $|Z| = R \to \infty$ where $|f(Z)| \leq M(R)$ and $0 \leq \arg Z \leq \pi$. In equation form, **Jordan's lemma** is

$$J = \lim_{R\to\infty} \int_{\Gamma} f(Z)e^{imZ}\,dZ \longrightarrow 0 \qquad (\text{for } m > 0). \tag{4.21}$$

By use of $Z = Re^{i\theta}$, we may write

$$J = \int_0^{\pi} f(Re^{i\theta})e^{imR(\cos\theta + i\sin\theta)} iRe^{i\theta}\,d\theta$$

and

$$|J| = \left| \int_0^\pi f(Re^{i\theta}) e^{imR(\cos\theta + i\sin\theta)} iRe^{i\theta}\, d\theta \right|$$
$$\leq \int_0^\pi \left| f(Re^{i\theta}) e^{imR(\cos\theta + i\sin\theta)} iRe^{i\theta} \right| d\theta$$
$$= M(R) \int_0^\pi Re^{-mR\sin\theta}\, d\theta \tag{4.22}$$

since

$$\left| \int_a^b f(Z)\, dZ \right| \leq \int_a^b |f(Z)|\, dZ$$

and

$$|e^{imR\cos\theta}| = 1.$$

In the range $0 \leq \theta \leq \pi/2$, $2\theta/\pi \leq \sin\theta \leq 0$ (see Problem 4.15) and

$$\int_0^\pi Re^{-mR\sin\theta}\, d\theta = 2\int_0^{\pi/2} Re^{-mR\sin\theta}\, d\theta$$

since $\sin\theta$ is symmetric about $\theta = \pi/2$. We may therefore write Eq. (4.22) in the form

$$|J| \leq 2M(R)R \int_0^{\pi/2} e^{-2mR\theta/\pi}\, d\theta$$
$$= \frac{M(R)\pi}{m}[1 - e^{-mR}].$$

Hence $|J| \to 0$ as $R \to \infty$ since $\lim_{R\to\infty} M(R) \to 0$, and the lemma is established.

4.6.4 Integrals of the Form: $\int_{-\infty}^{\infty} f(x)e^{imx}\, dx$

In this case, we have (see Fig. 4.3)

$$I = \int_{-\infty}^{\infty} f(x)e^{imx}\, dx + \lim_{R\to\infty} J$$
$$= 2\pi i \sum_{j=1}^{n} a_{-1Z_j} + \pi i \sum_{k=1}^{p} a_{-1x_k}$$

or

$$I_3 = \int_{-\infty}^{\infty} f(x)e^{imx}\, dx$$
$$= 2\pi i \sum_{j=1}^{n} a_{-1Z_j} + \pi i \sum_{k=1}^{p} a_{-1x_k} \tag{4.23}$$

since

$$\lim_{R\to\infty} J = 0$$

from Jordan's lemma. Simple poles on the real axis give rise to the second term on the right-hand side of Eq. (4.23). The above relation may be put in the following useful form:

$$\int_{-\infty}^{\infty} f(x)e^{imx}\,dx = 2\pi i \sum \text{Res}\,[f(Z)e^{imZ}]. \tag{4.24}$$

On equating the real and imaginary parts on both sides of Eq. (4.24), we obtain

$$\begin{aligned}\int_{-\infty}^{\infty} f(x)\cos mx\,dx = -2\pi \sum_{\substack{\text{enclosed}\\ \text{poles}}} \text{Im Res}\,[f(Z)e^{imZ}] \\ - \pi \sum_{\substack{\text{poles on}\\ \text{the path}}} \text{Im Res}\,[f(Z)^{imZ}]\end{aligned} \tag{4.25}$$

and

$$\begin{aligned}\int_{-\infty}^{\infty} f(x)\sin mx\,dx = 2\pi \sum_{\substack{\text{enclosed}\\ \text{poles}}} \text{Re Res}\,[f(Z)e^{imZ}] \\ + \pi \sum_{\substack{\text{poles on}\\ \text{the path}}} \text{Re Res}\,[f(Z)e^{imZ}].\end{aligned} \tag{4.26}$$

EXAMPLE 4.11 By use of the residue theorem, evaluate

$$I = \int_{-\infty}^{\infty} \frac{\sin x\,dx}{x}.$$

Solution Using Eq. (4.26), we write

$$\begin{aligned} I &= \int_{-\infty}^{\infty} \frac{e^{iZ}}{Z}\,dZ \\ &= 2\pi \sum_{\substack{\text{enclosed}\\ \text{poles}}} \text{Re Res}\left[\frac{e^{iZ}}{Z}\right] + \pi \sum_{\substack{\text{poles on}\\ \text{the path}}} \text{Re Res}\left[\frac{e^{iZ}}{Z}\right]. \end{aligned}$$

We have a simple pole at $Z = 0$. The corresponding residue is

$$\begin{aligned} a_{-1}\big|_{Z=0} &= \lim_{Z\to 0}(Z-0)f(Z) \\ &= \frac{Z\cdot e^{iZ}}{Z}\bigg|_{Z\to 0} = 1. \end{aligned}$$

The value of the integral is

$$\begin{aligned} I &= \int_{-\infty}^{\infty} \frac{\sin x}{x}\,dx \\ &= \pi \sum \text{Re Res}\left[\frac{e^{iZ}}{Z}\right] = \pi. \end{aligned}$$

4.7 DISPERSION RELATIONS

In mathematical physics, formulas expressing the real part of an analytic function in terms of its imaginary part, and the imaginary part in terms of its real part, are called **dispersion relations** (in mathematics such relations are called **Hilbert transforms**). Such relations are developed in this section. Consider the evaluation of the integral

$$\oint \frac{\chi(\omega)\,d\omega}{\omega - \omega_0}$$

around the contour shown in Fig. 4.4. We have

$$0 = \oint \frac{\chi(\omega)\,d\omega}{\omega - \omega_0} = \int_{-Q}^{\omega_0-\delta} \frac{\chi(\omega)\,d\omega}{\omega - \omega_0} + \int_{\omega_0+\delta}^{Q} \frac{\chi(\omega)\,d\omega}{\omega - \omega_0}$$
$$+ \int_{\pi}^{0} \frac{\chi(\omega_0 + \delta e^{i\theta}) i\delta e^{i\theta}\,d\theta}{\delta e^{i\theta}} + \int_{0}^{\pi} \frac{\chi(Qe^{i\theta}) iQe^{i\theta}\,d\theta}{Qe^{i\theta} - \omega_0}.$$

Note that

$$\lim_{\delta\to 0} i \int_{\pi}^{0} \chi(\omega_0 + \delta e^{i\theta})\,d\theta = -\pi i \chi(\omega_0).$$

If χ represents a physical quantity, we require $\chi(\infty) \longrightarrow 0$. Hence

$$\lim_{Q\to\infty} \int_{0}^{\pi} \frac{\chi(Qe^{i\theta}) iQe^{i\theta}\,d\theta}{Qe^{i\theta} - \omega_0} \longrightarrow 0$$

We therefore obtain

$$\pi i \chi(\omega_0) = \lim_{\substack{Q\to\infty \\ \delta\to 0}} \left[\int_{-Q}^{\omega_0-\delta} \frac{\chi(\omega)\,d\omega}{\omega - \omega_0} + \int_{\omega_0+\delta}^{Q} \frac{\chi(\omega)\,d\omega}{\omega - \omega_0} \right]$$
$$= \lim_{Q\to\infty} \left[\int_{-Q}^{\omega_0} \frac{\chi(\omega)\,d\omega}{\omega - \omega_0} + \int_{\omega_0}^{Q} \frac{\chi(\omega)\,d\omega}{\omega - \omega_0} \right]$$

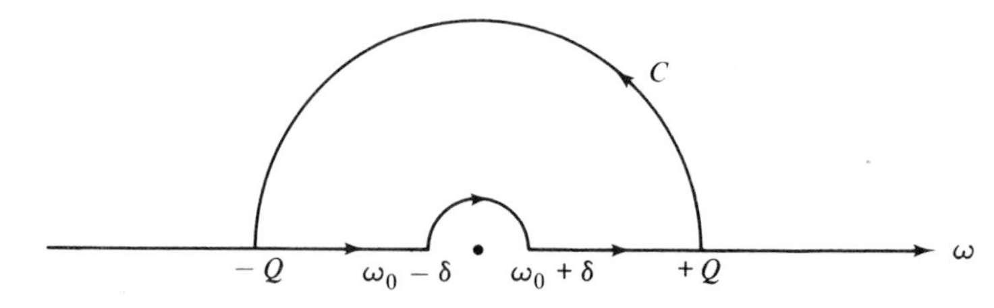

Figure 4.4

$$= \lim_{\Omega \to \infty} \int_{-\Omega}^{\Omega} \frac{\chi(\omega)\,d\omega}{\omega - \omega_0}$$

$$= P \int_{-\infty}^{\infty} \frac{\chi(\omega)\,d\omega}{\omega - \omega_0}$$

where P means the Cauchy principal value of the integral. We therefore obtain

$$-\chi(\omega_0) = \frac{i}{\pi} P \int_{-\infty}^{\infty} \frac{\chi(\omega)\,d\omega}{\omega - \omega_0}.$$

On substituting

$$\chi(\omega_0) = U(\omega_0) + iV(\omega_0) \quad \text{and} \quad \chi(\omega) = U(\omega) + iV(\omega)$$

into the preceding equation, we obtain

$$-[U(\omega_0) + iV(\omega_0)] = \frac{1}{\pi} P \int_{-\infty}^{\infty} \frac{[iU(\omega) - V(\omega)]\,d\omega}{\omega - \omega_0}$$

or

$$U(\omega_0) = \frac{1}{\pi} P \int_{-\infty}^{\infty} \frac{V(\omega)\,d\omega}{\omega - \omega_0} \tag{4.27}$$

and

$$V(\omega_0) = -\frac{1}{\pi} P \int_{-\infty}^{\infty} \frac{U(\omega)\,d\omega}{\omega - \omega_0}. \tag{4.28}$$

Equations (4.27) and (4.28) express one part of an analytic function in terms of an integral involving the other part. In mathematics, the functions U and V are referred to as the **Hilbert transform** of one another. In physics, these equations are usually written as

$$\text{Re}\,\chi(\omega_0) = \frac{1}{\pi} P \int_{-\infty}^{\infty} \frac{\text{Im}\,\chi(\omega)\,d\omega}{\omega - \omega_0} \tag{4.29}$$

and

$$\text{Im}\,\chi(\omega_0) = -\frac{1}{\pi} P \int_{-\infty}^{\infty} \frac{\text{Re}\,\chi(\omega)\,d\omega}{\omega - \omega_0}. \tag{4.30}$$

Equations (4.29) and (4.30) are called **dispersion relations.**

4.8 GEOMETRICAL REPRESENTATION

4.8.1 Introduction

In complex variable theory, we seek a graphical representation of $W = f(Z)$, where $Z = x + iy$ and $W = u(x, y) + iv(x, y)$. That is to say, each pair of values x and y corresponds to two values u and v. Hence we need a four-dimensional space to plot the real values u, v, x, and y. The mathematical subject of quaternions was developed by Hamilton (1805–1865) and Frobenius (1849–1917) to treat such systems. Limited applications of quaternions in quantum mechanics have been made in recent years, but it is not a widely discussed subject in mathematical methods of physics courses.

Riemann developed a mode of visualizing the relationship $W = f(Z)$ which uses two separate complex planes (Z-plane and W-plane). Riemann viewed a single-valued analytic function of a complex variable as indicating a specific one-to-one mapping between the points of two different two-dimensional surfaces. The relation $W = f(Z)$ establishes the connection between points of a given region Γ_Z in the Z-plane and the corresponding region Γ_W in the W-plane.

4.8.2 Conformal Transformation (Mapping)

For certain problems, complicated loci (figures) in the Z-plane can be transformed (mapped) into fairly simple loci (figures) in the W-plane. In these cases, the problem may be solved in the W-representation, and the result in the Z-representation can be obtained by use of the inverse transform, $Z = F(W)$. The converse of this illustration can also be performed.

On differentiating W with respect to Z, we obtain

$$\lim_{\Delta Z \to 0} \left(\frac{\Delta W}{\Delta Z}\right) = \frac{dW}{dZ} = f'(Z) = ae^{i\alpha} \tag{4.31}$$

where

$$a = |f'(Z)| \neq \begin{cases} \infty \\ 0. \end{cases} \tag{4.32}$$

Note that

$$\begin{aligned} \arg\left[\lim_{\Delta Z \to 0} \left(\frac{\Delta W}{\Delta Z}\right)\right] &= \arg[f'(Z)] \\ &= \lim_{\Delta Z \to 0} \left[\arg\left(\frac{\Delta W}{\Delta Z}\right)\right] \\ &= \lim_{\Delta Z \to 0} [\arg(\Delta W)] - \lim_{\Delta Z \to 0} [\arg(\Delta Z)]. \end{aligned} \tag{4.33}$$

We may therefore write

$$\alpha = \phi_W - \theta_Z \tag{4.34}$$

since

$$\arg f'(Z) = \alpha = \lim_{\Delta Z \to 0} [\arg (\Delta W)] - \lim_{\Delta Z \to 0} [\arg (\Delta Z)].$$

Thus, any line oriented at θ_Z in the Z-plane is rotated through α in the W-plane (see Fig. 4.5).

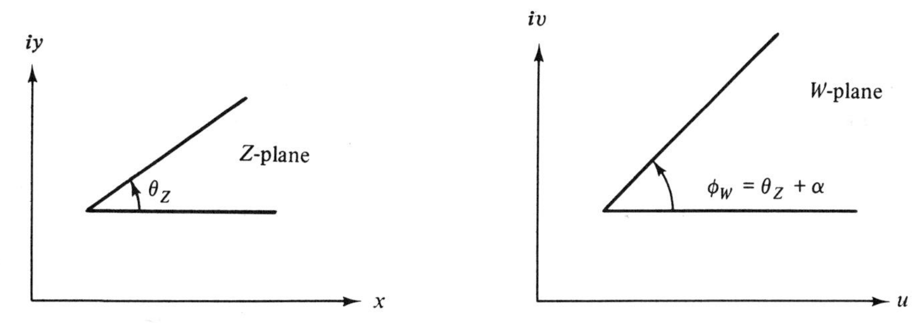

Figure 4.5

The angle between two lines L_W^1 and L_W^2 gives rise to

$$\phi_2 - \phi_1 = \theta_2 - \theta_1. \tag{4.35}$$

Hence the angle is preserved under an analytic transformation; this is referred to as a **conformal transformation (mapping)**. On taking the modulus of both sides of Eq. (4.31), we obtain

$$|dW| = |ae^{i\alpha}\, dZ| = a|dZ|. \tag{4.36}$$

This equation shows that an infinitesimal arc of Γ_W is a times the corresponding arc Γ_Z (magnification).

EXAMPLE 4.12 *Translation* Consider the transformation $W = Z + Z_0$. In terms of x and y, we have

$$\begin{aligned} W &= (x + iy) + (x_0 + iy_0) \\ &= (x + x_0) + i(y + y_0) \end{aligned}$$

or

$$u = x + x_0 \quad \text{and} \quad v = y + y_0 \qquad \text{(see Fig. 4.6).}$$

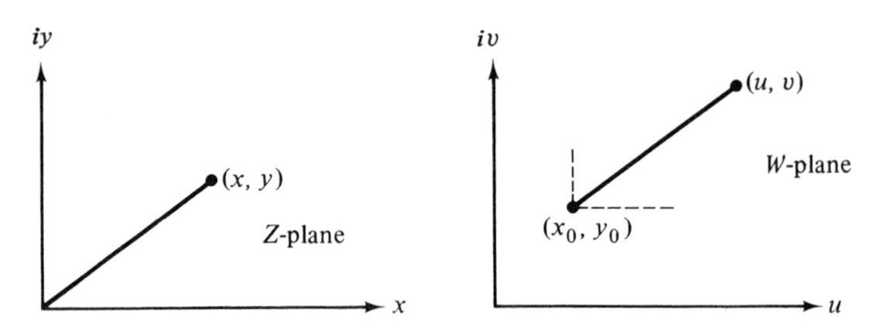

Figure 4.6

EXAMPLE 4.13 *Rotation* Consider the transformation $W = Z_0 Z$, where $Z = re^{i\theta}$, $Z_0 = r_0 e^{i\theta_0}$, and $W = \rho e^{i\phi}$. Here we may write

$$W = \rho e^{i\phi} = r_0 r e^{i(\theta + \theta_0)}.$$

Note that

$$\phi = \theta + \theta_0 \tag{4.37}$$

where θ_0 is the angle of rotation and r_0 is the corresponding modification of r ($\rho = r_0 r$).

EXAMPLE 4.14 *Inversion* Consider the transformation $W = 1/Z$, where $Z = re^{i\theta}$ and $W = \rho e^{i\phi}$. The quantity W becomes

$$W = \rho e^{i\phi} = \frac{1}{r} e^{-i\theta}.$$

In this case, we have

$$\rho = \frac{1}{r} \tag{4.38}$$

and

$$\phi = -\theta. \tag{4.39}$$

We now develop a scheme for transforming a locus in the Z-plane into the corresponding locus in the W-plane by inversion. Note that

$$\begin{aligned} W = u + iv &= \frac{1}{Z} \\ &= \frac{1}{x + iy} = \frac{x - iy}{x^2 + y^2}. \end{aligned}$$

In this case, we have

$$u = \frac{x}{x^2 + y^2}; \qquad x = \frac{u}{u^2 + v^2} \tag{4.40}$$

and

$$v = -\frac{y}{x^2 + y^2}; \qquad y = -\frac{v}{u^2 + v^2}. \tag{4.41}$$

EXAMPLE 4.15 Transform the equation of a circle centered at the origin in the Z-plane, $x^2 + y^2 = r^2$, into the corresponding locus in the W-plane (by inversion).

Solution In this example, we obtain

$$\frac{u^2}{(u^2 + v^2)^2} + \frac{v^2}{(u^2 + v^2)^2} = r^2$$

or

$$u^2 + v^2 = r^2(u^2 + v^2)^2 = \rho^2$$

since $u + iv = \rho e^{i\phi}$, $\rho^2 = u^2 + v^2$, and $\rho = 1/r$. The equation, $u^2 + v^2 = \rho^2$, is that of a circle in the W-plane centered at the origin with radius ρ. It is interesting to note that the interior points of the Z-plane circle are exterior points of the W-plane circle.

EXAMPLE 4.16 By inversion, transform the horizontal line $y = c$ into the corresponding locus in the W-plane.

Solution Here we have

$$\begin{aligned} y &= c \\ &= -\frac{v}{u^2 + v^2}. \end{aligned}$$

Simplifying this equation, we obtain

$$cu^2 + cv^2 + v = 0$$

or

$$(u - 0)^2 + \left(v + \frac{1}{2c}\right)^2 = \left(\frac{1}{2c}\right)^2.$$

The last equation is that of a circle centered at $(0, -1/2c)$ with radius $1/2c$ in the W-plane (see Fig. 4.7).

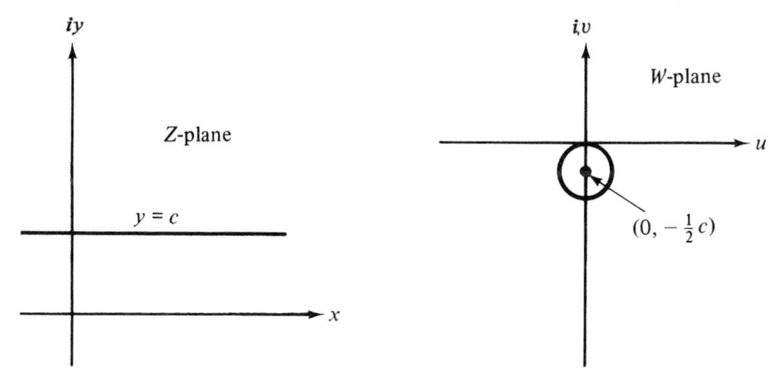

Figure 4.7

4.9 PROBLEMS

4.1 If $M\,dx + N\,dy$ is an exact differential $\left(\frac{\partial M}{\partial y} = \frac{\partial N}{\partial x}\right)$, show that

$$\oint (M\,dx + N\,dy) = 0.$$

4.2 Evaluate (select a convenient path)

(a) $\int_{0,0}^{1,1} Z^*\,dZ$

(b) $\oint \frac{dZ}{Z}$

(c) $\oint \frac{dZ}{Z^2}$.

4.3 (a) Show that

$$\int_{Z_0}^{Z} Z^n\,dZ = \frac{Z^{n+1} - Z_0^{n+1}}{n+1}$$

for all n except $n = -1$.

(b) Discuss the case for $n = -1$.

4.4 Show that

$$\oint_C (Z - Z_0)^n\,dZ = \begin{cases} 2\pi i & (n = -1) \\ 0 & (n \neq -1) \end{cases}$$

where Z_0 is within C.

4.5 Show that

$$\frac{1}{2\pi i}\oint_C Z^{m-n-1}\,dZ = \delta_{m,n}.$$

4.6 Locate and classify the singular point(s) of the following functions and evaluate the residue(s):

(a) $\frac{e^{1/Z}}{Z^2}$ (b) $\frac{1}{(Z^2 - 1)^2}$

(c) $\frac{\sin Z}{Z^2}$ (d) $\frac{e^Z}{Z}$

(e) $\frac{\sin Z}{(Z-1)^3}$ (f) $\frac{\cos(\sin Z)}{\sin Z}$.

4.7 Evaluate the residue(s) of the following functions:

(a) $\frac{Z^2 - 1}{Z^2 + 1}$ (b) $\frac{e^Z}{(Z-2)^3}$

(c) $\frac{1}{Z^2 + 4Z + 1}$ (d) $\frac{1}{(Z-1)^3}$

(e) $\dfrac{Ze^{iZ}}{Z^2 - a^2}$ (f) $\dfrac{1}{\cos Z}$.

4.8 By use of the residue theorem, evaluate:

(a) $\displaystyle\int_0^{2\pi} \frac{d\theta}{5 - 4\cos\theta}$

(b) $\displaystyle\int_0^{2\pi} \frac{d\theta}{a + b\cos\theta}$ for $|a| > |b|$

(c) $\displaystyle\int_0^{2\pi} \frac{d\theta}{3 - 2\cos\theta + \sin\theta}$.

4.9 By use of the residue theorem, evaluate:

(a) $\displaystyle\int_{|Z|=2} \frac{\cos Z\, dZ}{Z^3}$

(b) $\displaystyle\int_{|Z|=2} \frac{dZ}{1 - Z^2}$

(c) $\displaystyle\int_{|Z|=1} \frac{dZ}{Z + 2}$.

4.10 Show that:

(a) $\displaystyle\int_0^{\pi} \frac{d\theta}{a + \cos\theta} = \frac{\pi}{(a^2 + 1)^{1/2}}$ $(a > 1)$

(b) $\displaystyle P\int_{-\infty}^{\infty} \frac{e^{ix}}{x}\, dx = \pi i$

(c) $\displaystyle\int_0^{\infty} \frac{\sin^2 x}{x^2}\, dx = \frac{\pi}{2}$.

4.11 Refer to Fig. 4.8. Determine R_W if $W = Z + (1 - 2i)$. Draw the appropriate W-plane diagram.

4.12 Show that $W = Z^2$ is conformal except at $Z = 0$. Discuss the case for $Z = 0$.

4.13 For $W = Z^2$:

(a) Show that $u(x, y) = x^2 - y^2$.

(b) Sketch the Z-plane and W-plane areas bounded by $u = 1$, $v = 2$, and $u = 4$, $v = 8$.

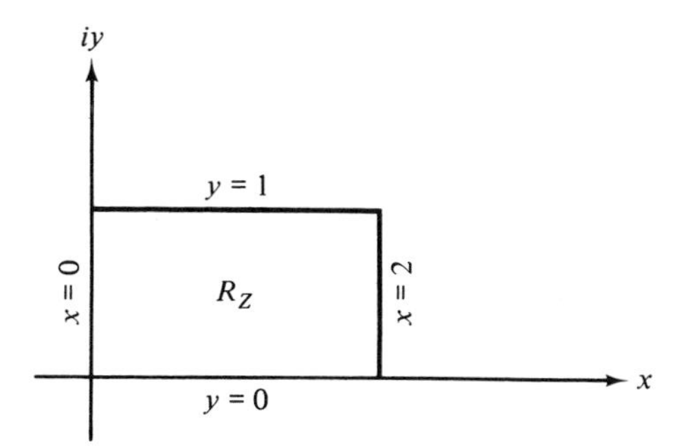

Figure 4.8

4.14 Of the following functions, determine the ones that satisfy the conditions in the statement of Jordan's lemma:

(a) $\dfrac{1}{Z^2+1}$ (b) $\sin Z$

(c) e^Z (d) $\dfrac{e^{iZ}}{Z^2+Z+1}$.

4.15 **Jordan's Inequality**: Show that

$$\frac{2\theta}{\pi} \leq \sin\theta \leq 0 \qquad \left(\text{for } 0 \leq \theta \leq \frac{\pi}{2}\right).$$

Hint: Use the fact that, since $\cos\theta$ decreases steadily as θ increases from 0 to $\pi/2$, the mean ordinate

$$\frac{1}{\theta}\int_0^\theta \cos x\, dx$$

of the graph of $y = \cos x$ over the range $0 \leq x \leq \theta$ also decreases steadily.

4.16 Evalute

$$P\int_{-1}^{2} \frac{dx}{x^3}$$

4.17 Show that $U(\omega_0) = \cos(k\omega_0)$ and $V(\omega_0) = \sin(k\omega_0)$ for positive k are a Hilbert transform pair.

4.18 For $\chi(\omega_0) = \chi(-\omega_0)$:

(a) Show that $U(\omega_0) = U(-\omega_0)$ and $V(\omega_0) = -V(-\omega_0)$—crossing relations for collision theory.

(b) Develop the corresponding equations for the Hilbert transform pair which involve integrations over the positive x-axis only.

5

Differential Equations

The set of mathematical techniques used to solve differential equations consists of a bag of tricks.

Physics is the study of the solutions of first- and second-order differential equations.

5.1 INTRODUCTION

A **differential equation** is an equation which contains derivatives, and it may be either an ordinary or a partial differential equation. **Ordinary differential equations** contain total derivatives (one independent variable), and **partial differential equations** contain partial derivatives with respect to two or more independent variables.

The **order** of a differential equation is the order of the highest derivative appearing in the equation. The **degree** of a differential equation is the power of the highest derivative after fractional powers of all derivatives have been removed. If the dependent variable and all of its derivatives occur in the first power without a product of the dependent variable and a derivative, the differential equation is said to be **linear**. We restrict this chapter to linear differential equations (or differential equations that are reducible to the linear form).

A physical process is described by a differential equation and the required boundary conditions in space and/or initial conditions in time. The boundary and/or initial conditions determine from the many possible solutions the one that describes the specific physical phenomenon involved.

Our main purpose here is the development of a solution of a differential equation which adequately describes the physical process under investigation. The mathematical subject of existence and uniqueness theorems for the solutions of differential equations is beyond the scope of this book.

An elementary introduction to the subject of differential equations, as it relates to the needs of students in the physical sciences, can be reduced to that of treating linear (or reducible to the linear form) first- and second-order differential equations. The emphasis in this chapter is on the construction of solutions for such differential equations.

5.2 ORDINARY DIFFERENTIAL EQUATIONS

First- and second-order linear ordinary differential equations have the following standard forms, respectively:

$$\frac{dy}{dx} + p(x)y = Q(x) \quad \text{or} \quad y' + p(x)y = Q(x) \tag{5.1}$$

and

$$\frac{d^2y}{dx^2} + p(x)\frac{dy}{dx} + q(x)y = f(x) \quad \text{or} \quad y'' + p(x)y' + q(x)y = f(x) \tag{5.2}$$

where $y'' \equiv d^2y/dx^2$ and $y' \equiv dy/dx$. When time, t, is the independent variable, we write $\ddot{y} \equiv d^2y/dt^2$ and $\dot{y} \equiv dy/dt$. If the right-hand sides of Eqs. (5.1) and (5.2) are equal to zero, the equations are called **homogeneous**. If the right-hand sides of Eqs. (5.1) and (5.2) are different from zero, the equations are classified as **nonhomogeneous (inhomogeneous)**. One should not confuse this nomenclature with that involving functions. A function $F(x, y)$ is said to be homogeneous of degree n if

$$F(tx, ty) = t^nF(x, y). \tag{5.3}$$

5.2.1 First-Order Homogeneous and Nonhomogeneous Equations with Variable Coefficients

A. Separable Equations Equations which can be put in the form

$$g(y)dy = f(x)dx \tag{5.4}$$

where the left-hand side is a function of y only and the right-hand side is a function of x only, are called **separable equations**. For $dy = f(x)\,dx$, the integral (solution) is

$$y = \int f(x)dx + C$$

where C is an arbitrary constant of integration. Since the general solution of a first-order differential equation results from one integration, it will contain one arbitrary constant. Similarly, the general solution of a second-order differential equation will contain two arbitrary constants. The values of the arbitrary constants will be determined from the physical boundary or initial conditions.

EXAMPLE 5.1 In the radioactive decay of nuclei, we find that the process is governed by the following equation:

$$\frac{dN(t)}{dt} = -\lambda N(t) \qquad (N(t=0) = N_0)$$

where $N(t)$ represents the number of parent atoms present at time t, λ is the decay constant which is characteristic of the particular element involved, and the negative sign is used to indicate that the number of parent atoms decreases with time.

Solution This equation may be written in the form

$$\frac{dN}{N} = -\lambda\, dt.$$

On integrating the above equation, we obtain

$$\ln N = -\lambda t + C_1$$

or

$$N(t) = C_2 e^{-\lambda t}$$

which is the required general solution. We now determine the value of the constant of integration by use of the initial condition, $N(t=0) = N_0$. We obtain $N(0) = N_0 = C_2$; hence the solution becomes

$$N(t) = N_0 e^{-\lambda t}.$$

B. Exact Differentials Consider a function of x and y with continuous first derivatives such that

$$F(x, y) = C; \qquad (C = \text{constant}). \tag{5.5}$$

The total differential of $F(x, y)$ is

$$dF = \frac{\partial F}{\partial x}\, dx + \frac{\partial F}{\partial y}\, dy = 0$$

or

$$M(x, y)\, dx + N(x, y)\, dy = 0 \tag{5.6}$$

where

$$M(x, y) = \frac{\partial F}{\partial x} \quad \text{and} \quad N(x, y) = \frac{\partial F}{\partial y}.$$

Since it is assumed that $F(x, y)$ has continuous first derivatives, we note that

$$\frac{\partial M}{\partial y} = \frac{\partial N}{\partial x}. \tag{5.7}$$

The differential equation involving M and N is called **exact** since the left-hand side of Eq. (5.6) can be obtained from $F(x, y)$ by differentiation only. The condition indicated in Eq. (5.7) is both necessary and sufficient for Eq. (5.6) to be an **exact differential equation.**

If the condition $\partial M/\partial y = \partial N/\partial x$ is not satisfied for the differential equation in the form of Eq. (5.6), it may be multiplied by a function, the **integrating factor**, so that the resulting equation is exact. However, there is no known method for finding the integrating factor in the general case.

EXAMPLE 5.2 Determine whether the following equation is exact, and find its solution if it is exact:

$$(4x^3 + 6xy + y^2)\frac{dx}{dy} = -(3x^2 + 2xy + 2).$$

Solution The standard form is

$$(4x^3 + 6xy + y^2)dx + (3x^2 + 2xy + 2)dy = 0$$

so that

$$\frac{\partial M}{\partial y} = \frac{\partial}{\partial y}(4x^3 + 6xy + y^2) = 6x + 2y$$

and

$$\frac{\partial N}{\partial x} = \frac{\partial}{\partial x}(3x^2 + 2xy + 2) = 6x + 2y.$$

The condition in Eq. (5.7) is satisfied; hence the differential equation is exact, and the solution has the form

$$F(x, y) = C$$

where

$$M = \frac{\partial F}{\partial x} \quad \text{and} \quad N = \frac{\partial F}{\partial y}.$$

For this equation, we have

$$\frac{\partial F}{\partial x} = 4x^3 + 6xy + y^2$$

or

$$F(x, y) = x^4 + 3x^2y + y^2x + f(y)$$

and

$$\frac{\partial F}{\partial y} = 3x^2 + 2xy + 2$$

or

$$F(x, y) = 3x^2y + y^2x + 2y + g(x)$$

where $f(y)$ and $g(x)$ are functions that arise from integrating $F(x, y)$ with respect to x and y, respectively. For consistency, we require that

$$f(y) = 2y \quad \text{and} \quad g(x) = x^4.$$

Therefore the required solution is

$$F(x, y) = x^4 + 3x^2y + xy^2 + 2y = C.$$

C. General First-Order Linear Differential Equation We seek a general solution for Eq. (5.1). The procedure for obtaining this solution is outlined by the following two steps:

Step 1 Solve the corresponding homogeneous equation ($Q = 0$).

$$y_h' + p(x)y_h = 0$$
$$\frac{dy_h}{y_h} = -p(x)\,dx$$

and

$$y_h = Ae^{-\int p(x)\,dx}.$$

The quantity $e^{\int p(x)\,dx}$ is referred to as the **integrating factor.**

Step 2 Multiply the original equation by the integrating factor.

On multiplying Eq. (5.1) by $e^{\int p(x)\,dx}$, we obtain

$$(y' + p(x)y)e^{\int p(x)\,dx} = Q(x)e^{\int p(x)\,dx}. \tag{5.8}$$

Since

$$\frac{d}{dx}(ye^{\int p(x)\,dx}) = (y' + p(x)y)e^{\int p(x)\,dx}$$

Eq. (5.8) may be written as

$$\frac{d}{dx}(ye^{\int p(x)\,dx}) = Q(x)e^{\int p(x)\,dx}. \tag{5.9}$$

Integrating Eq. (5.9), we get

$$ye^{\int p(x)\,dx} = \int Q(x)e^{\int p(x)\,dx}\,dx + C$$

or

$$y = e^{-\int p(x)\,dx}\int Q(x)e^{\int p(x)\,dx}\,dx + Ce^{-\int p(x)\,dx}. \tag{5.10}$$

Equation (5.10) is the general solution of Eq. (5.1). It is important to substitute the solution back into the original equation to see if a workable solution has been obtained.

EXAMPLE 5.3 Derive an expression for the current in the circuit in Fig. 5.1.

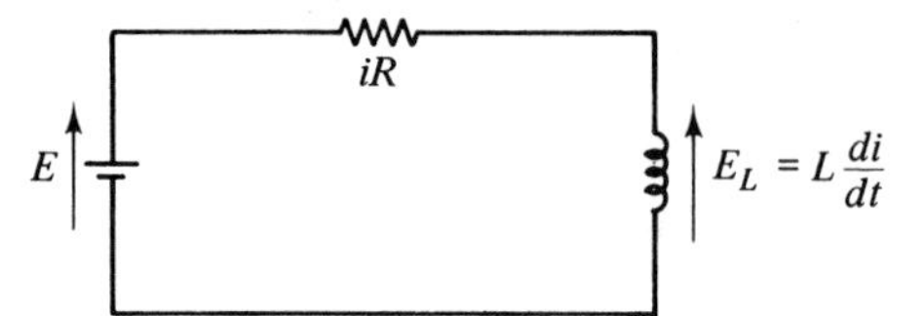

Figure 5.1

Solution The loop method gives

$$L\frac{di}{dt} + iR = E \qquad \text{(where } i(t = 0) = 0\text{)}.$$

The standard form of this differential equation is

$$\frac{di}{dt} + \frac{R}{L}i = \frac{E}{L}.$$

On applying Eq. (5.10) [$p(x) \to R/L$ and $Q(x) \to E/L$], we obtain

$$i(t) = e^{-Rt/L}\int \frac{E}{L}e^{Rt/L}\,dt + Ce^{-Rt/L}$$
$$= \frac{E}{R} + Ce^{-Rt/L}.$$

Note that

$$i(t = 0) = 0 = \frac{E}{R} + C.$$

Hence the required expression for the current is

$$i(t) = \frac{E}{R}(1 - e^{-Rt/L}).$$

D. The Jacob Bernoulli (1654–1705) Equation Bernoulli's equation has the form

$$y' + p(x)y = Q(x)y^n \qquad (n \neq 1). \tag{5.11}$$

Equation (5.11) is a nonlinear equation since it contains a nonlinear term in the dependent variable, y^n. This equation can be reduced to the linear form; it is for this reason that we consider solving this equation. The technique used here might be useful in treating other nonlinear equations. To reduce Eq. (5.11) to the linear form, we first make a change of variable. Let

$$z = y^{1-n}$$

so that

$$z' = (1 - n)y^{-n}y'.$$

On multiplying Eq. (5.11) by $(1 - n)y^{-n}$, we obtain

$$(1 - n)y^{-n}y' + (1 - n)y^{1-n}p(x) = (1 - n)Q(x).$$

In terms of the variable z, we write

$$z' + (1 - n)p(x)z = (1 - n)Q(x). \tag{5.12}$$

Equation (5.12) is the required linear form. It can be solved by use of the method outlined in Section 5.2.1C, Eq. (5.10). The solution of Eq. (5.12) is

$$z = e^{-(1-n)\int p(x)\,dx} \int (1 - n)Q(x)e^{(1-n)\int p(x)\,dx}\,dx + Ce^{-(1-n)\int p(x)\,dx}. \tag{5.13}$$

EXAMPLE 5.4 Solve the equation $y' + x = y/x$.

Solution In standard form, the equation becomes

$$y' - \frac{1}{x}y = -x.$$

$$\left(p(x) \longrightarrow -\frac{1}{x};\quad Q(x) \longrightarrow -x\right).$$

On applying Eq. (5.10), we obtain

$$
\begin{aligned}
y &= e^{-\int(-dx/x)} \int (-x) e^{\int(-dx/x)}\,dx + Ce^{-\int(-dx/x)} \\
&= -e^{\ln x} \int xe^{-\ln x}\,dx + Ce^{\ln x} \\
&= -x^2 + Cx.
\end{aligned}
$$

EXAMPLE 5.5 Solve the equation $y^3 y' + y^4/x = x$.

Solution The standard form is

$$
y' + \frac{1}{x} y = xy^{-3}.
$$

This is a differential equation of the Bernoulli form.

$$
p(x) \longrightarrow \frac{1}{x}; \qquad n \longrightarrow -3; \qquad Q(x) \longrightarrow x.
$$

On applying Eq. (5.13), we obtain

$$
\begin{aligned}
z &= e^{-4\int dx/x} \int 4xe^{4\int dx/x}\,dx + Ce^{-4\int dx/x} \\
&= 4e^{-4\ln x} \int xe^{4\ln x}\,dx + Ce^{-4\ln x} \\
&= \frac{4}{x^4} \int x^5\,dx + \frac{C}{x^4} \\
&= \frac{2}{3} x^2 + \frac{C}{x^4}.
\end{aligned}
$$

Since $z = y^{1-n}$, we have

$$
z = y^4 = \frac{2}{3} x^2 + \frac{C}{x^4}
$$

or

$$
y = \left(\frac{2}{3} x^2 + \frac{C}{x^4} \right)^{1/4}
$$

which is the required solution.

5.2.2 The Superposition Principle†

The **superposition of solutions principle**, stated in the form of the two theorems below, will be assumed valid and used extensively in solving second-order linear homogeneous differential equations.

†E. A. Coddington and N. Levison, *Theory of Ordinary Differential Equations*, McGraw-Hill Book Company, New York, 1955; Chapter 3.

THEOREM 1 The set of all solutions of an nth order linear homogeneous differential equation forms an n-dimensional vector space (see Section 2.2).

To illustrate Theorem 1, note that

$$y = c_1 y_1 + c_2 y_2$$

satisfies

$$y'' + p(x)y' + q(x)y = 0$$

if y_1 and y_2 are two linearly independent solutions of the above differential equation and c_1 and c_2 are two arbitrary constants.

THEOREM 2 A necessary and sufficient condition that solutions y_1 and y_2 of

$$y'' + p(x)y' + q(x)y = 0$$

be linearly independent is that

$$\begin{vmatrix} y_1 & y_2 \\ y_1' & y_2' \end{vmatrix} \neq 0.$$

The above determinant is called the **Wronskian** (Höne 1778–1853).

5.2.3 Second-Order Homogeneous Equations with Constant Coefficients

In this section, we outline the characteristic equation method for solving second-order homogeneous equations with constant coefficients. The general equation in standard form is

$$y'' + p_0 y' + q_0 y = 0 \tag{5.14}$$

where p_0 and q_0 are constants. Equation (5.14) may be written as

$$(D^2 + p_0 D + q_0)y = 0$$

where the linear operator D is defined by $D = d/dx$. The algebraic equation

$$D^2 + p_0 D + q_0 = 0 \tag{5.15}$$

is called the **auxiliary** or **characteristic equation**. Our procedure for solving Eq. (5.14) involves treating Eq. (5.15) algebraically. This clearly leads us to consider three cases which characterize the roots of a quadratic equation, that is, the roots may be (1) real and unequal, (2) real and equal, or (3) a complex-conjugate pair.

Case 1 The roots are real and unequal.

Suppose a and b are the roots of Eq. (5.15). We may then write

$$(D - a)(D - b)y = 0$$

or

$$(D - a)y_1 = 0 \quad \text{and} \quad (D - b)y_2 = 0.$$

By use of Eq. (5.10), we obtain

$$y_1 = C_1 e^{ax} \quad \text{and} \quad y_2 = C_2 e^{bx}. \tag{5.16}$$

The general solution of Eq. (5.14) is a linear combination of the two linearly independent solutions y_1 and y_2. That is to say,

$$\begin{aligned} y &= y_1 + y_2 \\ &= C_1 e^{ax} + C_2 e^{bx} \end{aligned} \tag{5.17}$$

where C_1 and C_2 are arbitrary constants. The Wronskian for y_1 and y_2 is (see Theorem 2 in Section 5.2.2)

$$\begin{vmatrix} C_1 e^{ax} & C_2 e^{bx} \\ C_1 a e^{ax} & C_2 b e^{bx} \end{vmatrix} \neq 0$$

since $a \neq b$ by hypothesis. Hence y_1 and y_2 are linearly independent.

Case 2 The roots are real and equal.

Since $a = b$, we have

$$(D - a)(D - a)y = 0. \tag{5.18}$$

One solution of Eq. (5.18) is

$$y_1 = C_3 e^{ax}.$$

We must now obtain a second linearly independent solution. Let

$$u = (D - a)y_2. \tag{5.19}$$

Equation (5.18) now becomes

$$(D - a)u = 0. \tag{5.20}$$

The solution of Eq. (5.20) is

$$u = C_4 e^{ax}.$$

Therefore Eq. (5.19) becomes

$$(D - a)y_2 = C_4 e^{ax}$$

or

$$y_2' - ay_2 = C_4 e^{ax}. \tag{5.21}$$

Equation (5.21) can be solved by the method outlined in Section 5.2.1C, Eq. (5.10). Its solution is

$$\begin{aligned} y_2 &= C_4 e^{ax} \int e^{ax} \cdot e^{-ax}\, dx + C_5 e^{ax} \\ &= (C_4 x + C_5)e^{ax}. \end{aligned}$$

In general, we write

$$y = (Ax + B)e^{ax}. \tag{5.22}$$

where A and B are arbitrary constants to be determined by the physical conditions (initial or boundary conditions).

Case 3 The roots are a complex-conjugate pair.

Here the problem is equivalent to that outlined in Case 1 if we set $b = a^*$ (a^* means the complex conjugate of a).

EXAMPLE 5.6 Solve $y'' + y' - 2y = 0$.

Solution The corresponding characteristic equation is

$$(D^2 + D - 2) = 0.$$

We may therefore write

$$(D - 1)(D + 2)y = 0.$$

Here we have

$$y_1 = C_1 e^x \qquad y_2 = C_2 e^{-2x}.$$

The general solution is therefore (by use of Case 1)

$$y = C_1 e^x + C_2 e^{-2x}.$$

EXAMPLE 5.7 Solve $y'' - 2y' + y = 0$.

Solution Here we write

$$(D - 1)(D - 1)y = 0.$$

Hence the solution is (by use of the development for Case 2)

$$y = (Ax + B)e^{x}.$$

EXAMPLE 5.8 Solve $y'' + 9y = 0$.

Solution Here we have

$$(D^2 + 9)y = 0.$$

The roots of the characteristic equation are $\pm 3i$. Therefore the solution is (by use of the analysis for Case 3)

$$y = Ae^{3ix} + Be^{-3ix}.$$

EXAMPLE 5.9 **Classical Harmonic Oscillator**. Consider the motion of a mass, m, attached to a spring with spring constant k (see Fig. 5.2). When the mass

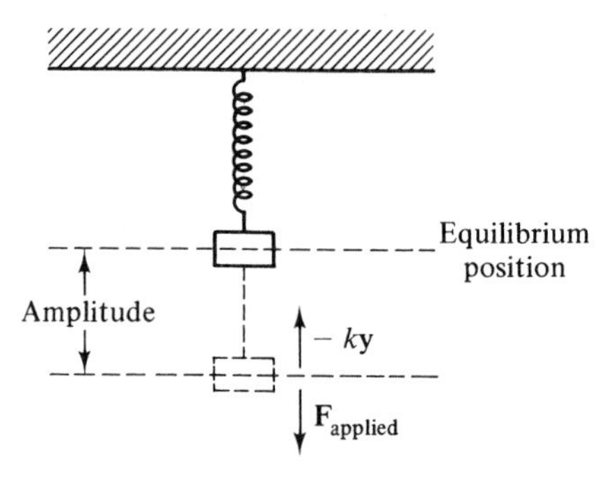

Figure 5.2

is released, the resulting motion is called **simple harmonic motion** since the acceleration is directly proportional to the displacement and in the opposite direction. The equation of motion is

$$F_y = -ky$$

or

$$m\frac{d^2y}{dt^2} = -ky.$$

The standard form of the above differential equation is

$$\frac{d^2y}{dt^2} + \omega^2 y = 0$$

where

$$\omega^2 = \frac{k}{m}.$$

Solution The general solution is

$$\begin{aligned} y(t) &= Ae^{i\omega t} + Be^{-i\omega t} \\ &= A(\cos \omega t + i \sin \omega t) + B(\cos \omega t - i \sin \omega t) \\ &= A' \cos \omega t + B' \sin \omega t \end{aligned}$$

where $A' = A + B$ and $B' = i(A - B)$. The arbitrary constants must be determined from initial conditions.

The **period** of the motion is determined by requiring that

$$y(t) = y(t + T)$$

where T is the period. Hence we have

$$\begin{aligned} A' \cos \omega t + B' \sin \omega t = A'(&\cos \omega t \cos \omega T - \sin \omega t \sin \omega T) \\ &+ B'(\sin \omega t \cos \omega T + \sin \omega T \cos \omega t). \end{aligned}$$

We now require that

$$\cos \omega T = 1 \quad \text{and} \quad \sin \omega T = 0$$

or

$$\omega T = 2\pi n \qquad (n = 0, 1, 2, \ldots).$$

Therefore the period is given by

$$T = \frac{2\pi n}{\omega} = 2\pi n \sqrt{\frac{m}{k}}.$$

5.2.4 Second-Order Nonhomogeneous Equations with Constant Coefficients

In this section, we consider equations of the form

$$y'' + p_0 y' + q_0 y = f(x). \tag{5.23}$$

The general solution of Eq. (5.23) may be written as

$$y = y_c + y_p$$

where $y_c = y_h$, **complementary solution,** is the solution of the corresponding homogeneous equation and y_p, **particular solution,** is any known solution of

Eq. (5.23). The complementary solution can be obtained by techniques discussed in Section 5.2.3. There are various methods for finding y_p which depend on the form of $f(x)$. In many cases, one may obtain y_p by inspection (guessing). However, we will develop a general procedure (**successive integration**) for solving Eq. (5.23) directly. Note that this method always yields a general solution, but it should be reserved for cases in which y_p cannot be obtained easily.

Consider Eq. (5.23) in the form

$$(D - a)(D - b)y = f(x) \tag{5.24}$$

where a and b are the roots of the corresponding characteristic equation. Let

$$u = (D - b)y. \tag{5.25}$$

Equation (5.24) now becomes

$$(D - a)u = f(x)$$

or

$$u' - au = f(x). \tag{5.26}$$

The solution of Eq. (5.26) can be obtained by applying Eq. (5.10); it is

$$u = e^{ax} \int f(x)e^{-ax}\,dx + C_1 e^{ax}.$$

Equation (5.25) now becomes

$$y' - by = e^{ax} \int f(x)e^{-ax}\,dx + C_1 e^{ax} \tag{5.27}$$

and the solution of Eq. (5.27) is

$$y = e^{bx} \int \bar{Q}(x)e^{-bx}\,dx + C_2 e^{bx} \tag{5.28}$$

where

$$\bar{Q}(x) = e^{ax} \int f(x)e^{-ax}\,dx + C_1 e^{ax}. \tag{5.29}$$

EXAMPLE 5.10 Solve $y'' - 2y' + y = 2 \cos x$.

Solution Here we have

$$(D^2 - 2D + 1)y = 2 \cos x$$

or

$$(D - 1)(D - 1)y = 2 \cos x.$$

The solution of the homogeneous equation is

$$y_c = (Ax + B)e^x.$$

By inspection, we observe that

$$y_p = -\sin x$$

since

$$y_p' = -\cos x \quad \text{and} \quad y_p'' = \sin x.$$

Therefore the solution is

$$\begin{aligned} y &= y_c + y_p \\ &= (Ax + B)e^x - \sin x. \end{aligned}$$

EXAMPLE 5.11 Solve $y'' - 2y' + y = 2\cos x$ by use of successive integration.

Solution The roots of the characteristic equation are real and equal, +1. On applying Eq. (5.28), we obtain

$$y = e^x \int \bar{Q}(x)e^{-x}\,dx + C_2e^x \tag{5.30}$$

where

$$\bar{Q}(x) = 2e^x \int e^{-x}\cos x\,dx + C_1e^x.$$

Note that

$$\int e^{mx}\cos nx\,dx = \frac{e^{mx}(n\sin nx + m\cos nx)}{m^2 + n^2} + C \tag{5.31}$$

and

$$\int e^{mx}\sin nx\,dx = \frac{e^{mx}(m\sin nx - n\cos nx)}{m^2 + n^2} + C. \tag{5.32}$$

By use of Eq. (5.31), we find that

$$\begin{aligned} \bar{Q}(x) &= \frac{2e^x[e^{-x}(\sin x - \cos x)]}{2} + C_1e^x \\ &= \sin x - \cos x + C_1e^x. \end{aligned}$$

Equation (5.30) now becomes [by use of Eq. (5.32)]

$$y = e^x\left[\int e^{-x} \sin x\, dx - \int e^{-x} \cos x\, dx + C_1 \int dx\right] + C_2 e^x$$

$$= e^x\left[\frac{e^{-x}(-\sin x - \cos x)}{2} - \frac{e^{-x}(\sin x - \cos x)}{2}\right] + C_1 x e^x + C_2 e^x$$

$$= -\sin x + (C_1 x + C_2)e^x.$$

Note that we obtained the same solution as in Example 5.10. However, more work was required using the method of successive integration.

5.2.5 Second-Order Nonhomogeneous Equations with Variable Coefficients

Consider equations of the form

$$y'' + p(x)y' + q(x)y = f(x). \tag{5.33}$$

The method of **variation of parameters** due to Lagrange (1736–1813) will be used to solve this general equation. Assume the following form for the solution:

$$y = v_1 y_1 + v_2 y_2 \tag{5.34}$$

where y_1 and y_2 are two independent solutions of the corresponding homogeneous equations

$$y_1'' + p(x)y_1' + q(x)y_1 = 0 \tag{5.35}$$

and

$$y_2'' + p(x)y_2' + q(x)y_2 = 0. \tag{5.36}$$

We will develop a general method for obtaining the required solution of the homogeneous equations in Section 5.2.6. The coefficients v_1 and v_2 are two unknown functions. It is also assumed that $f(x)$ is continuous in some interval $a \leq x \leq b$. Note that

$$\begin{aligned} y' &= v_1' y_1 + v_1 y_1' + v_2' y_2 + v_2 y_2' \\ &= (v_1 y_1' + v_2 y_2') + (v_1' y_1 + v_2' y_2). \end{aligned} \tag{5.37}$$

To simplify the calculation of y'', we choose v_1 and v_2 so that

$$v_1' y_1 + v_2' y_2 = 0. \tag{5.38}$$

The final solution obtained for Eq. (5.33) will naturally be subject to the condition in Eq. (5.38). Equation (5.37) becomes

$$y' = v_1 y_1' + v_2 y_2'. \tag{5.39}$$

On differentiating Eq. (5.39) with respect to x, we obtain

$$y'' = v_1'y_1' + v_1y_1'' + v_2'y_2' + v_2y_2''. \tag{5.40}$$

Substituting Eqs. (5.34), (5.39), and (5.40) into Eq. (5.33), we obtain

$$v_1'y_1' + v_1y_1'' + v_2'y_2' + v_2y_2'' + p(x)v_1y_1' + p(x)v_2y_2' + q(x)v_1y_1 + q(x)v_2y_2 = f(x)$$

or

$$v_1(y_1'' + p(x)y_1' + q(x)y_1) + v_2(y_2'' + p(x)y_2' + q(x)y_2) + v_1'y_1' + v_2'y_2' = f(x). \tag{5.41}$$

The two expressions in parentheses in Eq. (5.41) vanish since they represent solutions of the homogeneous equations. Hence Eq. (5.41) becomes

$$v_1'y_1' + v_2'y_2' = f(x). \tag{5.42}$$

Solving Eqs. (5.38) and (5.42) for v_1' and v_2', we obtain

$$v_1' = \frac{\begin{vmatrix} 0 & y_2 \\ f(x) & y_2' \end{vmatrix}}{\Delta} = -\frac{y_2 f(x)}{\Delta} \tag{5.43}$$

and

$$v_2' = \frac{\begin{vmatrix} y_1 & 0 \\ y_1' & f(x) \end{vmatrix}}{\Delta} = \frac{y_1 f(x)}{\Delta} \tag{5.44}$$

where

$$\Delta = \begin{vmatrix} y_1 & y_2 \\ y_1' & y_2' \end{vmatrix} = w(y_1, y_2).$$

The Δ quantity is just the Wronskian of y_1 and y_2. Note that $\Delta \neq 0$ since y_1 and y_2 are linearly independent by hypothesis. If we integrate Eqs. (5.43) and (5.44) and substitute the results into Eq. (5.34), the solution becomes

$$y = -y_1 \int \frac{f(x)y_2\,dx}{w(y_1, y_2)} + y_2 \int \frac{f(x)y_1\,dx}{w(y_1, y_2)}. \tag{5.45}$$

Equation (5.45) is the general solution of Eq. (5.33). When y_1 and y_2 cannot be easily obtained, the technique outlined in Section 5.2.6 should be used.

EXAMPLE 5.12 Solve $x^2y'' - 2xy' + 2y = x \ln x$; $(x \neq 0)$.

Solution The standard form for this equation is

$$y'' - \frac{2}{x}y' + \frac{2}{x^2}y = \frac{\ln x}{x}\left\{\begin{array}{l} p \longrightarrow \dfrac{-2}{x} \\ q \longrightarrow \dfrac{2}{x^2} \\ f(x) \longrightarrow \dfrac{\ln x}{x}. \end{array}\right.$$

Take $y_1 = x$ and $y_2 = x^2$. Hence we have

$$\Delta = \begin{vmatrix} y_1 & y_2 \\ y_1' & y_2' \end{vmatrix} = \begin{vmatrix} x & x^2 \\ 1 & 2x \end{vmatrix} = x^2.$$

By use of Eq. (5.45), we find that the solution is

$$\begin{aligned} y &= -x \int \frac{x^2\left(\dfrac{\ln x}{x}\right) dx}{x^2} + x^2 \int \frac{x\left(\dfrac{\ln x}{x}\right) dx}{x^2} \\ &= -x \int \frac{\ln x \, dx}{x} + x^2 \int \frac{\ln x \, dx}{x^2} \\ &= -x\left\{\frac{1}{2}(\ln x)^2 + C_1\right\} + x^2\left\{-\frac{\ln x}{x} - \frac{1}{x} + C_2\right\} \\ &= -x\left\{\frac{1}{2}(\ln x)^2 + \ln x + 1\right\} - C_1 x + C_2 x^2. \end{aligned}$$

5.2.6 Second-Order Homogeneous Equations with Variable Coefficients

The common method used to solve differential equations of the form

$$y'' + p(x)y' + q(x)y = 0 \tag{5.46}$$

is the **power series method** due to Frobenius (1849–1917) and Fuchs (1833–1902). Here we concentrate on developing a solution of Eq. (5.46) near the origin, $x = 0$. However, the Frobenius-Fuchs method can easily be extended to obtain solutions away from the origin.

The Frobenius-Fuchs theorem (whose proof is beyond the scope of this book) yields the following two kinds of information concerning the solution of Eq. (5.46): (1) the form of the solution as a result of the nature of $p(x)$ and $q(x)$, and (2) the form of the solution as indicated by the nature of the solutions of the indicial equation. (We will not make use of this second kind of information.) Information of the first kind is:

1. If $p(x)$ and $q(x)$ are regular at $x = 0$, then Eq. (5.46) possesses two distinct solutions of the form

$$y(x) = \sum_{\lambda=0}^{\infty} a_\lambda x^\lambda \qquad (a_0 \neq 0). \tag{5.47}$$

2. If $p(x)$ and $q(x)$ are singular at $x = 0$ but $xp(x)$ and $x^2q(x)$ are regular at $x = 0$, then there will always be at least one solution of Eq. (5.46) of the form

$$y(x) = \sum_{\lambda=0}^{\infty} a_\lambda x^{\lambda+k} \qquad (a_0 \neq 0). \tag{5.48}$$

3. If $p(x)$ and $q(x)$ are irregular singular points at $x = 0$ (that is, $xp(x)$ and $x^2q(x)$ are singular at $x = 0$), then regular solutions of Eq. (5.46) may not exist. In this case, no general method for solving the equation is known.

Since many problems in physics lead to equations in category 2 and equations under classification 1 are just special cases of 2, we concentrate on classification 2 only. Equation (5.46) may be rewritten in the form

$$y'' + \frac{f(x)}{x}y' + \frac{g(x)}{x^2}y = 0 \tag{5.49}$$

where $f(x)$ and $g(x)$ are regular at $x = 0$, and may therefore be expanded as

$$f(x) = \sum_{\beta=0}^{\infty} b_\beta x^\beta \qquad (b_0 \neq 0)$$

and

$$g(x) = \sum_{\gamma=0}^{\infty} c_\gamma x^\gamma \qquad (c_0 \neq 0). \tag{5.50}$$

On differentiating Eq. (5.48) twice with respect to x, we obtain

$$y' = \sum_{\lambda=0}^{\infty} a_\lambda(\lambda + k)x^{\lambda+k-1}$$

and

$$y'' = \sum_{\lambda=0}^{\infty} a_\lambda(\lambda + k)(\lambda + k - 1)x^{\lambda+k-2} \tag{5.51}$$

Substituting Eqs. (5.48), (5.50), and (5.51) into Eq. (5.49), we obtain†

$$\sum_{\lambda=0}^{\infty} a_\lambda(\lambda + k)(\lambda + k - 1)x^{\lambda+k-2} + \left(\sum_{\beta=0}^{\infty} b_\beta x^{\beta-1}\right)\left(\sum_{\lambda=0}^{\infty} a_\lambda(\lambda + k)x^{\lambda+k-1}\right) + \left(\sum_{\gamma=0}^{\infty} c_\gamma x^{\gamma-2}\right)\left(\sum_{\lambda=0}^{\infty} a_\lambda x^{\lambda+k}\right) = 0 \tag{5.52}$$

†In this development, we assume that term-by-term differentiations and multiplications of the series are valid.

The resulting equation for the lowest power of x is (**indicial equation**)

$$k(k-1)+b_0k+c_0=0$$

or

$$k^2+k(b_0-1)+c_0=0.$$

The remaining coefficients in Eq. (5.52) yield a **recursion** (one-after-another) **formula** for evaluating the coefficients of corresponding powers of x.

As in other cases, the series solution should be checked by substituting it back into the original equation. In addition, the domain of convergence of the series solution must be considered since it is possible for the solution to satisfy the original equation but not converge in the region of physical interest (as in the case of Legendre's equation in Chapter 6).

We now consider a simple example to illustrate the method. Additional applications of the power series method are given in Chapter 6 in connection with the development of certain special functions. The differential equation solved in Example 5.13 can obviously be solved by use of the simpler technique outlined in Section 5.2.3 of this chapter.

EXAMPLE 5.13 Solve $\ddot{y}+\omega^2y=0$.

Solution We assume that the solution is in the form

$$y(t)=\sum_{\lambda=0}^{\infty}a_\lambda t^{\lambda+k} \qquad (a_0 \neq 0) \tag{5.53}$$

then

$$\dot{y}(t)=\sum_{\lambda=0}^{\infty}a_\lambda(\lambda+k)t^{\lambda+k-1} \tag{5.54}$$

and

$$\ddot{y}=\sum_{\lambda=0}^{\infty}a_\lambda(\lambda+k)(\lambda+k-1)t^{\lambda+k-2}. \tag{5.55}$$

Substituting Eqs. (5.53) and (5.55) into the original differential equation, we obtain

$$\sum_{\lambda=0}^{\infty}a_\lambda(\lambda+k)(\lambda+k-1)t^{\lambda+k-2}+\omega^2\sum_{\lambda=0}^{\infty}a_\lambda t^{\lambda+k}=0$$

or

$$\begin{aligned}a_0k(k-1)t^{k-2}&+a_1k(k+1)t^{k-1}\\&+\sum_{\lambda=0}^{\infty}[a_{\lambda+2}(k+\lambda+2)(k+\lambda+1)+\omega^2a_\lambda]t^{k+\lambda}=0.\end{aligned} \tag{5.56}$$

Since Eq. (5.56) must be valid for all values of t, we require that

$$a_0 k(k-1) = 0 \qquad \text{(indicial equation)} \tag{5.57}$$
$$a_1 k(k+1) = 0$$

and

$$a_{\lambda+2}(k+\lambda+2)(k+\lambda+1) + \omega^2 a_\lambda = 0$$

or

$$a_{\lambda+2} = -\frac{\omega^2}{(k+\lambda+2)(k+\lambda+1)} a_\lambda. \tag{5.58}$$

Equation (5.58) is the recursion formula. The solutions of the indicial equation, Eq. (5.57), are $k = 0$ and $k = 1$. For $k = 0$, a_1 is arbitrary, and we obtain two independent series solutions from $k = 0$ since $a_0 \neq 0$ by hypothesis.

In this case, the recursion formula becomes

$$a_{\lambda+2} = -\frac{\omega^2}{(\lambda+2)(\lambda+1)} a_\lambda.$$

For λ even, we have:

$\lambda = 0$

$$a_2 = -\frac{\omega^2 a_0}{2 \cdot 1} = -\frac{\omega^2 a_0}{2!}$$

$\lambda = 2$

$$\begin{aligned} a_4 &= -\frac{\omega^2 a_2}{4 \cdot 3} \\ &= -\frac{\omega^2}{4 \cdot 3}\left(-\frac{\omega^2 a_0}{2!}\right) \\ &= \frac{\omega^4}{4!} a_0 \end{aligned}$$

$\lambda = 4$

$$\begin{aligned} a_6 &= -\frac{\omega^2}{6 \cdot 5} a_4 \\ &= -\frac{\omega^2}{6 \cdot 5}\left(\frac{\omega^4}{4!} a_0\right) \\ &= -\frac{\omega^6 a_0}{6!}. \end{aligned}$$

For λ odd, we have:

$\lambda = 1$

$$a_3 = -\frac{\omega^2}{3\cdot 2}a_1 = -\frac{\omega^2}{3!}a_1$$

$\lambda = 3$

$$\begin{aligned} a_5 &= -\frac{\omega^2}{5\cdot 4}a_3 \\ &= -\frac{\omega^2}{5\cdot 4}\left(-\frac{\omega^2 a_1}{3!}\right) = \frac{\omega^4}{5!}a_1 \end{aligned}$$

$\lambda = 5$

$$\begin{aligned} a_7 &= -\frac{\omega^2}{7\cdot 6}a_5 \\ &= -\frac{\omega^2}{7\cdot 6}\left(\frac{\omega^4 a_1}{5!}\right) \\ &= -\frac{\omega^6}{7!}a_1. \end{aligned}$$

The general solution now becomes

$$\begin{aligned} y(t) = a_0&\left[1 - \frac{(\omega t)^2}{2!} + \frac{(\omega t)^4}{4!} - \frac{(\omega t)^6}{6!} + \cdots\right] \\ &+ \frac{a_1}{\omega}\left[\omega t - \frac{(\omega t)^3}{3!} + \frac{(\omega t)^5}{5!} - \frac{(\omega t)^7}{7!} + \cdots\right] \end{aligned}$$

or

$$y(t) = a_0 \cos \omega t + \frac{a_1}{\omega} \sin \omega t.$$

This solution is the same as that obtained in Example 5.9.

5.3 PARTIAL DIFFERENTIAL EQUATIONS

5.3.1 Introduction

Questions of uniqueness and continuity involve a discussion of theoretical properties of partial differential equations such as characteristics, domains of dependence, and the maximum-minimum principle. The main purpose here is the construction of a solution of a partial differential equation subject to certain boundary and/or initial conditions. Construction techniques in this book will be (1) direct integration, (2) method of separation of variables, (3) Fourier series, and (4) Fourier transforms. It is important to note that partial differential equations whose solutions cannot be obtained by these four methods exist; constructing the solutions for such equations is beyond the scope of this book. However, many partial differential equations whose solutions have known physical interest can be solved by use of these four methods. In this section, we consider the methods of direct integration and separation of variables. The remaining two methods will be treated in Chapters 7 and 8.

The general linear second-order partial differential equation with two independent variables has the form

$$\begin{aligned} A(x, y)\frac{\partial^2\phi(x, y)}{\partial x^2} &+ B(x, y)\frac{\partial^2\phi(x, y)}{\partial y^2} + C(x, y)\frac{\partial\phi(x, y)}{\partial x} \\ &+ D(x, y)\frac{\partial\phi(x, y)}{\partial y} + E(x, y)\frac{\partial^2\phi(x, y)}{\partial x\,\partial y} + F(x, y)\phi(x, y) = G(x, y). \end{aligned} \tag{5.59}$$

Classification according to linear or nonlinear and homogeneous or nonhomogeneous is the same as that for ordinary differential equations. Equation (5.59) is classified as **elliptic, hyperbolic**, or **parabolic** corresponding to $E^2 - 4AB$ being less than, greater than, or equal to zero.

5.3.2 Some Important Partial Differential Equations in Physics

A. The Wave Equation in One Dimension Here we derive the one-dimensional wave equation for transverse waves in a flexible string of uniform mass per unit length, μ. The string is of length l, and it is fixed at the ends. It is stretched by a constant tension T which is so great that the gravitational force (the weight of the string) on the string can be neglected in comparison with the tension.

Suppose the string is distorted at a time $t = 0$, released, and allowed to vibrate in the xy-plane (see Fig. 5.3). The main problem is to characterize the vibration of the string. This motion is represented by $u(x, t)$ for $t > 0$.

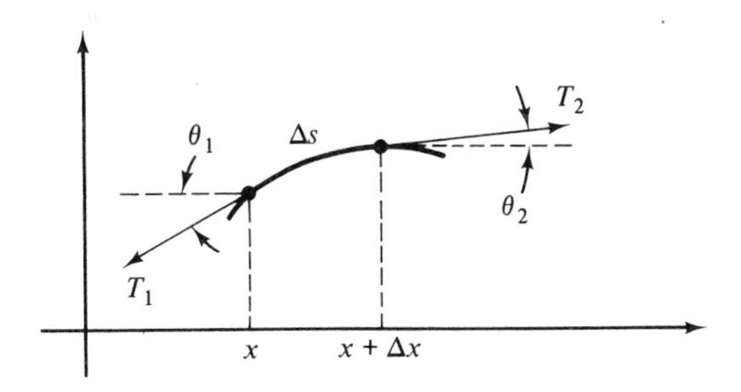

Figure 5.3

For equilibrium, we require that

(x-components—no motion) $T_1 \cos\theta_1 = T_2 \cos\theta_2.$

(y-components—net force equals mass times acceleration—Newton's second law) $T_2 \sin\theta_2 - T_1 \sin\theta_1 = \mu\,\Delta s \dfrac{\partial^2 u(x, t)}{\partial t^2}.$

For small oscillations, we may use the small-angle approximation [$\sin\theta \approx \tan\theta \approx \theta$ (in radians); hence $T_1 \approx T_2$] and obtain

$$\tan\theta_2 - \tan\theta_1 = \frac{\mu\,\Delta x}{T}\frac{\partial^2 u(x, t)}{\partial t^2}$$

or

$$\frac{1}{\Delta x}\left[\left(\frac{\partial u(x, t)}{\partial x}\right)_{x+\Delta x} - \left(\frac{\partial u(x, t)}{\partial x}\right)_x\right] = \frac{\mu}{T}\frac{\partial^2 u(x, t)}{\partial t^2}$$

In the limit as $\Delta x \to 0$, we find that

$$\frac{\partial^2 u(x, t)}{\partial x^2} = \frac{1}{v^2}\frac{\partial^2 u(x, t)}{\partial t^2}. \tag{5.60}$$

where $v = \sqrt{T/\mu}$ which has the dimensions of speed. Equation (5.60) is the required one-dimensional wave equation. The three-dimensional wave equation is

$$\nabla^2 u(x, y, z, t) = \frac{1}{v^2}\frac{\partial^2 u(x, y, z, t)}{\partial t^2}. \tag{5.61}$$

B. Poisson's and Laplace's Equations In terms of the scalar electrical potential and charge density, we have

$$\nabla^2\phi(x, y, z) = \begin{cases} -\dfrac{\rho}{\epsilon_0} & \text{(Poisson's equation)} \\ 0 & \text{(Laplace's equation).} \end{cases} \tag{5.62}$$

This type of equation occurs in other areas of physics. For example, $\phi =$ steady-state temperature or $\phi =$ the gravitational potential.

C. Heat Conduction (or Diffusion) Equation The heat conduction equation is given by (see Example 1.11)

$$\nabla^2 T(x, y, z, t) = \frac{1}{\sigma}\frac{\partial T(x, y, z, t)}{\partial t} \tag{5.63}$$

where $T = T(x, y, z, t)$ is the temperature at position (x, y, z) and σ is called the diffusivity which equals $k/c\rho$ (for thermal conductivity k, specific heat c, and density ρ).

D. The Schrödinger Wave Equation The **classical Hamiltonian** (total energy) of a particle of mass m and momentum p is given by

$$E = \frac{p^2}{2m} + V(x, y, z) \tag{5.64}$$

where $V(x, y, z)$ is the potential energy. The corresponding **quantum mechanical Hamiltonian** is obtained by use of the following procedure:

$$E \longrightarrow i\hbar \frac{\partial}{\partial t} \quad \text{and} \quad p \longrightarrow -i\hbar \nabla \tag{5.65}$$

where $\hbar = h/2\pi$ and h is called **Planck's constant**; its value is 6.625×10^{-34} Joule·second. The resulting operator equation for the total energy is

$$i\hbar \frac{\partial}{\partial t} = -\frac{\hbar^2}{2m} \nabla^2 + V(x, y, z). \tag{5.66}$$

If we operate on the function $\Psi(x, y, z, t)$ (the wave function), we obtain **Schrödinger's wave equation**

$$i\hbar \frac{\partial \Psi(x, y, z, t)}{\partial t} = \left[-\frac{\hbar^2}{2m} \nabla^2 + V(x, y, z)\right] \Psi(x, y, z, t). \tag{5.67}$$

The corresponding **time-independent equation** is

$$E\psi(x, y, z) = \left[-\frac{\hbar^2}{2m} \nabla^2 + V(x, y, z)\right] \psi(x, y, z). \tag{5.68}$$

E. Helmholtz's Equation The **Helmholtz equation** (see Chapter 6) has the form

$$\nabla^2 u(x, y, z) + k^2 u(x, y, z) = 0. \tag{5.69}$$

5.3.3 An Illustration of the Method of Direct Integration

Solve the equation

$$\frac{\partial^2 z(x, y)}{\partial x \, \partial y} = x^2 y \tag{5.70}$$

subject to the conditions

$$z(x, 0) = x^2$$

$$\text{and} \quad z(1, y) = \cos y.$$

On integrating Eq. (5.70) with respect to x, we obtain

$$\frac{\partial}{\partial x}\left(\frac{\partial z}{\partial y}\right) = x^2 y$$

$$\frac{\partial z}{\partial y} = \frac{1}{3}x^3 y + f(y). \tag{5.71}$$

Integrating Eq. (5.71) with respect to y, we obtain

$$\begin{aligned} z(x, y) &= \tfrac{1}{6}x^3y^2 + \int f(y)\,dy + g(x) \\ &= \tfrac{1}{6}x^3y^2 + F(y) + g(x) \end{aligned} \tag{5.72}$$

where

$$F(y) = \int f(y)\,dy.$$

Condition 1 $z(x, 0) = x^2$.
Equation (5.72) becomes

$$z(x, 0) = x^2 = F(0) + g(x).$$

Hence

$$z(x, y) = \tfrac{1}{6}x^3y^2 + F(y) + x^2 - F(0). \tag{5.73}$$

Condition 2 $z(1, y) = \cos y$.
Equation (5.73) becomes

$$z(1, y) = \cos y = \tfrac{1}{6}y^2 + F(y) + 1 - F(0).$$

Solving for $F(y)$, we obtain

$$F(y) = \cos y - \tfrac{1}{6}y^2 - 1 + F(0).$$

The required solution is

$$\begin{aligned} z(x, y) &= \tfrac{1}{6}x^3y^2 + \cos y - \tfrac{1}{6}y^2 - 1 + F(0) + x^2 - F(0) \\ &= \tfrac{1}{6}x^3y^2 + \cos y - \tfrac{1}{6}y^2 - 1 + x^2. \end{aligned}$$

5.3.4 Method of Separation of Variables

The method of **separation of variables** was introduced and developed by d'Alembert, D. Bernoulli, and Euler during the middle of the eighteenth century. It is the oldest systematic technique (and still the most useful) for solving partial differential equations. To illustrate the separation of variables method, we solve the two-dimensional Laplace equation

$$\frac{\partial^2\phi}{\partial x^2}+\frac{\partial^2\phi}{\partial y^2}=0 \quad \text{or} \quad \phi_{xx}+\phi_{yy}=0. \tag{5.74}$$

Assume that the solution $\phi(x, y)$ can be written as

Step 1
$$\phi(x, y)=X(x)Y(y) \tag{5.75}$$

where X is a function of x only and Y is a function of y only. Substituting Eq. (5.75) into Eq. (5.74), we obtain

Step 2
$$X''Y+XY''=0. \tag{5.76}$$

Dividing Eq. (5.76) by the right-hand side of Eq. (5.75), we obtain

Step 3
$$\frac{X''}{X}=-\frac{Y''}{Y}. \tag{5.77}$$

Here the left-hand side of Eq. (5.77) is a function of x only and the right-hand side is a function of y only. Since

$$\frac{d}{dx}\left(\frac{X''}{X}\right)=\frac{d}{dx}\left(\frac{Y''}{Y}\right)=0$$

and

$$\frac{d}{dy}\left(\frac{X''}{X}\right)=\frac{d}{dy}\left(\frac{Y''}{Y}\right)=0$$

it follows that

$$\frac{X''}{X}=-\lambda \quad \text{and} \quad \frac{Y''}{Y}=\lambda$$

or

$$X''+\lambda X=0 \tag{5.78}$$

and

$$Y''-\lambda Y=0 \tag{5.79}$$

where λ must be independent of x and y; λ is called the **separation constant.** Equations (5.78) and (5.79) are ordinary differential equations which can be solved by use of techniques developed in previous sections of this chapter. Additional illustrations of the separation of variables method are given in Chapters 6, 7, and 8.

In general, the separation of variables method reduces a partial differential equation with n independent variables to n ordinary differential equations involving $n - 1$ separation constants.

5.4 PROBLEMS

5.1 Classify (order and degree; homogeneous or nonhomogeneous; linear or nonlinear) the following differential equations:

(a) $(l + y^2)y'' + xy' + x = e^x$

(b) $\ddot{\theta} + \frac{g}{l} \sin \theta = 0$

(c) $y' + xy^2 = 0$

(d) $y'' + (y')^2 + xy = 0$

(e) $y'' + (y')^{1/2} + xy = 0.$

5.2 Verify that the first equation in each set is a solution of the corresponding differential equation.

(a) $x^2 = 2y^2 \ln y;\ y' = \dfrac{xy}{x^2 + y^2}$

(b) $x^2 + y = xy;\ x^3y' - x^2y + y^2 = 0$

(c) $y = e^x + e^{2x};\ y'' - 3y' + 2y$

(d) $y = x - x \ln x;\ xy' + x - y = 0$

(e) $y = x \tan (x + c);\ xy' = x^2 + y^3 + y.$

5.3 What is the degree of homogeneity of

$$x + y \sin \frac{y}{x}?$$

5.4 Test the following differential equations for exactness, and solve the exact equations:

(a) $(2x^3 - 3x^2y + y^3)\dfrac{dy}{dx} = 2x^3 - 6x^2y + 3xy^2$

(b) $2xy \dfrac{dy}{dx} + y^2 = a + 2bx$

(c) $x(x^2 + 2y^2)\, dx + y(2x^2 + y^2)\, dy = 0$

(d) $m \dfrac{dv}{dt} + V \dfrac{dm}{dt} = 0$

(e) $x(y^2 + 2)\dfrac{dy}{dx} = (x + 1)(y^2 + 1).$

5.5 Find the general solution for each of the following differential equations:

(a) $y' = ky$

(b) $y' = ax + by$

(c) $xy' + 2y = 12x$

(d) $y - y' - xy^2 = 0$

(e) $y' + 3y - x - e^{-2x} = 0.$

5.6 Find the solution of each of the following differential equations subject to the indicated condition:

(a) $y' = y^2$; $y(0) = 1$

(b) $y' = e^y$; $y(0) = 0$

(c) $e^y(y' + 1) = 1$; $y(0) = 1$

(d) $m\dot{v} + kv = g$; $v(0) = v_0$

(e) $y' + y = y^2 e^{2x}$; $y(0) = 1$.

5.7 If a rocket is projected vertically upward under gravity, the equation describing its motion (the equation of motion) is

$$m\frac{dv}{dt} + V\frac{dm}{dt} = -mg$$

where V is the speed of the exhaust gas relative to the rocket. Find $v(t)$ for $v(0) = 0$ and $m(0) = M$ (The other symbols in the equation of motion have their usual meanings, and other external forces are neglected.)

5.8 Assume that a spherical drop evaporates at a rate proportional to its surface area. Find an expression for the radius of the drop as a function of time if $r(0) = 3$ mm and $r(1 \text{ hr}) = 2$ mm.

5.9 If the interest on \$500 is continuously compounded at 5% per year, how long will it take for the amount to become \$700?

5.10 In a certain chemical reaction, a given substance is converted into another substance at a rate that is proportional to the amount of the unconverted substance. If one-fourth of the original amount has been transformed in four minutes, how much time will be required to transform one-half of the substance.

5.11 A certain radioactive material has a half-life of 2 hr. Find the time required for a given amount of this material to decay to one-tenth of its original amount. (**Half-life** is the time required for one-half of the original amount to decay.)

5.12 Five percent of a radioactive substance is lost in 100 yr. How much of the original amount will be present after 250 yr?

5.13 (a) Write the appropriate differential equation for the circuit in Fig. 5.4. (Here I, the current, is the dependent variable, and time is the independent variable.)

(b) Solve the differential equation in part (a) for $I(0) = 0$ and constant E, R, and L

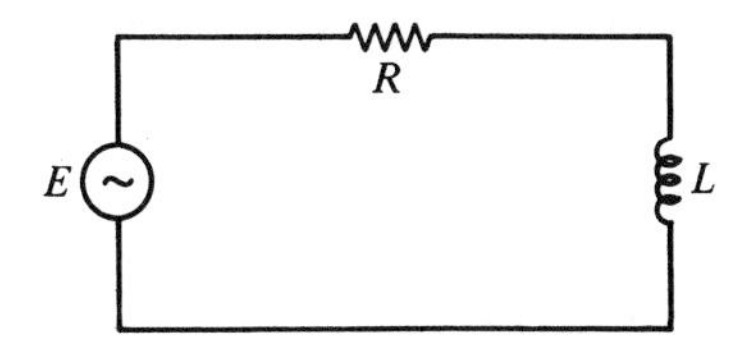

Figure 5.4

(c) Take $E(t) = E \sin \omega t$ in part (a) and solve the resulting differential equation with constant E and ω.

5.14 A metal ball having a temperature of 80°C is placed into m grams of ice water at 0°C. After ten minutes, the temperatures of the ball and water are 60°C and 20°C, respectively. The ball is now transferred to another m grams of ice water at 0°C. Calculate the temperature of the ball at the end of another (a second) ten-minute interval if the only exchange of heat is between the ball and the water. (*Hint:* Use Newton's law of cooling.)

5.15 A raindrop falls from rest at a place where the air resistance is proportional to the speed, kv.

(a) Derive the (particular) expression for the speed of the drop as a function of time.

(b) When ($t \to \infty$), the expression for the **terminal speed** results. Derive the expression for the terminal speed of the drop.

(c) Derive the (particular) expression for the position of the drop as a function of time.

5.16 Solve the following differential equations:

(a) $\frac{d}{dr}\left(r^2 \frac{dv}{dr}\right) = 0$; $v(a) = v_a$ and $v(b) = v_b$

(b) $y'' + 7y' + 12y = 0$

(c) $y'' + 6y' + 9y = 0$

(d) $y'' + 2y' + 2y = 0$

(e) $\ddot{y} + 6\dot{y} + 13y = 0$; $y(0) = 2$ and $\dot{y}(0) = 1$.

5.17 Solve the following differential equations:

(a) $y'' + y' - 6y = 12e^{-x}$

(b) $y'' + 3y' + 2y = e^{-x}$

(c) $y'' + 3y' = 10 \sin x$; $y(0) = y'(0) = 0$

(d) $y'' - y' - 2y = x^3$

(e) $y'' + y = x$.

5.18 If $y_1 = x$ and $y_2 = xe^x$ are two solutions of the homogeneous differential equation associated with

$$x^2y'' - (x^2 + 2x)y' + (x + 2)y = x^3$$

(a) show that y_1 and y_2 are linearly independent, (b) find the general solution of the differential equation.

5.19 **Euler's differential equation** (sometimes called **Cauchy's differential equation**),

$$x^n \frac{d^n y}{dx^n} + a_1 x^{n-1} \frac{d^{n-1} y}{dx^{n-1}} + \cdots + a_{n-1} x \frac{dy}{dx} + a_n y = f(x)$$

sometimes occurs in physical problems. It can be reduced to an nth-order linear differential equation with constant coefficients by use of a change of variable, $x = e^z$.

(a) Show that

$$x^2 \frac{d^2 y}{dx^2} + x \frac{dy}{dx} - y = \ln x$$

can be reduced to

$$\frac{d^2 y}{dz^2} - y = z.$$

(b) By first solving the reduced differential equation, find the solution of the original differential equation.

5.20 Consider the following differential equation:

$$\ddot{y} + 2\delta \dot{y} + k^2 y = 0 \qquad (\text{for } k > 0).$$

Discuss the behavior of the solution for each of the following cases:

(a) $\delta = 0$
(b) $\delta < k$ (light or under-damping)
(c) $\delta = k$ (critical damping)
(d) $\delta > k$ (heavy or over-damping).

5.21 The **Riccati equation**

$$\frac{dy}{dx} = P(x)y^2 + Q(x)y + R(x)$$

is a nonlinear differential equation which is of considerable importance in particle dynamics. By use of a change of the dependent variable

$$\left(y = -\frac{1}{P(x)z}\frac{dz}{dx}\right)$$

show that the Riccati differential equation becomes

$$\frac{d^2 z}{dx^2} - \left(Q + \frac{1}{P}\frac{dP}{dx}\right)\frac{dz}{dx} + PRz = 0$$

which is a second-order linear differential equation.

5.22 (a) By use of

$$y = ze^{-1/2 \int^x p(t)\, dt}$$

show that

$$y'' + p(x)y' + q(x)y = 0$$

transforms into

$$z'' + f(x)z = 0.$$

(b) What is the expression for $f(x)$? Note that **Leibnitz's** (1646–1716) **formula** for the derivative of an integral is useful in solving this problem,

$$\frac{d}{d\alpha}\int_{g(\alpha)}^{h(\alpha)} f(x,\alpha)\,dx = \int_{g(\alpha)}^{h(\alpha)} \frac{\partial f(x,\alpha)}{\partial \alpha}\,dx + f(h(\alpha),\alpha)\frac{dh(\alpha)}{d\alpha} - f(g(\alpha),\alpha)\frac{dg(\alpha)}{d\alpha}.$$

5.23 The vertical motion of a particle of mass m on a spring with spring constant k is described by the following differential equation:

$$m\ddot{y} = -ky + mg \qquad (y(0) = y_0 \text{ and } \dot{y}(0) = 0).$$

Solve this equation for the position of the particle as a function of time.

5.24 A particle of mass m is projected vertically upward with a speed of v_0. The particle is subject to gravity and a force of air resistance that is proportional to v. Find the expression for the maximum height reached by the particle.

5.25 The steady flow of heat in a uniform rod of length l is described by the equation,

$$\frac{d^2T}{dx^2} - \gamma T = 0 \qquad (T(0) = T_1 \text{ and } T(l) = T_2)$$

where γ is a constant which characterizes the material. Find the solution of this differential equation subject to the indicated boundary conditions.

5.26 If a thin wire of length l is heated by an electric current, the differential equation for its temperature distribution has the form

$$\frac{d^2T}{dx^2} + b^2T = -k \qquad (T(0) = 0 \text{ and } T(l) = 0)$$

where $k > 0$ and the coefficient b characterizes the material. Discuss the solution of this differential equation for the following three cases:

(a) $b^2 > 0$

(b) $b^2 = 0$

(c) $b^2 < 0$.

5.27 A force of $F_0 \sin \omega t$ is applied to a damped harmonic oscillator, $\ddot{x} + 2\alpha\dot{x} + n^2x = F(t)/m$, with critical damping, $\alpha = n$. Derive the expression for a particular solution and show that it can be put into the following form:

$$x(t) = \frac{F_0 \sin(\omega t - 2\phi)}{m(n^2 + \omega^2)}; \qquad \phi = \tan^{-1}\left(\frac{\omega}{n}\right).$$

5.28 The equation of motion for a simple pendulum is $ml\ddot{\theta} = -mg \sin \theta$, where l is the length, m is the mass of the bob, and g is the acceleration due to gravity.

(a) By use of the small-angle approximation [$\sin \theta \approx \tan \theta \approx \theta$ (in radians)], find $\theta(t)$ and the expression for the period of the motion.

(b) Show that the original equation of motion may be written as

$$\frac{1}{2}\frac{d}{d\theta}(\dot{\theta}^2) = -n^2 \sin\theta \qquad \left(\text{where } n^2 = \frac{g}{l}\right).$$

(c) Show that the following integral (elliptic integral) must be evaluated to obtain the solution of the equation in part (b):

$$-\int_{\theta_0}^{\theta} \frac{d\theta}{(2\cos\theta - 2\cos\theta_0)^{1/2}} = nt \qquad (\text{where } \theta(0) = \theta_0 \text{ and } \dot{\theta}(0) = 0).$$

The expression for the period in this general case is

$$T = \frac{4}{n}\left[\frac{\pi}{2} + \frac{1}{2}\frac{\pi}{4}\sin^2\left(\frac{\theta_0}{2}\right) + \frac{3}{8}\frac{3\pi}{16}\sin^4\left(\frac{\theta_0}{2}\right) + \cdots\right]$$
$$\approx \frac{2\pi}{n}\left(1 + \frac{\theta_0^2}{16}\right).$$

5.29 Show that $u = \cos x \cos at$ is a solution of $u_{tt} = a^2 u_{xx}$.

5.30 Solve

$$\frac{\partial^2 u}{\partial x\,\partial y} = 0.$$

5.31 By use of the separation of variables method, solve

$$u_x = 4u_y; \qquad u(0, y) = 8e^{-3y}.$$

5.32 Consider the one-dimensional heat equation,

$$\frac{\partial^2 T}{\partial x^2} = \frac{1}{\sigma}\frac{\partial T}{\partial t}.$$

If b^2 is the separation constant, discuss the nature of the solution for

(a) $b^2 > 0$

(b) $b^2 = 0$

(c) $b^2 < 0$.

(d) Find the particular solution for

$$T(0, t) = T(L, t) = 0 \quad \text{and} \quad T(x, 0) = T_0 \sin\frac{n\pi x}{L} \qquad (\text{for } 0 < x < L)$$

if the general solution is

$$T(x, t) = Ce^{-b^2\sigma t}(A\cos bx + B\sin bx).$$

5.33 (a) Solve

$$\frac{\partial^2 I}{\partial x^2} - \frac{2}{k}\frac{\partial I}{\partial t} + I = 0$$

for

$$I(l, t) = 0 \quad \text{and} \quad \left.\frac{\partial I}{\partial x}\right|_{x=0} = -ae^{-kt}.$$

(b) Solve

$$\frac{\partial^2 T}{\partial x^2} = \frac{1}{\sigma}\frac{\partial T}{\partial t}$$

for

$$T(x, t \longrightarrow 0) \longrightarrow \text{finite}, \qquad T(0, t) = 0, \quad \text{and} \quad \left.\frac{\partial T}{\partial x}\right|_{x=l} = 0.$$

5.34 The motion of a transverse wave in a string is characterized by

$$\frac{\partial^2 \varphi}{\partial x^2} = \frac{1}{v^2}\frac{\partial^2 \varphi}{\partial t^2}.$$

(a) Discuss the nature of the solution for $k^2 > 0$, $k^2 = 0$, and $k^2 < 0$, where k^2 is the separation constant.
(b) Find the general periodic motion solution of the wave equation.
(c) For

$$\varphi(0, t) = \varphi(L, t) = \left.\frac{\partial \varphi}{\partial t}\right|_{t=0} = 0$$

and

$$\varphi(x, 0) = Y$$

show that the solution for periodic motion is

$$\varphi(x, t) = \sum_{n=1}^{\infty} Y_n \sin\left(\frac{n\pi x}{L}\right) \cos\left(\frac{nv\pi t}{L}\right).$$

5.35 Consider the two-dimensional Laplace equation,

$$\frac{\partial^2 \phi}{\partial x^2} + \frac{\partial^2 \phi}{\partial y^2} = 0.$$

(a) Show that

$$\phi(x, y) = (A \cos bx + B \sin bx)(C \cosh by + D \sinh by)$$

where b^2 is the separation constant.

(b) For

$$\phi(x, 0) = \phi(0, y) = \phi(\omega, y) = 0$$

and

$$\phi(x, h) = \phi_0 \sin \frac{n\pi x}{\omega}$$

find $\phi(x, y)$.

5.36 By use of the generalized power series method, solve

$$xy'' + 2y' + xy = 0 \qquad (y(0) = 1 \text{ and } y'(0) = 0).$$

5.37 Solve the **Euler-Cauchy differential equation**, $x^2y'' + 2xy' = 0$, by use of the generalized power series method.

5.38 Develop the solution of the three-dimensional Laplace equation in Cartesian coordinates, and show that it can be written as

$$\phi(x, y, z) = \sum_{j,k,l=1}^{2} c_{jkl} e^{(-1)^j \sqrt{a}\, x} e^{(-1)^k \sqrt{b}\, y} e^{(-1)^l i \sqrt{a+b}\, z}$$

where a and b are separation constants.

5.39 By separating the variables in Laplace's equation in cylindrical coordinates, show that

$$u(\rho, \phi, z) = P(\rho)\Phi(\phi)Z(z)$$

where

$$Z = Ae^{\sqrt{a}\, z} + Be^{-\sqrt{a}\, z}, \qquad \Phi = C \cos n\phi + D \sin n\phi$$

and $P(\rho)$ is the solution of

$$\frac{d^2P}{d\rho^2} + \frac{1}{\rho}\frac{dP}{d\rho} + \left(a - \frac{n^2}{\rho^2}\right)P = 0.$$

5.40 Show that the solution of Laplace's equation in spherical polar coordinates is

$$u = R(r)\Theta(\theta)\Phi(\phi)$$

where

$$\Phi = A \cos m\phi + B \sin m\phi, \qquad R(r) = Cr^l + Dr^{-l-1};$$

$$\frac{1}{R}\frac{d}{dr}\left(r^2 \frac{dR}{dr}\right) = l(l + 1)$$

and Θ is a solution of

$$[l(l + 1) \sin^2 \theta - m^2]\,\Theta + \sin \theta \frac{d}{d\theta}\left(\sin \theta \frac{d\Theta}{d\theta}\right) = 0.$$

5.41 The so-called **Sturm-Liouville problem**: Consider the equation $y''(x) + \lambda^2 y(x) = 0$. Find the eigenvalues (the restricted values of λ_n) and the corresponding eigenfunctions, $y_n(x)$, of this differential equation for (a) $y'(0) = 0$ and $y'(\pi) = 0$, and (b) $y(0) = 0$ and $y'(\pi) = 0$.

5.42 The **WKB-approximation**: Consider solving the differential equation

$$\frac{d^2\psi(x)}{dx^2} + \frac{2m}{\hbar^2}[E - V(x)]\psi(x) = 0$$

by first making the substitution

$$\psi(x) = R(x)e^{iS(x)/\hbar}$$

and separating (equating corresponding real and imaginary parts) the resulting differential equation into two differential equations. Now show that the solution of the original differential equation is

$$\psi(x) = R \exp\left\{i \int_{x_1}^{x} dy \left(\frac{2m}{\hbar^2}[E - V(y)]\right)^{1/2}\right\} \qquad (E > V(x)).$$

Hints: (a) Solve the equation resulting from equating imaginary parts,

$$\frac{d}{dx}\left[ln\left(\frac{dS}{dx}R^2\right)\right] = 0$$

and substitute the result into the differential equation obtained by equating real parts,

$$\frac{d^2R}{dx^2} - \frac{R}{\hbar^2}\left(\frac{dS}{dx}\right)^2 + \frac{2m}{\hbar^2}[E - V(x)]R = 0.$$

(b) Then use the following approximation (Wentzel-Kramers-Brillouin, WKB):

$$\frac{1}{R}\frac{d^2R}{dx^2} \ll \frac{(\text{constant})^2}{\hbar^2}\frac{1}{R^4} = \frac{1}{\hbar^2}\left(\frac{dS}{dx}\right)^2.$$

6

Special Functions of Mathematical Physics

6.1 INTRODUCTION

In this chapter, we treat some of the so-called "special functions" that have known physical applications. Primary emphasis will be devoted to (1) Hermite polynomials, (2) Legendre and associated Legendre polynomials, (3) Laguerre and associated Laguerre polynomials, and (4) Bessel functions. The order of the list does not reflect the degree of importance. At the end of this chapter we list other special functions that are also used in physics.

A careful study of the list of important differential equations in mathematical physics (see Chapter 5) reveals that Laplace's equation, Poisson's equation, the time-independent diffusion or heat equation, the time-independent wave equation, and the time-independent Schrödinger equation are all mathematical special cases of Helmholtz's equation [see Eq. (5.69)] with the proper choice of k^2. Hence our development centers around Helmholtz's equation. The solutions of Helmholtz's equation for various k^2 and in various coordinate systems lead to special functions. Since Helmholtz's equation is the general case of important equations in mathematical physics, the need for studying certain special functions is established.

6.2 THE HERMITE POLYNOMIALS

6.2.1 Basic Equations of Motion in Mechanics

The laws of classical mechanics are valid for systems whose dimensions are greater than the atomic size (10^{-8} cm). The basic equation of motion of a particle or the constituent particles of a classical system is the differential equation form of Newton's (1642–1727) second law (or the various generalizations of this law). The laws of special relativity must also be introduced when the speed approaches the speed of light.

The laws of quantum mechanics are valid for systems whose dimensions are of atomic size (10^{-8} cm). Here the basic equation of motion of a particle or the constituent particles of such a system is Schrödinger's wave equation. If the speed approaches the speed of light, the laws of special relativity must again be introduced.

Solution of Schrödinger's Equation for the Harmonic Oscillator

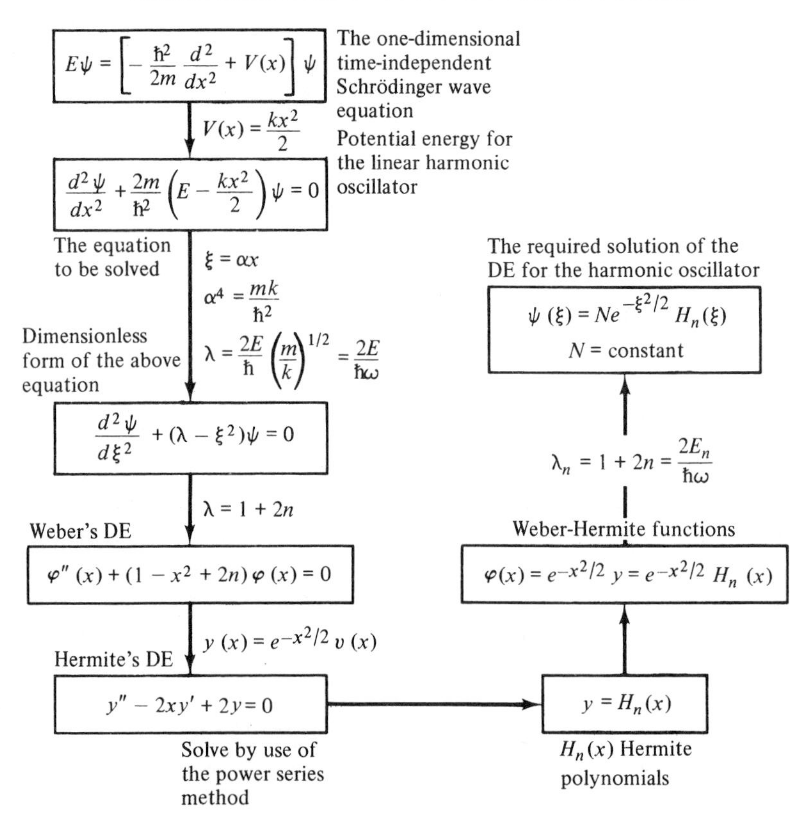

The fundamental equation of motion for systems whose dimensions are of nuclear size (10^{-13} cm) is unknown.

The many interesting and important physical concepts of quantum physics are the subject of modern physics and quantum mechanics courses. Our purpose here is to consider some of the prerequisite mathematical details related to the quantum mechanical solutions of the linear harmonic oscillator and the central force problem.

6.2.2 The One-Dimensional Linear Harmonic Oscillator

The one-dimensional motion of a mass, m, attached to a spring is governed by the force equation

$$F_x = -kx \qquad (k = \text{spring constant})$$

with potential energy

$$V(x) = \frac{k}{2}x^2.$$

This system provides us with one of the most fundamental problems in physics (the analysis of the **linear harmonic oscillator**). Its study enables us to analyze certain complicated systems in terms of normal modes of motion (whenever the interparticle forces are linear functions of the relative displacements of the particles from their equilibrium positions). These **normal modes** are formally equivalent to harmonic oscillators.

The linear harmonic oscillator problem will now be resolved by solving the one-dimensional time-independent Schrödinger wave equation. That is, we must solve

$$\frac{-\hbar^2}{2m}\frac{d^2\psi(x)}{dx^2} + \frac{1}{2}kx^2\psi(x) = E\psi(x). \tag{6.1}$$

Note that this equation is just the one-dimensional Helmholtz (1821–1894) equation with

$$k^2(x) = \frac{2m}{\hbar^2}\left(E - \frac{kx^2}{2}\right).$$

By use of the following transformation, we can put Eq. (6.1) in a simpler (dimensionless) form:

$$\xi = \alpha x \tag{6.2}$$

$$\alpha^4 = \frac{mk}{\hbar^2} \tag{6.3}$$

$$\lambda = \frac{2E}{\hbar}\left(\frac{m}{k}\right)^{1/2} = \frac{2E}{\hbar\omega} \tag{6.4}$$

whence

$$\frac{d^2}{dx^2} = \frac{d}{dx}\left(\frac{d}{d\xi}\frac{d\xi}{dx}\right) = \alpha^2\frac{d^2}{d\xi^2}.$$

Equation (6.1) now becomes

$$\frac{-\hbar^2\alpha^2}{2m}\frac{d^2\psi(\xi)}{d\xi^2} + \frac{1}{2}\frac{k\xi^2}{\alpha^2}\psi(\xi) - E\psi(\xi) = 0$$

or

$$\frac{d^2\psi(\xi)}{d\xi^2} + (\lambda - \xi^2)\psi(\xi) = 0. \tag{6.5}$$

Equation (6.5) is the differential equation that we must solve, but the solution is not readily obtainable by direct use of the power series method. However, Eq. (6.5) reduces to **Weber's** (1842–1913) **differential equation,**

$$\varphi'' + (1 - x^2 + 2n)\varphi = 0 \tag{6.6}$$

when

$$\lambda \equiv 1 + 2n. \tag{6.7}$$

Equation (6.6), as will be shown later, can be reduced to the famous **Hermite** (1822–1901) **differential equation**

$$y'' - 2xy' + 2ny = 0 \qquad (n = \text{constant}). \tag{6.8}$$

Our plan is to develop the solution of Hermite's differential equation by use of the generalized power series method of Frobenius and Fuchs and deduce from this solution the solution of Eq. (6.5).

6.2.3 The Solution of Hermite's Differential Equation

Assume that the solution of Hermite's equation, Eq. (6.8), has the form

$$y(x) = \sum_{r=0}^{\infty} a_r x^{k+r} \qquad (a_0 \neq 0) \tag{6.9}$$

so that

$$y'(x) = \sum_{r=0}^{\infty} (k + r)a_r x^{k+r-1} \tag{6.10}$$

and

$$y''(x) = \sum_{r=0}^{\infty} (k + r)(k + r - 1)a_r x^{k+r-2}. \tag{6.11}$$

On substituting Eqs. (6.9), (6.10), and (6.11) into Eq. (6.8) and rearranging terms, we find that

$$k(k-1)a_0x^{k-2} + k(k+1)a_1x^{k-1} + \sum_{r=0}^{\infty} [(k+r+2)(k+r+1)a_{r+2} - 2(k+r-n)a_r]x^{k+r} = 0. \tag{6.12}$$

By requiring that Eq. (6.12) hold for all values of x, we obtain

$$k(k-1)a_0 = 0 \qquad \text{(indicial equation)}$$

or

$$k = 0 \quad \text{and} \quad k = 1 \tag{6.13}$$

$$k(k+1)a_1 = 0 \tag{6.14}$$

and

$$a_{r+2} = \frac{2(k+r-n)}{(k+r+2)(k+r+1)}a_r \qquad \text{(recursion formula).} \tag{6.15}$$

When $k = 0$, a_1 is arbitrary because of Eq. (6.14), and we obtain two independent series solutions for $k = 0$. The recursion formula, in this case, becomes

$$a_{r+2} = -\frac{2(n-r)a_r}{(r+2)(r+1)}. \tag{6.16}$$

For r even, we have:

$r = 0$

$$a_2 = -\frac{2na_0}{2\cdot 1}$$

$r = 2$

$$a_4 = -\frac{2(n-2)a_2}{4\cdot 3} = \frac{2^2n(n-2)a_0}{4!}$$

$r = 4$

$$a_6 = -\frac{2(n-4)a_4}{6\cdot 5} = -\frac{2^3n(n-2)(n-4)a_0}{6!}.$$

The general term is

$$a_{2j} = \frac{(-2)^j n(n-2)\cdots(n-2j+2)a_0}{(2j)!} \tag{6.17}$$

where $j = 1, 2, 3, \ldots$.

For r odd, we have:

$r = 1$

$$a_3 = -\frac{2(n-1)a_1}{3\cdot 2}$$

$r = 3$

$$a_5 = -\frac{2(n-3)a_3}{5\cdot 4}$$

$$= \frac{2^2(n-1)(n-3)a_1}{5!}$$

$r = 5$

$$a_7 = -\frac{2(n-5)a_5}{7\cdot 6}$$

$$= -\frac{2^3(n-1)(n-3)(n-5)a_1}{7!}.$$

The general term is

$$a_{2j+1} = \frac{(-2)^j(n-1)(n-3)\cdots(n-2j+1)a_1}{(2j+1)!} \tag{6.18}$$

where $j = 1, 2, 3, \ldots$.

The solution of Eq. (6.8) is†

$$\begin{aligned} y(x) = a_0\left[1 + \sum_{j=1}^{\infty}\frac{(-2)^j n(n-2)\cdots(n-2j+2)x^{2j}}{(2j)!}\right] \\ + a_1 x\left[1 + \sum_{j=1}^{\infty}\frac{(-2)^j(n-1)\cdots(n-2j+1)x^{2j}}{(2j+1)!}\right]. \end{aligned} \tag{6.19}$$

The first infinite series in Eq. (6.19) becomes a finite and even polynomial of degree n when $n - 2j + 2 = 0$ (n is equal to a positive even integer). The second infinite series in Eq. (6.19) becomes a finite and odd polynomial of degree n when $n - 2j + 1 = 0$ (n is equal to a positive odd integer). We may obtain a single series, in descending powers of x, valid for both even and odd

†A Physical requirement on the wave function, $\psi(\xi) \longrightarrow 0$ as $\xi \longrightarrow \infty$, makes this infinite series solution unsatisfactory.

n. The first term of this series is $a_n x^n$ and subsequent terms are given by the recursion formula

$$a_r = -\frac{(r+2)(r+1)a_{r+2}}{2(n-r)} \tag{6.20}$$

where

$$a_{n-2} = -\frac{n(n-1)a_n}{2 \cdot 2}$$

and

$$\begin{aligned} a_{n-4} &= -\frac{(n-2)(n-3)a_{n-2}}{2 \cdot 4} \\ &= \frac{n(n-1)(n-2)(n-3)a_n}{2^2 \cdot 2 \cdot 4}. \end{aligned}$$

The general term is

$$a_{n-2j} = \frac{(-1)^j n(n-1)(n-2) \cdots (n-2j+1)a_n}{2^j \cdot 2 \cdot 4 \cdots 2j}.$$

Since

$$\begin{aligned} n(n-1) \cdots (n-2j+1) &= n(n-1) \cdots (n-2j+1) \\ &\quad \cdot \frac{(n-2j)(n-2j-1) \cdots 3 \cdot 2 \cdot 1}{(n-2j)(n-2j-1) \cdots 3 \cdot 2 \cdot 1} \\ &= \frac{n!}{(n-2j)!} \end{aligned}$$

and

$$\begin{aligned} 2 \cdot 4 \cdots 2 &= (2 \cdot 1)(2 \cdot 2)(2 \cdot 3) \cdots (2 \cdot j) \\ &= 2^j j! \end{aligned}$$

the general term may be written as

$$a_{n-2j} = \frac{(-1)^j n! a_n}{j! 2^{2j} (n-2j)!}. \tag{6.21}$$

The required finite solution is

$$y(x) = a_n \sum_{j=0}^{N} \frac{(-1)^j n! x^{n-2j}}{2^{2j} j! (n-2j)!} \tag{6.22}$$

where

$$N = \begin{cases} \dfrac{n}{2} & \text{(if } n \text{ is even)} \\ \dfrac{n-1}{2} & \text{(if } n \text{ is odd).} \end{cases}$$

The standard particular solution is obtained when a_n is defined as 2^n. This solution is

$$y_n(x) \equiv H_n(x) = \sum_{j=0}^{N} \frac{(-1)^j n!(2x)^{n-2j}}{j!(n-2j)!} \tag{6.23}$$

where $H_n(x)$ are known as **Hermite polynomials** of degree n.

The **generating function** for a polynomial is a function $W(t, x)$ which has a series expansion in the variable t of the form

$$W(t, x) = \sum_{n=0}^{\infty} \phi_n(x)t^n \tag{6.24}$$

where $\phi_n(x)$ is the required polynomial. The generating function for the Hermite polynomials is

$$e^{2t\xi - t^2} = \sum_{n=0}^{\infty} \frac{H_n(\xi)t^n}{n!} \tag{6.25}$$

and the **Rodrigues** (1794–1851) **formula** is

$$H_n(\xi) = (-1)^n e^{\xi^2} \frac{d^n}{d\xi^n}(e^{-\xi^2}). \tag{6.26}$$

Note that

$$H_0(\xi) = 1 \quad \text{and} \quad H_3(\xi) = 8\xi^3 - 12\xi. \tag{6.27}$$

The following transformation

$$\varphi(x) = e^{-x^2/2}y(x)$$

reduces Weber's differential equation,

$$\varphi''(x) + (1 - x^2 + 2n)\varphi(x) = 0 \tag{6.6$'$}$$

to

$$y'' - 2xy' + 2ny = 0 \tag{6.8$'$}$$

which is Hermite's equation. Hence the solution of Eq. (6.6′) is

$$\varphi_n(x) = e^{-x^2/2}H_n(x). \tag{6.28}$$

Here $\varphi_n(x)$ are called **Weber-Hermite functions**. The substitution in Eq. (6.7), $\lambda_n = 1 + 2n$, reduces Eq. (6.6′) to the equation we set out to solve, Eq. (6.5). Therefore the solution of Eq. (6.5) is

$$\psi_n(\xi) = Ne^{-\xi^2/2}H_n(\xi) \tag{6.29}$$

where

$$\lambda_n = 1 + 2n = \frac{2E_n}{\hbar\omega} \qquad (n = 0, 1, 2, \ldots).$$

Therefore

$$E_n = \hbar\omega\left(n + \frac{1}{2}\right). \tag{6.30}$$

Equation (6.30) represents the eigenenergy of the linear harmonic oscillator. The **principal quantum number** is denoted by n. When $n = 0$, $E_0 = \hbar\omega/2$; this is the **zero-point energy** which has no classical analog. The behavior of $\psi_n(\xi)$ for the first six n values is illustrated in the graphs in Fig. 6.1.

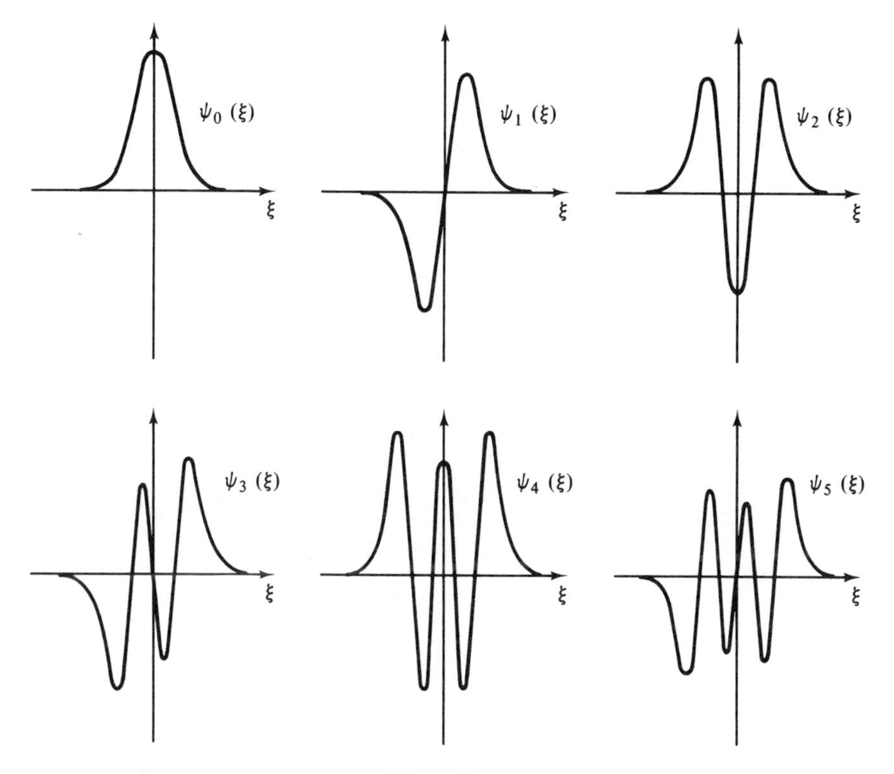

Figure 6.1

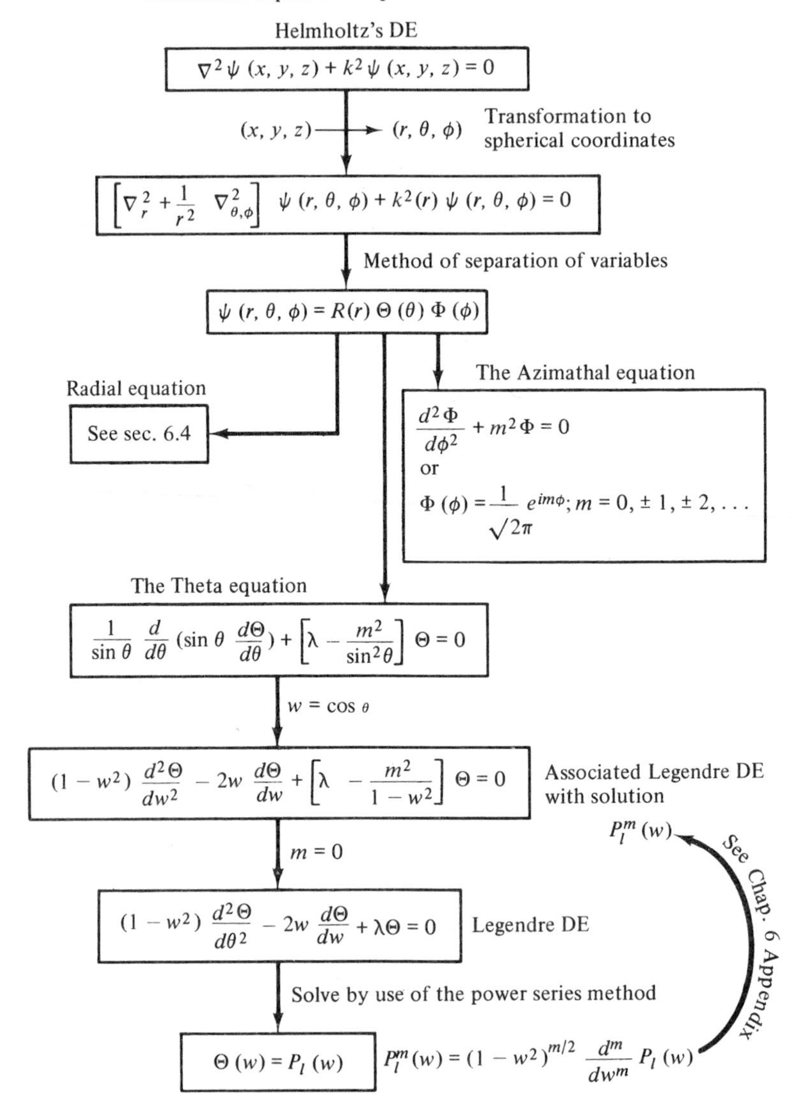

6.3 LEGENDRE AND ASSOCIATED LEGENDRE POLYNOMIALS

6.3.1 Spherical Harmonics

The Laplacian in spherical polar coordinates is

$$\nabla^2 = \nabla_r^2 + \frac{1}{r^2} \nabla_{\theta,\phi}^2$$

where

$$\nabla_r^2 = \frac{1}{r^2}\frac{\partial}{\partial r}\left(r^2\frac{\partial}{\partial r}\right)$$

and

$$\nabla_{\theta,\phi}^2 = \frac{1}{\sin\theta}\frac{\partial}{\partial\theta}\left(\sin\theta\frac{\partial}{\partial\theta}\right) + \frac{1}{\sin^2\theta}\frac{\partial^2}{\partial\phi^2}.$$

In spherical polar coordinates, Helmholtz's equation becomes

$$\left[\nabla_r^2 + \frac{1}{r^2}\nabla_{\theta,\phi}^2\right]\psi(r,\theta,\phi) + k^2(r)\psi(r,\theta,\phi) = 0. \tag{6.31}$$

The reason for making k^2 a function of r will become clear later.

The method of separation of variables will now be used to reduce the partial differential in Eq. (6.31) to three ordinary differential equations. It is convenient to proceed using two steps. First, assume that the desired solution can be represented as a product of the radial and angular parts, that is,

$$\psi(r,\theta,\phi) = R(r)Y(\theta,\phi) \tag{6.32}$$

Substituting Eq. (6.32) into Eq. (6.31) and multiplying by r^2/RY, we obtain

$$\frac{r^2}{R}\nabla_r^2 R + r^2k^2(r) = -\frac{\nabla_{\theta,\phi}^2 Y}{Y}. \tag{6.33}$$

Since the left-hand side of Eq. (6.33) is a function of r only and the right-hand side is a function of θ and ϕ only, this equation can be satisfied only if both sides are separately equal to a constant. We denote this separation constant by λ and obtain

$$\nabla_r^2 R(r) + \left[k^2(r) - \frac{\lambda}{r^2}\right]R(r) = 0$$

and

$$\nabla_{\theta,\phi}^2 Y(\theta,\phi) + \lambda Y(\theta,\phi) = 0. \tag{6.34}$$

The solutions of Eq. (6.34) are called **spherical harmonics.**

The second step involves separating the variables in Eq. (6.34). Here we assume that

$$Y(\theta,\phi) = \Theta(\theta)\Phi(\phi). \tag{6.35}$$

From Eqs. (6.34) and (6.35), we obtain

$$\sin^2\theta\frac{\nabla_\theta^2\Theta}{\Theta} + \lambda\sin^2\theta = -\frac{\nabla_\phi^2\Phi}{\Phi} \equiv m^2$$

where

$$\nabla_\theta^2 = \frac{1}{\sin\theta}\frac{d}{d\theta}\left(\sin\theta\,\frac{d}{d\theta}\right)$$

$$\nabla_\phi^2 = \frac{d^2}{d\phi^2}$$

and m^2 is the separation constant. In summary, we have developed the following three ordinary differential equations from Eq. (6.31):

1. *Radial equation:*

$$\frac{1}{r^2}\frac{d}{dr}\left(r^2\frac{dR}{dr}\right) + \left[k^2(r) - \frac{\lambda}{r^2}\right]R = 0. \tag{6.36}$$

2. *Theta equation:*

$$\frac{1}{\sin\theta}\frac{d}{d\theta}\left(\sin\theta\,\frac{d\Theta}{d\theta}\right) + \left[\lambda - \frac{m^2}{\sin^2\theta}\right]\Theta = 0. \tag{6.37}$$

3. *Azimuthal equation:*

$$\frac{d^2\Phi}{d\phi^2} + m^2\Phi = 0. \tag{6.38}$$

6.3.2 The Azimuthal Equation

The specific form of $k^2(r)$ is required in order to obtain a solution of the radial equation, and we will return to this equation in later sections on Laguerre and associated Laguerre polynomials and Bessel functions. It is important to note that the theta and the azimuthal equations are independent of physical parameters which characterize the particular dynamical system under investigation. These parameters are contained in the radial equation through k^2, since k^2 is expressed in terms of the kinetic and potential energies and the mass (or reduced mass) of the particle involved.

We begin the solution of the remaining two equations with Eq. (6.38) since its solution is relatively easy to obtain. The general solution of the azimuthal equation is

$$\Phi(\phi) = c_1 e^{im\phi} + c_2 e^{-im\phi}.$$

For physical problems, we require that $\Phi(\phi)$ be a single-valued function. In other words, we require that

$$\Phi(\phi) = \Phi(\phi + 2\pi)$$

or

$$e^{2\pi im} = 1 \qquad (m = 0, \pm 1, \pm 2, \ldots).$$

The integer m is called the **magnetic quantum number** when k^2 is chosen so that the time-independent Schrödinger wave equation is obtained.

Normalization of $\Phi(\phi)$ requires that

$$\int_0^{2\pi} \Phi^*\Phi \, d\phi = 1.$$

Hence the normalized form of $\Phi(\phi)$ is

$$\Phi(\phi) = \frac{1}{\sqrt{2\pi}} e^{im\phi} \qquad (m = 0, \pm 1, \pm 2, \ldots). \tag{6.39}$$

6.3.3 Legendre (1752–1833) Polynomials

We may now proceed to solve the theta equation since m is known from the solution of the azimuthal equation. It is convenient to introduce a new variable, W, which is defined by

$$W = \cos\theta \tag{6.40}$$

so that

$$\frac{dW}{d\theta} = -\sin\theta \quad \text{and} \quad \sin^2\theta = 1 - W^2.$$

In terms of W, the theta equation becomes

$$\frac{d}{dW}\left[(1 - W^2)\frac{d\Theta}{dW}\right] + \left[\lambda - \frac{m^2}{1 - W^2}\right]\Theta = 0$$

or

$$(1 - W^2)\frac{d^2\Theta}{dW^2} - 2W\frac{d\Theta}{dW} + \left[\lambda - \frac{m^2}{1 - W^2}\right]\Theta = 0. \tag{6.41}$$

Equation (6.41) is the **associated Legendre differential equation**. In the special case of $m = 0$, Eq. (6.41) becomes the **Legendre differential equation**. Our plan is to develop a solution of Legendre's equation by use of the Frobenius-Fuchs power series method. The corresponding solution for the associated Legendre differential equation will be generated from the solution of Legendre's differential equation. Legendre's equation is

$$(1 - W^2)\frac{d^2\Theta}{dW^2} - 2W\frac{d\Theta}{dW} + \lambda\Theta = 0. \tag{6.42}$$

We assume that the form of the solution is

$$\Theta(W) = \sum_{n=0}^{\infty} a_n W^{k+n} \qquad (a_0 \neq 0). \tag{6.43}$$

Substituting Eq. (6.43) into Eq. (6.42) and rearranging terms, we obtain

$$k(k-1)a_0W^{k-2} + k(k+1)a_1W^{k-1} + \sum_{n=0}^{\infty}[(k+n+2)(k+n+1)a_{n+2} - (k+n)(k+n-1)a_n - 2(k+n)a_n + \lambda a_n]W^{k+n} = 0. \quad (6.44)$$

By requiring that Eq. (6.44) hold for all values of W, we obtain

$$k(k-1)a_0 = 0 \qquad \text{(indicial equation)}$$

or

$$k = 0 \quad \text{and} \quad k = 1 \quad (6.45)$$

$$k(k+1)a_1 = 0 \quad (6.46)$$

and

$$a_{n+2} = \frac{[(k+n)(k+n-1) + 2(k+n) - \lambda]a_n}{(k+n+2)(k+n+1)}. \quad (6.47)$$

Equation (6.47) is the required recursion formula.

The roots of the indicial equation are 0 and 1 with a_0 arbitrary. For $k = 0$, a_1 is arbitrary from Eq. (6.46). The root $k = 0$ therefore gives rise to the general solution in the form

$$\Theta(W) = a_0 \sum_{n=\text{even}}^{\infty} (\cdots)W^n + a_1 \sum_{n=\text{odd}}^{\infty} (\cdots)W^n.$$

The recursion formula for $k = 0$ is

$$a_{n+2} = \frac{[n(n-1) + 2n - \lambda]a_n}{(n+2)(n+1)}$$

$$= \frac{[n(n+1) - \lambda]a_n}{(n+2)(n+1)}. \quad (6.48)$$

For n even, we have:

$n = 0$

$$a_2 = -\frac{\lambda a_0}{2!}$$

$n = 2$

$$a_4 = \frac{[2(3) - \lambda]a_2}{4 \cdot 3}$$

$$= -\frac{\lambda(6 - \lambda)a_0}{4!}$$

$n = 4$

$$a_6 = \frac{[4(5) - \lambda]a_4}{6 \cdot 5}$$

$$= -\frac{\lambda(6 - \lambda)(20 - \lambda)a_0}{6!}.$$

For n odd, we have:

$n = 1$

$$a_3 = \frac{[1(2) - \lambda]a_1}{3 \cdot 2}$$

$$= \frac{(2 - \lambda)a_1}{3!}$$

$n = 3$

$$a_5 = \frac{[3(4) - \lambda]a_3}{5 \cdot 4}$$

$$= \frac{(2 - \lambda)(12 - \lambda)a_1}{5!}$$

$n = 5$

$$a_7 = \frac{[5(6) - \lambda]a_5}{7 \cdot 6}$$

$$= \frac{(2 - \lambda)(12 - \lambda)(30 - \lambda)a_1}{7!}.$$

The required solution is

$$\begin{aligned}\Theta(W) = a_0\Big[1 - \frac{\lambda}{2!}W^2 - \frac{\lambda(6 - \lambda)}{4!}W^4 \\ - \frac{\lambda(6 - \lambda)(20 - \lambda)}{6!}W^6 + \cdots\Big] \\ + a_1\Big[W + \frac{(2 - \lambda)}{3!}W^3 + \frac{(2 - \lambda)(12 - \lambda)W^5}{5!} \\ + \frac{(2 - \lambda)(12 - \lambda)(30 - \lambda)W^7}{7!} + \cdots\Big].\end{aligned} \tag{6.49}$$

The domain of convergence of $\Theta(W)$ in Eq. (6.49) is $-1 < W < 1$; this can be seen from the power series analysis in Chapter 3. It can be shown that $\Theta(W)$ diverges for $W = \pm 1$. Since $W = \cos\theta$, it is important to construct a solution valid for $-1 \leq W \leq 1$. This is achieved as follows: (1) replace λ

by the always positive integer $l(l+1)$, and (2) rewrite the coefficients of $\Theta(W)$ in terms of $l(l+1)$. The recursion formula, Eq. (6.48), now becomes

$$a_{n+2} = \frac{[n(n+1) - l(l+1)]a_n}{(n+2)(n+1)}$$
$$= -\frac{(l-n)(l+n+1)a_n}{(n+2)(n+1)}. \tag{6.50}$$

For n even, we have:

$n = 0$

$$a_2 = -\frac{l(l+1)a_0}{2!}$$

$n = 2$

$$a_4 = \frac{l(l-2)(l+1)(l+3)a_0}{4!}.$$

The general term is

$$a_{2j} = \frac{(-1)^j l(l-2)(l-4)\cdots(l-2j+2)(l+1)(l+3)\cdots(l+2j-1)a_0}{(2j)!} \tag{6.51}$$

where $j = 1, 2, 3, \ldots$.

For n odd, we have:

$n = 1$

$$a_3 = -\frac{(l-1)(l+2)a_1}{3!}$$

$n = 3$

$$a_5 = \frac{(l-1)(l-3)(l+2)(l+4)a_1}{5!}.$$

The general term is

$$a_{2j+1} = \frac{(-1)^j(l-1)(l-3)\cdots(l-2j+1)(l+2)(l+4)\cdots(l+2j)a_1}{(2j+1)!}. \tag{6.52}$$

The solution in terms of l is

$$\Theta_l(W) = a_0\Theta_l^{\text{even}}(W) + a_1\Theta_l^{\text{odd}}(W)$$

where

$$\Theta_l^{\text{even}} = 1 + \sum_{j=1}^{\infty} \frac{(-1)^j l(l-2)\cdots(l-2j+2)(l+1)\cdots(l+2j-1)W^{2j}}{(2j)!}$$

and

$$\Theta_l^{\text{odd}} = W + \sum_{j=1}^{\infty} \frac{(-1)^j(l-1)\cdots(l-2j+1)(l+2)\cdots(l+2j)W^{2j+1}}{(2j+1)!}.$$

For $l = 2j$ (an even positive integer), $\Theta_l^{\text{even}}(W)$ is a finite series (an even polynomial of degree l) since $a_{2j} \neq 0$ but $a_{2j+2} = 0$ where $l - (2j+2) + 2 = 0$. Similarly, $\Theta_l^{\text{odd}}(W)$ becomes an odd polynomial of degree l for $l = 2j + 1$ (an odd positive integer) since $a_{2j+1} \neq 0$ but $a_{2j+3} = 0$. A single finite series (polynomial), in descending powers of W, valid for both even and odd l will now be obtained. The first term of this polynomial is $a_l W^l$, and subsequent terms are given by the recursion formula

$$a_{j+2} = \frac{[j(j+1) - l(l+1)]a_j}{(j+2)(j+1)}$$
$$= -\frac{(l-j)(l+j+1)a_j}{(j+2)(j+1)}$$

or

$$a_j = -\frac{(j+2)(j+1)a_{j+2}}{(l-j)(l+j+1)}. \tag{6.53}$$

If $j = l - 2$ in Eq. (6.53), then

$$a_{l-2} = -\frac{l(l-1)a_l}{2(2l-1)}. \tag{6.54}$$

If $j = l - 4$ in Eq. (6.53), then

$$a_{l-4} = -\frac{(l-2)(l-3)a_{l-2}}{4(2l-3)}. \tag{6.55}$$

On substituting Eq. (6.54) into Eq. (6.55), we obtain

$$a_{l-4} = \frac{l(l-1)(l-2)(l-3)a_l}{2\cdot4(2l-1)(2l-3)}.$$

The general term is

$$a_{l-2r} = \frac{(-1)^r l(l-1)(l-2)\cdots(l-2r+1)a_l}{2\cdot4\cdots2r(2l-1)\cdots(2l-2r+1)}. \tag{6.56}$$

Equation (6.56) can be written in a compact form if we note that

$$l(l-1)\cdots(l-2r+1) = l(l-1)\cdots(l-2r+1) \cdot \frac{(l-2r)(l-2r-1)\cdots3\cdot2\cdot1}{(l-2r)(l-2r-1)\cdots3\cdot2\cdot1}$$
$$= \frac{l!}{(l-2r)!} \tag{6.57}$$

$$2 \cdot 4 \cdot 6 \cdots 2r = (2 \cdot 1)(2 \cdot 2)(2 \cdot 3) \cdots (2 \cdot r)$$
$$= 2^r r! \tag{6.58}$$

and

$$(2l - 1)(2l - 3) \cdots (2l - 2r + 1)$$
$$= \frac{2l(2l-1)(2l-2)(2l-3)\cdots(2l-2r+1)}{2l(2l-2)(2l-4)\cdots(2l-2r+2)} \cdot \frac{(2l-2r)!}{(2l-2r)!}$$
$$= \frac{(2l)!}{2^r l(l-1)\cdots(l-r+1)(2l-2r)!}$$
$$= \frac{(2l)!(l-r)!}{2^r(2l-2r)!l!}. \tag{6.59}$$

By use of Eqs. (6.57), (6.58), and (6.59), Eq. (6.56) becomes

$$a_{l-2r} = (-1)^r \left[\frac{(l!)^2(2l-2r)!}{(2l)!r!(l-2r)!(l-r)!} \right] a_l.$$

The solution in terms of l, the **angular momentum quantum number,**

$$a_0 \Theta_l^{\text{even}}(W) + a_1 \Theta_l^{\text{odd}}(W)$$

reduces to

$$\Theta_l(W) = a_l \sum_{r=0}^{N} \frac{(-1)^r (l!)^2 (2l-2r)! W^{l-2r}}{r!(2l)!(l-r)!(l-2r)!} \tag{6.60}$$

where

$$N = \begin{cases} \dfrac{l}{2} & \text{(if } l \text{ is even)} \\[2ex] \dfrac{l-1}{2} & \text{(if } l \text{ is odd).} \end{cases}$$

If we choose a_l to be

$$a_l = \frac{(2l)!}{2^l (l!)^2}$$

then the solution valid for $-1 \leq W \leq 1$ is

$$\Theta_l(W) \equiv P_l(W) = \sum_{r=0}^{N} \frac{(-1)^r (2l-2r)! W^{l-2r}}{2^l r!(l-r)!(l-2r)!} \tag{6.61}$$

where $P_l(W)$ are called the **Legendre polynomials of the first kind** of order l. The general solution of Legendre's equation is

$$\Theta_l(W) = P_l(W) + Q_l(W)$$

where $Q_l(W)$ are known as **Legendre functions of the second kind** and are seldom used in solving physical problems.

The generating function for the Legendre polynomials is

$$\frac{1}{\sqrt{1-2tW+t^2}} = \sum_{l=0}^{\infty} P_l(W)t^l \tag{6.62}$$

and the Rodrigues formula is

$$P_l(W) = \frac{1}{2^l l!}\frac{d^l}{dW^l}(W^2-1)^l. \tag{6.63}$$

Note that

$$P_0(W) = 1$$

$$P_1(W) = W$$

$$P_2(W) = \frac{1}{2}(3W^2-1)$$

and

$$P_3(W) = \frac{1}{2}(5W^3-3W).$$

The orthogonality property is given by

$$\int_{-1}^{1} P_l(W)P_m(W)\,dW = \begin{cases} 0 & (l \neq m) \\ \dfrac{2}{2l+1} & (l = m). \end{cases}$$

The recursion relation is

$$(2l+1)WP_l(W) = (l+1)P_{l+1}(W) + lP_{l-1}(W) \qquad (l = 1, 2, 3, \ldots).$$

Graphical illustrations of Legendre polynomials are given in Fig. 6.2.

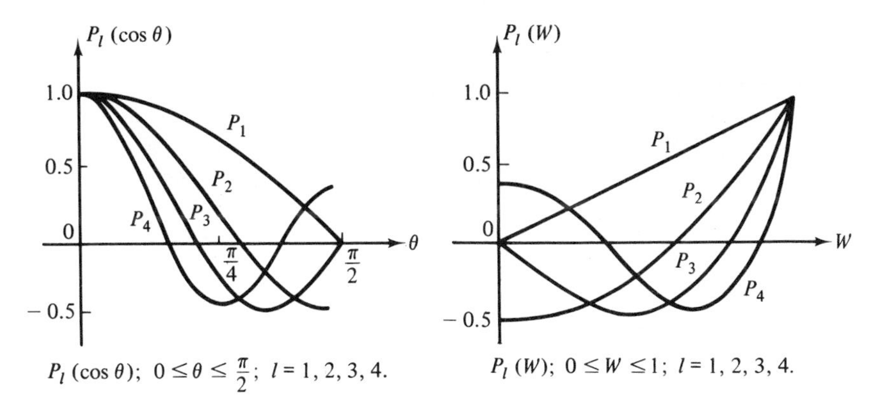

Figure 6.2

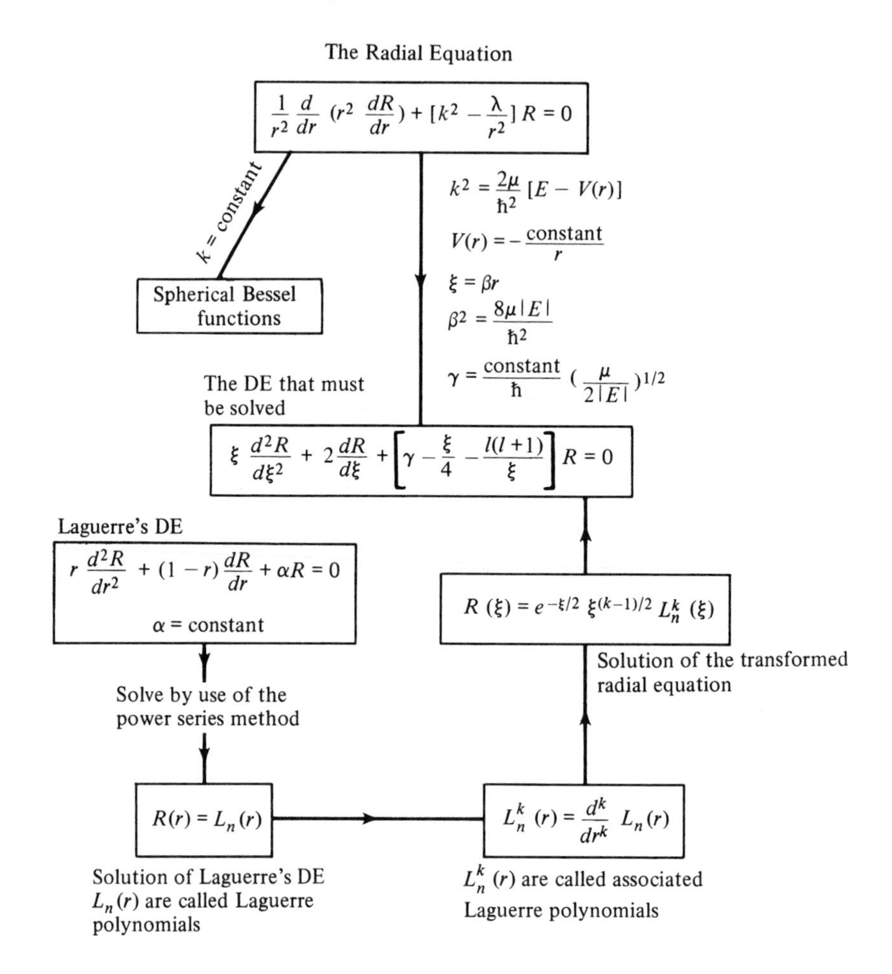

6.4 THE CENTRAL FORCE PROBLEM

6.4.1 Introduction

The quantum-mechanical problem of the motion of a particle in a central force field (field in which the potential depends on the distance alone) is of great importance in physics. This problem provides the basis for the quantum theory of the rigid rotator which is of considerable importance in the study of the spectra of diatomic molecules, the theory of the hydrogen atom, the nonrelativistic theory of the deuteron, and many other fundamental physical problems.

Let $k^2(r)$ in the radial equation [Eq. (6.36)] be given by

$$k^2(r) = \frac{2\mu}{\hbar^2}[E - V(r)]$$

where the potential is $V(r) = -\text{constant}/r$ and $\lambda = l(l + 1)$. The radial equation becomes

$$\frac{1}{r^2}\frac{d}{dr}\left(r^2\frac{dR}{dr}\right) + \left\{\frac{2\mu}{\hbar^2}\left[E + \frac{A}{r}\right] - \frac{l(l+1)}{r^2}\right\}R = 0.$$

We now write this equation in a dimensionless form. Let $\xi = \beta r$,

$$\beta^2 = \frac{8\mu|E|}{\hbar^2} \quad \text{and} \quad \gamma = \frac{2\mu A}{\beta\hbar^2} = \frac{A}{\hbar}\left(\frac{\mu}{2|E|}\right)^{1/2}$$

where $A = $ constant. In terms of ξ, the radial equation becomes

$$\xi\frac{d^2R}{d\xi^2} + 2\frac{dR}{d\xi} + \left[\gamma - \frac{\xi}{4} - \frac{l(l+1)}{\xi}\right]R = 0. \tag{6.64}$$

6.4.2 The Laguerre (1834–1866) Polynomials

The plan for solving Eq. (6.64) is to show that its solution can be generated from the solution of **Laguerre's equation,**

$$r\frac{d^2R}{dr^2} + (1 - r)\frac{dR}{dr} + \alpha R = 0 \qquad (\alpha = \text{constant}). \tag{6.65}$$

We assume that the solution of Eq. (6.65) has the form

$$R(r) = \sum_{\lambda=0}^{\infty} a_\lambda r^{k+\lambda} \qquad (a_0 \neq 0) \tag{6.66}$$

so that

$$\begin{aligned} R'(r) &= \sum_{\lambda=0}^{\infty} (k + \lambda)a_\lambda r^{k+\lambda-1} \\ R''(r) &= \sum_{\lambda=0}^{\infty} (k + \lambda)(k + \lambda - 1)a_\lambda r^{k+\lambda-2}. \end{aligned} \tag{6.67}$$

Substituting Eq. (6.66) and Eq. (6.67) into Eq. (6.65) and collecting terms, we obtain

$$\begin{aligned} [k(k - 1) + k]a_0 r^{k-1} + \sum_{\lambda=0}^{\infty} \{[(k + \lambda + 1)(k + \lambda) + k \\ + \lambda + 1]a_{\lambda+1} + (\alpha - k - \lambda)a_\lambda\}r^{k+\lambda} = 0. \end{aligned} \tag{6.68}$$

By requiring that Eq. (6.68) be valid for all values of r, we find that

$$[k(k - 1) + k]a_0 = 0 \tag{6.69}$$

and

$$a_{\lambda+1} = \frac{(k + \lambda - \alpha)a_\lambda}{(k + \lambda + 1)(k + \lambda) + k + \lambda + 1}. \tag{6.70}$$

The indicial equation and recursion formula are given, respectively, by Eqs. (6.69) and (6.70).

The indicial equation has a double root, $k = 0$; this results from the fact that Laguerre's equation, Eq. (6.65), has a nonessential singularity at the origin ($r = 0$). Hence the power series method will yield only one solution that is finite for all values of r. However, this solution is extremely important in physics. For $k = 0$, the recursion formula is

$$a_{\lambda+1} = -\frac{(\alpha - \lambda)a_\lambda}{(\lambda + 1)^2} \tag{6.71}$$

$\lambda = 0$

$$a_1 = -\frac{\alpha a_0}{1^2}$$

$\lambda = 1$

$$a_2 = -\frac{(\alpha - 1)a_1}{2^2}$$
$$= \frac{\alpha(\alpha - 1)a_0}{(2!)^2}$$

$\lambda = 2$

$$a_3 = -\frac{(\alpha - 2)a_2}{3^2}$$
$$= -\frac{\alpha(\alpha - 1)(\alpha - 2)}{(3!)^2}.$$

The solution is therefore

$$R(r) = a_0\left[1 - \frac{\alpha}{1^2}r + \frac{\alpha(\alpha - 1)}{(2!)^2}r^2 + \cdots + \frac{(-1)^j\alpha(\alpha - 1)\cdots(\alpha - j + 1)r^j}{(j!)^2} + \cdots\right] \tag{6.72}$$

where $j = 1, 2, 3, \ldots$. The right-hand side of Eq. (6.72) becomes finite (a polynomial of degree n) when $\alpha - j + 1 = 0$ or α equals a positive integer. The required finite solution is

$$\begin{aligned} R(r) &= a_0 \sum_{\lambda=0}^{n} \frac{(-1)^\lambda \alpha(\alpha - 1)\cdots(\alpha - \lambda + 1)r^\lambda}{(\lambda!)^2} \\ &= a_0 \sum_{\lambda=0}^{n} \frac{(-1)^\lambda \alpha! r^\lambda}{(\alpha - \lambda)!(\lambda!)^2} \\ &\equiv L_n(r) \end{aligned} \tag{6.73}$$

Helmholtz's Equation in Cylindrical Coordinates

Helmholtz's DE

$$\nabla^2 \psi(x, y, z) + k^2 \psi(x, y, z) = 0$$

$(x, y, z) \longrightarrow (\rho, \phi, z)$ Transformation to cylindrical coordinates

$$\rho \frac{\partial^2 \psi}{\partial \rho^2} + \frac{\partial \psi}{\partial \rho} + \frac{1}{\rho} \frac{\partial^2 \psi}{\partial \phi^2} + \rho \frac{\partial^2 \psi}{\partial z^2} + k^2 \rho \psi = 0$$

Use of the method of separation of variables

$$\psi(\rho, \phi, z) = P(\rho)\, \Phi(\phi)\, Z(z)$$

$$-\frac{1}{Z} \frac{d^2 Z}{dz^2} = \lambda^2 \text{ (separation constant)}$$

or

$$Z(z) = A \cos \lambda z + B \sin \lambda z$$

$$-\frac{1}{\Phi} \frac{d^2 \Phi}{d\phi^2} = n^2 \text{ (separation constant)}$$

or

$$\Phi(\phi) = C \cos n\phi + D \sin n\phi$$

$\xi = \alpha\rho;\; k^2 - \lambda^2 \equiv \alpha^2$

Bessel's DE

$$\xi^2 \frac{d^2 P}{d\xi^2} + \xi \frac{dP}{d\xi} + (\xi^2 - n^2) P = 0$$

n = constant; $P = P(\xi)$

Solve by use of the power series method

$P(\xi) = J_n(\xi)$ = Bessel functions of the first kind

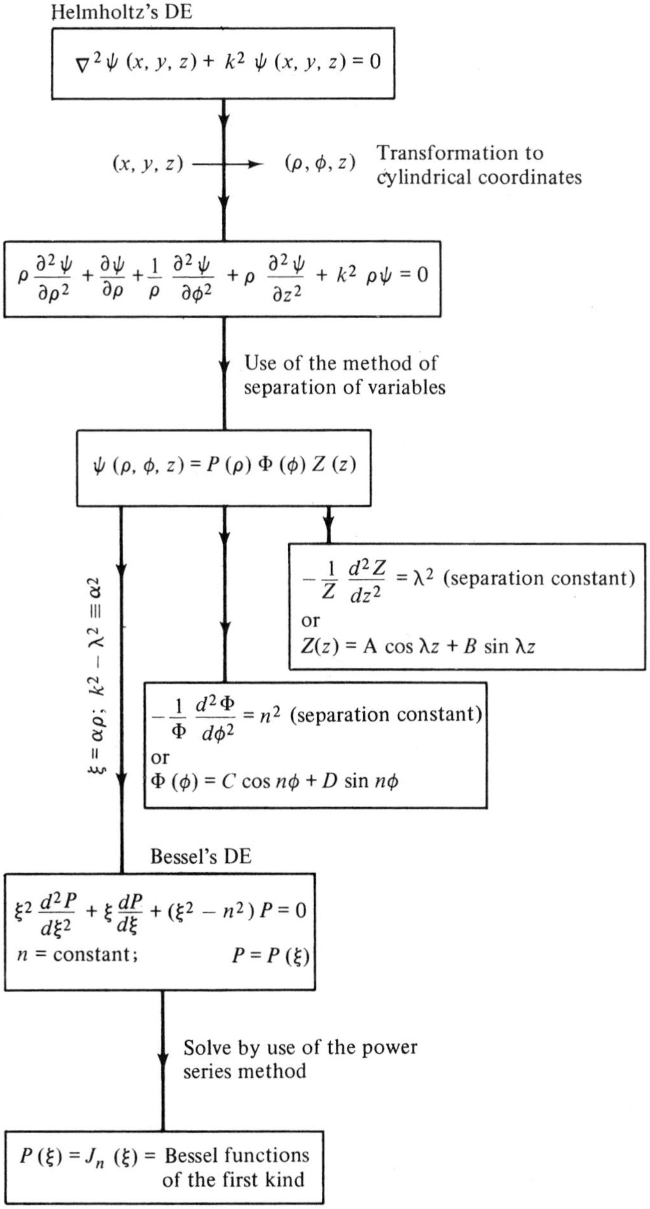

$$\frac{1}{\rho}\left[\frac{\partial}{\partial \rho}\left(\rho \frac{\partial \psi}{\partial \rho}\right) + \frac{\partial}{\partial \phi}\left(\frac{1}{\rho}\frac{\partial \psi}{\partial \phi}\right) + \frac{\partial}{\partial z}\left(\rho \frac{\partial \psi}{\partial z}\right)\right] + k^2(\rho)\psi = 0$$

which reduces to

$$\rho\frac{\partial^2 \psi}{\partial \rho^2} + \frac{\partial \psi}{\partial \rho} + \frac{1}{\rho}\frac{\partial^2 \psi}{\partial \phi^2} + \rho\frac{\partial^2 \psi}{\partial z^2} + k^2(\rho)\rho\psi = 0. \tag{6.85}$$

Equation (6.85) will now be solved by the method of separation of variables,

$$\psi = P(\rho)\Phi(\phi)Z(z). \tag{6.86}$$

Substituting Eq. (6.86) into Eq. (6.85) and dividing by $P\Phi Z\rho$, we obtain

$$\frac{1}{P}\frac{d^2P}{d\rho^2} + \frac{1}{\rho P}\frac{dP}{d\rho} + k^2(\rho) + \frac{1}{\rho^2\Phi}\frac{d^2\Phi}{d\phi^2} = -\frac{1}{Z}\frac{d^2Z}{dz^2}. \tag{6.87}$$

Since the left-hand side of Eq. (6.87) is a function of ρ and ϕ and the right-hand side is a function of z only, we may write

$$-\frac{1}{Z}\frac{d^2Z}{dz^2} = \lambda^2$$

with solution

$$Z(z) = A\cos\lambda z + B\sin\lambda z \tag{6.88}$$

and

$$\frac{\rho^2}{P}\frac{d^2P}{d\rho^2} + \frac{\rho}{P}\frac{dP}{d\rho} + \rho^2[k^2 - \lambda^2] = -\frac{1}{\Phi}\frac{d^2\Phi}{d\phi^2}. \tag{6.89}$$

On separating the variables in Eq. (6.89), we obtain

$$\frac{1}{\Phi}\frac{d^2\Phi}{d\phi^2} = -n^2$$

with solution

$$\Phi(\phi) = C\cos n\phi + D\sin n\phi \tag{6.90}$$

and

$$\rho^2\frac{d^2P}{d\rho^2} + \rho\frac{dP}{d\rho} + \{\rho^2[k^2 - \lambda^2] - n^2\}P = 0. \tag{6.91}$$

For

$$k^2 - \lambda^2 = \alpha^2$$

where $k^2 =$ constant, Eq. (6.91) becomes

$$\rho^2 \frac{d^2P}{d\rho^2} + \rho \frac{dP}{d\rho} + \{\rho^2\alpha^2 - n^2\}P = 0. \tag{6.92}$$

We now make the following change of variable:

$$\xi = \alpha\rho$$

so that

$$\frac{d}{d\rho} = \frac{d}{d\xi}\frac{d\xi}{d\rho} = \alpha \frac{d}{d\xi}$$

$$\frac{d^2}{d\rho^2} = \alpha^2 \frac{d^2}{d\xi^2}.$$

By use of this change of variable, Eq. (6.92) reduces to

$$\xi^2 \frac{d^2P}{d\xi^2} + \xi \frac{dP}{d\xi} + (\xi^2 - n^2)P = 0 \tag{6.93}$$

which is **Bessel's equation**, and its solutions are called **Bessel** (1784–1846) **functions** (or **cylindrical functions**).

6.5.2 The Solution of Bessel's Equation

We now proceed to solve Bessel's equation by the generalized power series method. The form of the solution is assumed to be

$$P(\xi) = \sum_{\lambda=0}^{\infty} a_\lambda \xi^{k+\lambda} \qquad (a_0 \neq 0) \tag{6.94}$$

so that

$$P'(\xi) = \sum_{\lambda=0}^{\infty} (k + \lambda) a_\lambda \xi^{k+\lambda-1}$$

and

$$P''(\xi) = \sum_{\lambda=0}^{\infty} (k + \lambda)(k + \lambda - 1) a_\lambda \xi^{k+\lambda-2}. \tag{6.95}$$

Substituting Eqs. (6.94) and (6.95) into Eq. (6.93) and rearranging terms, we obtain

$$(k^2 - n^2)a_0\xi^k + [(k + 1)^2 - n^2]a_1\xi^{k+1} + \sum_{\lambda=2}^{\infty} \{[(k + \lambda)^2 - n^2]a_\lambda + a_{\lambda-2}\}\xi^{k+\lambda} = 0. \tag{6.96}$$

By requiring that Eq. (6.96) hold for all values of ξ, we find that

$$(k^2 - n^2)a_0 = 0$$

or

$$k = \pm n \tag{6.97}$$

$$[(k+1)^2 - n^2]a_1 = 0 \tag{6.98}$$

and

$$[(k+\lambda)^2 - n^2]a_\lambda + a_{\lambda-2} = 0 \qquad (\lambda = 2, 3, \ldots)$$

or

$$a_\lambda = -\frac{a_{\lambda-2}}{(k+\lambda)^2 - n^2} \qquad (\lambda \geq 2). \tag{6.99}$$

Since $k^2 = n^2$ from the indicial equation, a_1 must equal zero for $n =$ integer in Eq. (6.98).

Case 1 $k = n$; $a_0 \neq 0$; $a_1 = 0$.

Here the recursion formula is

$$a_\lambda = -\frac{a_{\lambda-2}}{(n+\lambda)^2 - n^2}$$
$$= -\frac{a_{\lambda-2}}{\lambda(2n+\lambda)} \qquad (\lambda \geq 2) \tag{6.100}$$

$\lambda = 2$

$$a_2 = -\frac{a_0}{2(2n+2)}$$
$$= -\frac{a_0}{2^2 \cdot 1(n+1)}$$

$\lambda = 4$

$$a_4 = -\frac{a_2}{4(2n+4)}$$
$$= \frac{a_0}{2^4 \cdot 2!(n+1)(n+2)}$$

$\lambda = 6$

$$a_6 = -\frac{a_4}{6(2n+6)}$$
$$= -\frac{a_0}{2^6 \cdot 3!(n+1)(n+2)(n+3)}.$$

The general coefficient is

$$a_{2j} = (-1)^j \frac{a_0}{2^{2j} j!(n+1)(n+2) \cdots (n+j)} \tag{6.101}$$

where $j = 0, 1, 2, \ldots$. In terms of gamma functions, the general coefficient reduces to

$$a_{2j} = (-1)^j \frac{\Gamma(n+1)a_0}{2^{2j} j! \Gamma(n+j+1)} \tag{6.102}$$

since

$$\begin{aligned}(n+1)(n+2) \cdots (n+j) &= (n+j)(n+j-1) \cdots (n+2)(n+1) \\ &\quad \cdot \frac{\Gamma(n+1)}{\Gamma(n+1)} \\ &= \frac{\Gamma(n+j+1)}{\Gamma(n+1)}.\end{aligned} \tag{6.103}$$

If we choose a_0 to be

$$a_0 = \frac{1}{2^n \Gamma(n+1)}$$

the solution for $k = n$, where $n =$ integer, becomes

$$\begin{aligned}P(\xi) \equiv J_n(\xi) &= \sum_{j=0}^{\infty} \frac{(-1)^j \xi^{2j+n}}{2^{2j+n} j! \Gamma(n+j+1)} \\ &= \sum_{j=0}^{\infty} \frac{(-1)^j (\xi/2)^{2j+n}}{j! \Gamma(n+j+1)}\end{aligned} \tag{6.104}$$

where $J_n(\xi)$, usually written in the misleading† form of $J_n(x)$, are called **Bessel functions of the first kind of order** n.

The **gamma function,** Γ, is a generalization (to the case of nonintegers) of the factorial,

$$\begin{aligned}n! &= n(n-1) \cdots 2 \cdot 1 \\ &= \int_0^{\infty} e^{-t} t^n \, dt \qquad \text{(for } n \text{ integer).}\end{aligned}$$

Note that $0! = 1$ and $n! = \pm\infty$ if n is a negative integer. The general definition of the gamma function is

$$\Gamma(p) = \int_0^{\infty} e^{-t} t^{p-1} \, dt \qquad \text{(for any } p > 0\text{).} \tag{6.105}$$

†Since x is normally used as a Cartesian coordinate.

If p is a positive integer in Eq. (6.105), we obtain

$$\Gamma(n+1) = \int_0^\infty e^{-t} t^n \, dt$$
$$= n! \tag{6.106}$$

On evaluating the integral in Eq. (6.106) by parts, we obtain

$$\Gamma(n+1) = \int_0^\infty e^{-t} t^n \, dt = n\Gamma(n)$$

which is the recursion relation for the gamma function.

Case 2 $k = -n; a_0 \neq 0; a_1 = 0.$

The solution in this case is obtained by replacing n by $-n$ in all of the equations for Case 1. The result is that

$$J_{-n}(\xi) = \sum_{j=0}^{\infty} \frac{(-1)^j (\xi/2)^{2j-n}}{j!\Gamma(-n+j+1)}. \tag{6.107}$$

The **generating function** for the Bessel functions is

$$\exp\left[\frac{1}{2}\xi\left(t - \frac{1}{t}\right)\right] = \sum_{n=-\infty}^{\infty} t^n J_n(\xi). \tag{6.108}$$

6.5.3 Analysis of the Various Solutions of Bessel's Equation

A. n = Integer Here

$$J_n(\xi) = \sum_{j=0}^{\infty} \frac{(-1)^j (\xi/2)^{2j+n}}{j!(n+j)!}$$

and

$$J_{-n}(\xi) = \sum_{j=0}^{\infty} \frac{(-1)^j (\xi/2)^{2j-n}}{j!(j-n)!}.$$

Since $P! = \pm\infty$ for negative P, all of the terms up to $j = n$ in the $J_{-n}(\xi)$ equation vanish. Hence the series for $J_{-n}(\xi)$ may be considered to start with $j = n$.

$$J_{-n}(\xi) = \sum_{j=n}^{\infty} \frac{(-1)^j (\xi/2)^{2j-n}}{j!(j-n)!}$$
$$= \sum_{m=0}^{\infty} \frac{(-1)^{m+n} (\xi/2)^{2m+n}}{(m+n)!m!}$$
$$= (-1)^n \sum_{m=0}^{\infty} \frac{(-1)^m (\xi/2)^{2m+n}}{m!(m+n)!}$$
$$= (-1)^n J_n(\xi) \tag{6.109}$$

where $m = j - n$. Hence we have only one independent solution when n is an integer.

B. $n =$ ***Odd-Half Integer*** Note that

$$J_n(\xi) = \sum_{j=0}^{\infty} \frac{(-1)^j(\xi/2)^{2j+n}}{\Gamma(j+1)\Gamma(j+n+1)}.$$

For $n = \frac{1}{2}$, we obtain

$$J_{1/2}(\xi) = \left(\frac{\xi}{2}\right)^{1/2} \sum_{j=0}^{\infty} \frac{(-1)^j \xi^{2j}}{2^{2j} j! \Gamma(3/2 + j)}. \tag{6.110}$$

In general, we find that

$$J_{\nu+1/2}^{(\xi)} = \sum_{j=0}^{\infty} \frac{(-1)^j(\xi/2)^{\nu+1/2+2j}}{j!\Gamma(\nu + j + 3/2)}$$

where $2n = 2\nu + 1$. Here ν is any positive integer or zero.

6.5.4 Neumann (1832–1925) Functions

For $n =$ integer, it was shown that only one independent solution is obtained. The **Neumann functions** give us the remaining solution for $n =$ integer. They are defined by†

$$N_n(\xi) = Y_n(\xi) = \frac{J_n(\xi)\cos n\pi - J_{-n}(\xi)}{\sin n\pi}. \tag{6.111}$$

Hence the most general solution for $n =$ integer is

$$P(\xi) = AJ_n(\xi) + BN_n(\xi). \tag{6.112}$$

Neumann functions are often referred to as **Bessel functions of the second kind.** The general solution for $n \neq$ integer is

$$P(\xi) = CJ_n(\xi) + DJ_{-n}(\xi). \tag{6.113}$$

6.5.5 Hankel (1839–1873) Functions

The two kinds of **Hankel functions** are defined by

$$\begin{aligned} H_n^{(1)}(\xi) &= \frac{i}{\sin n\pi}[e^{-n\pi i}J_n(\xi) - J_{-n}(\xi)] \\ &= J_n(\xi) + iN_n(\xi) \end{aligned} \tag{6.114}$$

†L'Hospital's rule should be used to evaluate $N_n(\xi)$ for $n =$ integer.

and

$$H_n^{(2)}(\xi) = -\frac{i}{\sin n\pi}[e^{n\pi i}J_n(\xi) - J_{-n}(\xi)]$$
$$= J_n(\xi) - iN_n(\xi). \tag{6.115}$$

The Hankel functions are independent solutions of Bessel's equation. Their usefulness is connected with their behavior for large ξ since they are infinite at $\xi = 0$. They are often called **Bessel functions of the third kind.**

6.5.6 Modified Bessel Functions

By replacing ξ with it in Eq. (6.93), we obtain the **modified Bessel equation** which can be written in the form

$$t^2\frac{d^2P}{dt^2} + t\frac{dP}{dt} - (t^2 + n^2)P = 0 \tag{6.116}$$

where $P_n = P_n(it)$ are called **modified Bessel functions of the first kind** and are denoted by $I_n(\xi)$; they are given by

$$I_n(\xi) = i^{-n}J_n(i\xi) = \sum_{\lambda=0}^{\infty}\frac{(\xi/2)^{2\lambda+n}}{\lambda!(\lambda+n)!}. \tag{6.117}$$

When n is a noninteger, $I_n(\xi)$ and $I_{-n}(\xi)$ are independent solutions of the modified Bessel equation. When n is an integer, $I_n(\xi) = I_{-n}(\xi)$. The **modified Bessel functions of the second kind,** $K_n(\xi)$, are defined by

$$K_n(\xi) = \frac{\pi}{2}\left[\frac{I_{-n}(\xi) - I_n(\xi)}{\sin n\pi}\right] \tag{6.118}$$

and are well behaved for all values of n.

6.5.7 Spherical Bessel Functions

The radial equation, Eq. (6.36), for $k^2 =$ constant is

$$r^2\frac{d^2R}{dr^2} + 2r\frac{dR}{dr} + [k^2r^2 - l(l+1)]R = 0. \tag{6.119}$$

Equation (6.119) strongly resembles Bessel's equation. Its solution can be expressed in terms of Bessel functions in the following way:

$$R(r) = Aj_n(\zeta) = \sqrt{\frac{\pi}{2\zeta}}J_{n+1/2}(\zeta); \qquad \zeta = kr \tag{6.120}$$

where $j_n(\zeta)$ are called **spherical Bessel functions** and $A =$ constant.

The **spherical Neumann functions** are defined by

$$n_n(\zeta) = \sqrt{\frac{\pi}{2\zeta}} N_{n+1/2}(\zeta). \tag{6.121}$$

The **spherical Hankel functions** are defined by

$$h_n^{(1)}(\zeta) = j_n(\zeta) + i n_n(\zeta) \tag{6.122}$$

and

$$h_n^{(2)}(\zeta) = j_n(\zeta) - i n_n(\zeta). \tag{6.123}$$

The **spherical modified Bessel functions** are defined by

$$i_n(\zeta) = \sqrt{\frac{\pi}{2\zeta}} I_{n+1/2}(\zeta) \tag{6.124}$$

and

$$k_n(\zeta) = \sqrt{\frac{2}{\pi\zeta}} K_{n+1/2}(\zeta) \tag{6.125}$$

6.5.8 The Characteristics of the Various Bessel Functions

The Bessel functions are tabulated for various values of n in many books.† The $J_{n\neq 0}(\xi)$ functions all behave like $\sin \xi$ with decreasing amplitude. The $J_0(\xi)$ function equals 1 at $\xi = 0$ and looks like a damped cosine. All $N_n(\xi)$ are $\pm\infty$ at the origin, but they oscillate with decreasing amplitude away from the origin. The Hankel functions behave like imaginary exponentials, and the modified Bessel functions, $I_n(\xi)$ and $K_n(\xi)$, go into positive and negative exponentials, respectively.

As $\xi \to \infty$, we have

$$J_n(\xi) \sim \sqrt{\frac{2}{\pi\xi}} \cos\left[\xi - \left(n + \frac{1}{2}\right)\frac{\pi}{2}\right]$$

$$N_n(\xi) \sim \sqrt{\frac{2}{\pi\xi}} \sin\left[\xi - \left(n + \frac{1}{2}\right)\frac{\pi}{2}\right]$$

$$H_n^{(1)}(\xi) \sim \sqrt{\frac{2}{\pi\xi}} \exp\left[i\left\{\xi - \left(n + \frac{1}{2}\right)\frac{\pi}{2}\right\}\right]$$

$$H_n^{(2)}(\xi) \sim \sqrt{\frac{2}{\pi\xi}} \exp\left[-i\left\{\xi - \left(n + \frac{1}{2}\right)\frac{\pi}{2}\right\}\right]$$

†See for example, M. Abramowitz and I. A. Stegun, *Handbook of Mathematical Functions with Formulas, Graphs, and Mathematical Tables*, New York, Dover, 1969.

$$I_n(\xi) \sim \frac{1}{\sqrt{2\pi\xi}} e^{\xi}$$

$$K_n(\xi) \sim \sqrt{\frac{\pi}{2\xi}} e^{-\xi}$$

$$j_n(\xi) \sim \frac{1}{\xi} \sin\left(\xi - \frac{n\pi}{2}\right)$$

$$n_n(\xi) \sim -\frac{1}{\xi} \cos\left(\xi - \frac{n\pi}{2}\right)$$

$$h_n^{(1)}(\xi) \sim -\frac{i}{\xi} \exp\left[i\left(\xi - \frac{n\pi}{2}\right)\right]$$

$$h_n^{(2)}(\xi) \sim \frac{i}{\xi} \exp\left[-i\left(\xi - \frac{n\pi}{2}\right)\right].$$

As $\xi \to 0$, we have

$$J_n(\xi) \sim \frac{1}{\Gamma(n+1)}\left(\frac{\xi}{2}\right)^n$$

$$N_n(\xi) \sim \begin{cases} -\dfrac{1}{\pi}\Gamma(n)\left(\dfrac{2}{\xi}\right)^n & (n > 0) \\ \dfrac{2}{\pi} \ln \xi & (n = 0) \end{cases}$$

$$H_n^{(1)}(\xi) \sim iN_n(\xi) + 0(\xi^n)$$

$$H_n^{(2)}(\xi) \sim -iN_n(\xi) + 0(\xi^n)$$

$$I_n(\xi) \quad \text{behaves like} \quad J_n(\xi)$$

$$K_n(\xi) = \begin{cases} -\ln \xi + 0(1) & (n = 0) \\ \dfrac{1}{2}\Gamma(n)\left(\dfrac{2}{\xi}\right)^n + 0\left(\dfrac{1}{\xi^{n-1}}\right) & (n > 0) \end{cases}$$

$$j_n(\xi) \sim \frac{\xi^n}{(2n+1)!!} \qquad (n \text{ integer})$$

$$n_n(\xi) \sim -\frac{(2n-1)!!}{\xi^{n+1}} \qquad (n \text{ integer})$$

where

$$n!! = n(n-2)(n-4)\cdots \begin{cases} 5\cdot 3\cdot 1 & (\text{if } n \text{ is odd}) \\ 6\cdot 4\cdot 2 & (\text{if } n \text{ is even}). \end{cases}$$

Graphs indicating the behavior of the various Bessel functions are given in Figs. 6.3 through 6.8.

Numerical values for certain Bessel functions are given in Tables 6.1, 6.2, and 6.3.

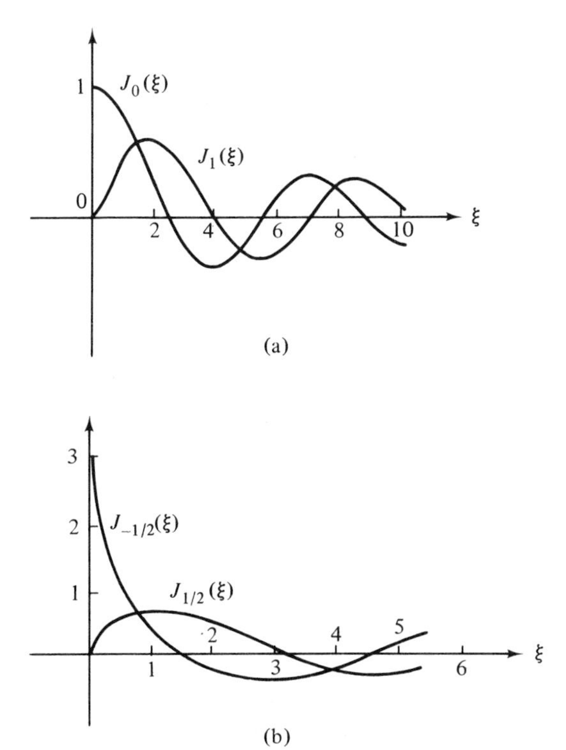

(a)

(b)

Figure 6.3

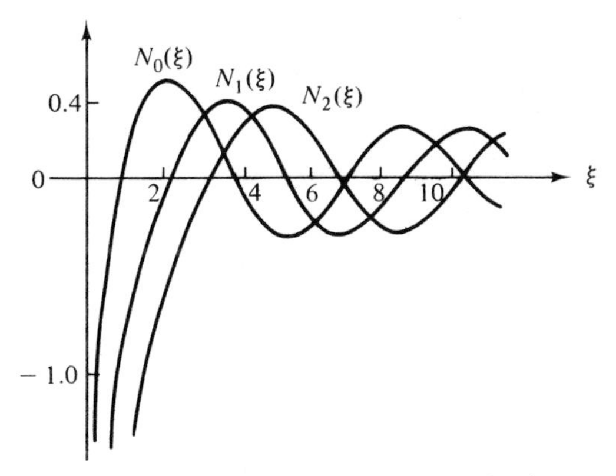

Neumann functions, $N_0(\xi)$, $N_1(\xi)$, and $N_2(\xi)$

Figure 6.4

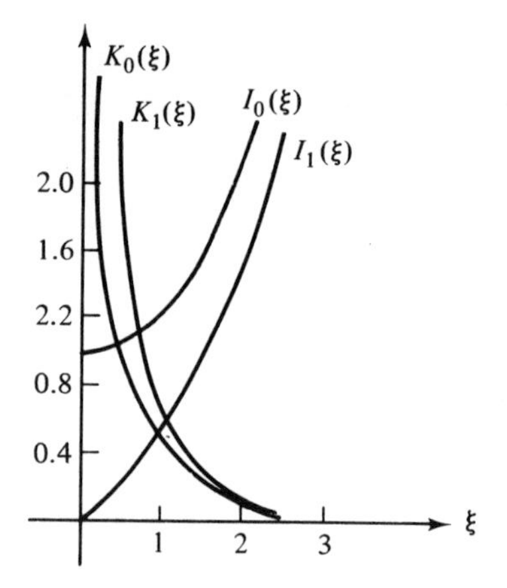

Modified Bessel functions **Figure 6.5**

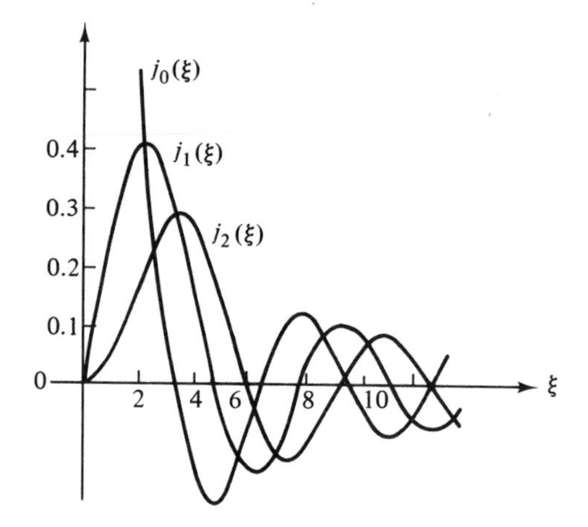

Spherical Bessel functions **Figure 6.6**

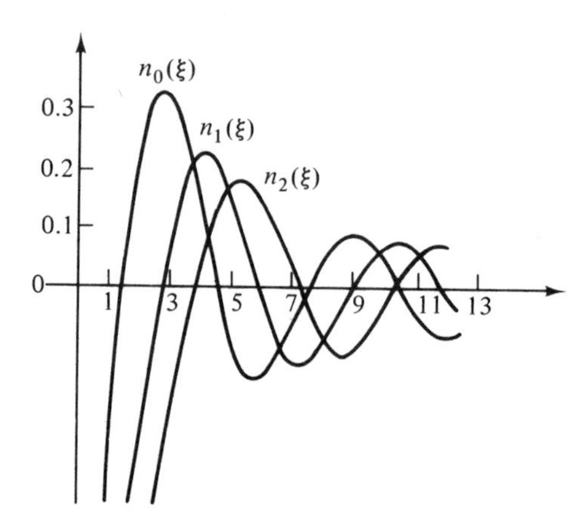

Spherical Neumann functions **Figure 6.7**

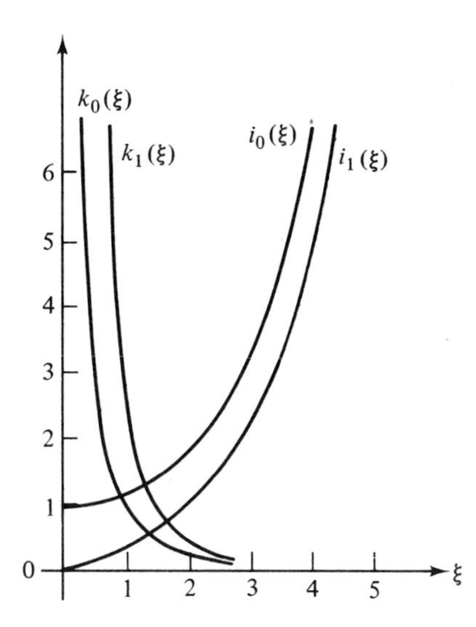

Figure 6.8 Spherical modified Bessel functions

TABLE 6.1 SOME VALUES OF $J_0(\xi)$ AND $J_1(\xi)$

ξ	$J_0(\xi)$	$J_1(\xi)$	ξ	$J_0(\xi)$	$J_1(\xi)$	ξ	$J_0(\xi)$	$J_1(\xi)$
0.0	1.0000	0.0000	3.0	−0.2601	0.3391	6.0	0.1506	−0.2767
0.1	0.9975	0.0499	3.1	−0.2921	0.3009	6.1	0.1773	−0.2552
0.2	0.9900	0.0995	3.2	−0.3202	0.2613	6.2	0.2017	−0.2329
0.3	0.9776	0.1483	3.3	−0.3443	0.2207	6.3	0.2238	−0.2081
0.4	0.9604	0.1960	3.4	−0.3643	0.1792	6.4	0.2433	−0.1816
0.5	0.9385	0.2423	3.5	−0.3801	0.1374	6.5	0.2601	−0.1538
0.6	0.9120	0.2867	3.6	−0.3918	0.0955	6.6	0.2740	−0.1250
0.7	0.8812	0.3290	3.7	−0.3992	0.0538	6.7	0.2851	−0.0953
0.8	0.8463	0.3688	3.8	−0.4026	0.0128	6.8	0.2931	−0.0652
0.9	0.8075	0.4059	3.9	−0.4018	−0.0272	6.9	0.2981	−0.0349

TABLE 6.2 SOME VALUES OF $Y_0(\xi)$ AND $Y_1(\xi)$

ξ	$Y_0(\xi)$	$Y_1(\xi)$	ξ	$Y_0(\xi)$	$Y_1(\xi)$	ξ	$Y_0(\xi)$	$Y_1(\xi)$
0.0	$(-\infty)$	$(-\infty)$	2.5	0.498	0.146	5.0	−0.309	0.148
0.5	−0.445	−1.471	3.0	0.377	0.325	5.5	−0.340	−0.024
1.0	0.088	−0.781	3.5	0.189	0.410	6.0	−0.288	−0.175
1.5	0.382	−0.412	4.0	−0.017	0.398	6.5	−0.173	−0.274
2.0	0.510	−0.107	4.5	−0.195	0.301	7.0	−0.026	−0.303

TABLE 6.3 SOME VALUES FOR WHICH $J_{...}(\xi) = 0$

Number of zero	$J_0(\xi) = 0$ ξ	$J_1(\xi) = 0$ ξ	$J_2(\xi) = 0$ ξ	$J_3(\xi) =$ ξ	$J_4(\xi) = 0$ ξ	$J_5(\xi) = 0$ ξ
1	2.4048	3.8317	5.1356	6.3802	7.5883	8.7715
2	5.5201	7.0156	8.4172	9.7610	11.0647	12.3386
3	8.6537	10.1735	11.6198	13.0152	14.3725	15.7002
4	11.7915	13.3237	14.7960	16.2235	17.6160	18.9801
5	14.9309	16.4706	17.9598	19.4094	20.8269	22.2178

6.5.9 Certain Other Special Functions

Our solution of the equation for the classical harmonic oscillator in Chapter 5 was in terms of sines and cosines. These functions may be called "special functions." In general, no precise definition of a special function can be given. Hence the number of special functions is arbitrary.

The following is a list of other "special functions" that are also used in mathematical physics: (1) hypergeometric functions, (2) confluent hypergeometric (Kummer's) functions, (3) Chebyshev (Tschebycheff) polynomials, (4) beta functions, (5) the error function, (6) Riemann's zeta function, (7) Debye's function, (8) Mathieu's functions, (9) Lamé functions, (10) theta functions, (11) Kelvin functions, and (12) Gegenbauer functions.

EXAMPLE The equation which characterizes the motion of a vibrating membrane is

$$\frac{\partial^2 u}{\partial x^2} + \frac{\partial^2 u}{\partial y^2} = \frac{1}{v^2}\frac{\partial^2 u}{\partial t^2}$$

where $v = \sqrt{T/\mu}$, μ is the mass per unit area of the membrane, T is the tension, and $u = u(x, y, t)$ is the displacement of the membrane at (x, y) and time t. Develop a solution of this differential equation for a circular membrane (a drum).

Solution Here it is convenient to use cylindrical coordinates $(\rho, \phi, z = 0)$. In this case, the equation of motion becomes

$$\frac{1}{\rho}\left[\frac{\partial}{\partial \rho}\left(\rho \frac{\partial u}{\partial \rho}\right) + \frac{\partial}{\partial \phi}\left(\frac{1}{\rho}\frac{\partial u}{\partial \phi}\right)\right] = \frac{1}{v^2}\frac{\partial^2 u}{\partial t^2}$$

or

$$\frac{\partial^2 u}{\partial \rho^2} + \frac{1}{\rho}\frac{\partial u}{\partial \rho} + \frac{1}{\rho^2}\frac{\partial^2 u}{\partial \phi^2} = \frac{1}{v^2}\frac{\partial^2 u}{\partial t^2}.$$

We now use the method of separation of variables and assume that the solution of the above partial differential equation has the form

$$u(\rho, \phi, t) = R(\rho)\Phi(\phi)T(t).$$

On substituting the right-hand side of the above equation into the original partial differential equation to be solved and then dividing the resulting equation by the right-hand side of the above equation, we obtain

$$\begin{aligned} \frac{R''}{R} + \frac{1}{\rho}\frac{R'}{R} + \frac{1}{\rho^2}\frac{\Phi''}{\Phi} &= \frac{1}{v^2}\frac{\ddot{T}}{T} \\ &= -\alpha^2 \end{aligned} \tag{6.124}$$

where

$$\frac{d^2T}{dt^2} + \omega^2 T = 0, \qquad T(t) = A\cos\omega t + B\sin\omega t$$

and $\omega^2 = v^2\alpha^2$. On multiplying Eq. (6.124) by ρ^2, we find that

$$\rho^2\frac{R''}{R} + \rho\frac{R'}{R} + \rho^2\alpha^2 = -\frac{\Phi''}{\Phi} = n^2$$

or

$$\frac{d^2\Phi}{d\phi^2} + n^2\Phi = 0, \qquad \Phi(\phi) = C\cos n\phi + D\sin n\phi$$

and

$$\rho^2\frac{d^2R}{d\rho^2} + \rho\frac{dR}{d\rho} + (\rho^2\alpha^2 - n^2)R = 0. \tag{6.125}$$

By means of the change of variable $\xi = \alpha\rho$, Eq. (6.125) becomes

$$\xi^2\frac{d^2R}{d\xi^2} + \xi\frac{dR}{d\xi} + (\xi^2 - n^2)R = 0.$$

This is Bessel's differential equation with solution

$$R(\xi) = EJ_n(\xi) + FN_n(\xi)$$

since $n =$ integer as determined from requiring that $\Phi(\phi)$ be single-valued.

The solution of the equation of motion of the vibrating membrane is therefore

$$u(\xi, \phi, t) = (A\cos\omega t + B\sin\omega t)(C\cos n\phi + D\sin n\phi)EJ_n(\xi)$$

since we require that $u(\rho, \phi, t)$ be finite at $\rho = \xi\alpha = 0$ $[N_n(\xi = 0) \to \infty]$.

At the edge of the membrane $\rho = r$ (radius), $u(\rho, \phi, t) = 0$, or $R(\xi) = EJ_n(\xi) = 0$ (a circular node). By use of the properties of Bessel functions, we know that there are nodes at $\alpha\rho = \xi_k$ where ξ_k are the values of ξ for which $J_n(\xi)$ has a zero. Hence a single term in the solution corresponds to a standing wave whose nodes are concentric circles. The complete solution is obtained by summing over all such modes of vibration.

6.6 PROBLEMS

6.1 By differentiating both sides of the equation for the generating function for $H_n(\xi)$ with respect to t and equating the coefficients of like powers of t, show that

$$H_{n+1}(\xi) = 2\xi H_n(\xi) - 2nH_{n-1}(\xi).$$

6.2 By differentiating both sides of the equation for the generating function for $H_n(\xi)$ with respect to ξ and equating the coefficients of like powers of t, show that

$$H'_n(\xi) = 2nH_{n-1}(\xi).$$

6.3 Show that

$$\int_{-\infty}^{\infty} e^{-\xi^2} H_n(\xi) H_m(\xi)\, d\xi = \begin{cases} \sqrt{\pi}\, 2^n n! & (n = m) \\ 0 & (n \neq m). \end{cases}$$

6.4 Evaluate

$$\int_{-\infty}^{\infty} \{H_0(\xi)\}^2 \xi e^{-\xi^2}\, d\xi.$$

6.5 Find (a) $J_0(0)$, (b) $P_1(W)$, and (c) $H_2(\xi)$.

6.6 By use of the Rodrigues formula for $P_n(W)$, find P_0, P_1, P_2, P_3, and P_4. Use Leibnitz's formula† for the differentiation of a product to obtain P_3 and P_4.

6.7 Show that $P_l(-W) = (-1)^l P_l(W)$. *Hint:* Consider replacing W by $-W$ in the Rodrigues formula for $P_l(W)$.

6.8 Show that

$$\int_{-1}^{1} P_n(W) P_m(W)\, dW = \begin{cases} 0 & (n \neq m) \\ \dfrac{2}{2n + 1} & (n = m). \end{cases}$$

6.9 When $W(t, x) = (1 - 2xt + t^2)^{-1/2}$ is expanded in a power series of the form

$$W(t, x) = \phi_0(x) + \phi_1(x)t + \phi_2(x)t^2 + \cdots + \phi_n(x)t^n + \cdots$$

†See the Appendix at the end of Chapter 6.

show that (a) $\phi_n(x)$ is a polynomial of degree n, (b) $\phi_0(x) = 1$, and (c) $\phi_n(x)$ is a solution of Legendre's differential equation [i.e., $\phi_n(x) \to P_n(x)$]. (*Hint:* Develop the binomial expansion for $(1 - u)^{-1/2}$ where $u = 2xt - t^2$.)

6.10 Show that $L_n''(0) = \frac{1}{2}n(n - 1)$.

6.11 By use of the analysis in Section 6.4, develop the explicit expression for the eigenenergy for the hydrogen atom.

6.12 Show that (a) $\Gamma(1) = 1$ and (b) $\Gamma(n) = (n - 1)!$.

6.13 Show that $2 \cdot 4 \cdot 6 \cdots 2n = 2^n\Gamma(n + 1)$.

6.14 Show that

$$1 \cdot 3 \cdot 5 \cdots (2n - 1) = \frac{2^{1-n}\Gamma(2n)}{\Gamma(n)}.$$

6.15 Show that

$$\Gamma(x + n) = (x + n - 1)(x + n - 2) \cdots (x + 1)x\Gamma(x).$$

6.16 For n positive and $x - n \neq 0, -1, -2, \ldots$, show that

$$\frac{\Gamma(x + n)}{\Gamma(x - n)} = (x + n - 1)(x + n - 2) \cdots (x - n).$$

6.17 By use of the definition for the **beta function,**

$$B(x, y) \equiv \int_0^1 t^{x-1}(1 - t)^{y-1}\, dt \qquad (x \text{ and } y > 0)$$

show that $B(x, y) = B(y, x)$. (Let $t = 1 - z$.)

6.18 Show that $B(x, y)$ can be expressed as an integral with limits from zero to infinity. [Let $t = u/(1 + u)$.]

6.19 Show that

$$B(x, y) = \frac{\Gamma(x)\Gamma(y)}{\Gamma(x + y)}.$$

6.20 Show that $J_n(-\xi) = J_n(\xi)$ for n even and $J_n(\xi) = -J_n(-\xi)$ for n odd.

6.21 Show that $N_n(\xi)$ and $J_n(\xi)$ are linearly independent for n = integer.

6.22 Show that $N_n(\xi)$ is a solution of Bessel's differential equation for n = integer.

6.23 Show that $J_1(\xi) = -J_0'(\xi)$.

6.24 Show that $H_n^{(1)}(\xi)$ and $H_n^{(2)}(\xi)$ satisfy Bessel's differential equation.

6.25 By differentiating both sides of the equation for the generating function for $J_n(\xi)$ with respect to t and equating the coefficients of like powers of t, show that

$$J_{n-1}(\xi) + J_{n+1}(\xi) = \frac{2n}{\xi}J_n(\xi).$$

6.26 By differentiating both sides of the equation for the generating function for $J_n(\xi)$ with respect to ξ and equating the coefficients of like powers of t, show that

$$J_{n-1}(\xi) - J_{n+1}(\xi) = 2J'_n(\xi).$$

6.27 Consider a finite heat-conducting cylinder where $0 \le \rho \le a$ and $0 < z \le l$. Find the steady-state temperature distribution $T(\rho, z)$ when the surface $z = 0$ is kept at temperature T_0 and all other surfaces are kept at zero temperature.

6.28 Solve the two-dimensional Laplace equation in cylindrical coordinates for the electrostatic potential, $\phi(\rho, z)$.

6.29 The error function is defined by

$$\operatorname{erf}(x) = \frac{2}{\sqrt{\pi}} \int_0^x e^{-t^2}\, dt.$$

(a) Show that the error function is an odd function, $\operatorname{erf}(x) = -\operatorname{erf}(-x)$.

(b) Show that $\operatorname{erf}(\infty) = 1$.

6.30 Show that

$$n! = \int_0^\infty e^{-t} t^n \, dt$$
$$\sim n^n e^{-n} \sqrt{2\pi n} \qquad (n \to \infty) \qquad \text{(Stirling's formula)}$$

Hint: Make the change of variable, $t = ny$, and use the fact that $\ln(1 + y) \sim y - \frac{1}{2}y^2$.

6.31 By use of the generating function for the Hermite polynomials, evaluate

$$\int_{-\infty}^{\infty} e^{-\xi^2} H_n(\xi) \xi H_m(\xi)\, d\xi.$$

APPENDIX: THE RELATION BETWEEN $P_l(W)$ AND $P_l^m(W)$

Here we prove the following theorem: If $P_l(W)$ is a solution of Legendre's equation,

$$(1 - W^2)\frac{d^2\Theta}{dW^2} - 2W\frac{d\Theta}{dW} + l(l + 1)\Theta = 0$$

then

$$(1 - W^2)^{m/2} \frac{d^m P_l(W)}{dW^m} \quad \text{for} \quad m \ge 0$$

is a solution of the associated Legendre equation,

$$(1 - W^2)\frac{d^2\Theta}{dW^2} - 2W\frac{d\Theta}{dW} + \left[l(l + 1) - \frac{m^2}{1 - W^2}\right]\Theta = 0.$$

On differentiating

$$(1 - W^2)\frac{d^2P_l(W)}{dW^2} - 2W\frac{dP_l(W)}{dW} + l(l+1)P_l(W) = 0$$

m times with respect to W, we obtain

$$\frac{d^m}{dW^m}\left[(1 - W^2)\frac{d^2P_l}{dW^2}\right] - 2\frac{d^m}{dW^m}\left[W\frac{dP_l}{dW}\right] + l(l+1)\frac{d^mP_l}{dW^m} = 0.$$

By use of **Leibnitz's formula** for the differentiation of a product,

$$\frac{d^n}{dx^n}[A(x)B(x)] = \sum_{s=0}^{n}\left[\frac{n!}{(n-s)!s!}\frac{d^{n-s}}{dx^{n-s}}A(x)\frac{d^s}{dx^s}B(x)\right]$$

we obtain

$$\sum_{s=0}^{m}\left[\frac{m!}{(m-s)!s!}\frac{d^{m-s}(1 - W^2)}{dW^{m-s}}\frac{d^s}{dW^s}\left(\frac{d^2P_l}{dW^2}\right)\right]$$
$$- 2\sum_{s=0}^{m}\left[\frac{m!}{(m-s)!s!}\frac{d^{m-s}W}{dW^{m-s}}\frac{d^s}{dW^s}\left(\frac{dP_l}{dW}\right)\right] + l(l+1)\frac{d^mP_l}{dW^m} = 0. \qquad \text{(A6.1)}$$

The first and second terms in Eq. (A6.1) reduce, respectively, to

$$\sum_{s=0}^{m}\left[\frac{m!}{(m-s)!s!}\frac{d^{m-s}}{dW^{m-s}}(1 - W^2)\frac{d^s}{dW^s}\left(\frac{d^2P_l}{dW^2}\right)\right]$$
$$= -m(m-1)\frac{d^mP_l}{dW^m} - 2mW\frac{d^{m+1}P_l}{dW^{m+1}} + (1 - W^2)\frac{d^{m+2}P_l}{dW^{m+2}} \qquad \text{(A6.2)}$$

and

$$-2\sum_{s=0}^{m}\left[\frac{m!}{(m-s)!s!}\frac{d^{m-s}}{Wd^{m-s}}\frac{d^s}{dW^s}\left(\frac{dP_l}{dW}\right)\right] = -2\left[m\frac{d^mP_l}{dW^m} + W\frac{d^{m+1}P_l}{dW^{m+1}}\right]. \qquad \text{(A6.3)}$$

On substituting Eqs. (A6.2) and (A6.3) into Eq. (A6.1) and collecting terms, we obtain

$$(1 - W^2)\frac{d^{m+2}P_l}{dW^{m+2}} - 2W(m+1)\frac{d^{m+1}P_l}{dW^{m+1}} + [l(l+1) - m(m+1)]\frac{d^mP_l}{dW^m} = 0. \qquad \text{(A6.4)}$$

Now let

$$P_l^m(W) = (1 - W^2)^{m/2}\frac{d^mP_l(W)}{dW^m}.$$

Solving the above equation for $d^m P_l(W)/dW^m$ and differentiating twice with respect to W, we obtain

$$\frac{d^m P_l(W)}{dW^m} = (1 - W)^{-m/2} P_l^m(W) \tag{A6.5}$$

$$\frac{d^{m+1} P_l(W)}{dW^{m+1}} = (1 - W^2)^{-m/2}\left[\frac{dP_l^m(W)}{dW} + \frac{mWP_l^m(W)}{1 - W^2}\right] \tag{A6.6}$$

and

$$\frac{d^{m+2} P_l(W)}{dW^{m+2}} = (1 - W^2)^{-m/2}\left\{\frac{d^2 P_l^m(W)}{dW^2} + \frac{2mW}{1 - W^2}\frac{dP_l^m(W)}{dW} + \left[\frac{m}{1 - W^2} + \frac{m(m+2)W^2}{(1 - W^2)^2}\right]P_l^m(W)\right\}. \tag{A6.7}$$

On substituting Eqs. (A6.5), (A6.6), and (A6.7) into Eq. (A6.4), we obtain

$$(1 - W^2)\frac{d^2 P_l^m(W)}{dW^2} - 2W\frac{dP_l^m(W)}{dW} + \left[l(l+1) - \frac{m^2}{1 - W^2}\right]P_l^m(W) = 0 \tag{A6.8}$$

which is the associated Legendre equation where $P_l^m(W)$ are called the **associated Legendre functions**.

The Rodrigues formula for $P_l^m(W)$ is

$$P_l^m(W) = \frac{1}{2^l l!}(1 - W^2)^{m/2}\frac{d^{l+m}(W^2 - 1)^l}{dW^{l+m}}.$$

Note that

$$P_l^0(W) = P_l(W) \quad \text{and} \quad P_l^m(W) = 0 \qquad (\text{if } m > l).$$

7

Fourier Series

7.1 INTRODUCTION

The theory of the representation of a function of a real variable by means of series of sines and cosines is an indispensable technique in mathematical physics. Fourier analysis is such a theory. We begin the discussion of the Fourier series with a statement of Fourier's theorem. In 1807, Fourier (1768–1830) stated without proof and used, in developing a solution of the heat equation, the following theorem:

FOURIER'S THEOREM Any single-valued function $f(x)$ defined on the interval† $[-\pi, \pi]$ may be represented over this interval by the **trigonometric series**

$$\frac{a_0}{2} + \sum_{n=1}^{\infty} [a_n \cos nx + b_n \sin nx] \tag{7.1}$$

provided the **expansion coefficients**, a_n and b_n, are determined by use of **Euler's formulas,**

†Throughout this chapter, reference is made to whole (closed) intervals (the end points are included in the intervals).

$$a_n = \frac{1}{\pi} \int_{-\pi}^{\pi} f(x) \cos nx \, dx \qquad (n = 0, 1, 2, \dots) \tag{7.2}$$

and

$$b_n = \frac{1}{\pi} \int_{-\pi}^{\pi} f(x) \sin nx \, dx \qquad (n = 1, 2, 3, \dots). \tag{7.3}$$

For historical reasons, it is important to note that d'Alembert (1717–1783), Euler, D. Bernoulli (1700–1782), and Lagrange had previously made fruitful use of the trigonometric series in solving the wave equation for a vibrating string. However, the trigonometric series associated with a function $f(x)$ by means of Euler's formulas is known as a **Fourier series.**

Fourier investigated many special cases of the above theorem, but he was unable to develop a logical proof of it.† The first step toward a logical proof was made by Dirichlet in 1829. Dirichlet (1805–1859) proved the following theorem.

DIRICHLET'S THEOREM If for the interval $[-\pi, \pi]$ the function $f(x)$ (1) is single-valued, (2) is bounded, (3) has at most a finite number of maxima and minima, and (4) has only a finite number of discontinuities—piecewise continuous, and if (5) $f(x + 2\pi) = f(x)$ for values of x outside of $[-\pi, \pi]$, then

$$s_p(x) = \frac{a_0}{2} + \sum_{n=1}^{p} [a_n \cos nx + b_n \sin nx] \tag{7.4}$$

converges to $f(x)$ as $p \to \infty$ at values of x for which $f(x)$ is continuous and to $\frac{1}{2}[f(x + 0) + f(x - 0)]$ at points of discontinuity. The quantities $f(x + 0)$ and $f(x - 0)$ refer to the limits from the right and left, respectively. It is assumed that the coefficients in Eq. (7.4) are given by Euler's formulas. A function $f(x)$ is said to be bounded if the inequality $|f(x)| \leq M$ holds for some constant M for all values of x.

The conditions imposed on $f(x)$ by Dirichlet's theorem are called **Dirichlet's conditions.** From a rigorous mathematical point of view, the Dirichlet conditions are sufficient but not necessary. However, they are satisfied by most functions representing the solutions of physical problems, and we will work within the constraints of Dirichlet's theorem. Although a rigorous proof of Dirichlet's theorem is beyond the scope of this book, we will discuss the convergence properties of the Fourier series in Section 7.3.

7.1.1 The Fourier Cosine and Sine Series

If $f(x)$ is an even function, $[f(x) = f(-x)]$, in Eqs. (7.2) and (7.3), then

$$f(x) = \frac{a_0}{2} + \sum_{n=1}^{\infty} a_n \cos nx \qquad (f(x) \text{ even}) \tag{7.5}$$

†In fact, the theorem is not valid as stated.

since $b_n = 0$ for even $f(x)$. Similarly, $a_n = 0$ and

$$f(x) = \sum_{n=1}^{\infty} b_n \sin nx \qquad (f(x) \text{ odd}) \tag{7.6}$$

for odd $f(x)$, $[f(x) = -f(-x)]$. Equations (7.5) and (7.6) are called the **Fourier cosine** and **Fourier sine series**, respectively; the expansion coefficients, a_n and b_n, are given in Eqs. (7.2) and (7.3).

7.1.2 Change of Interval

Thus far, the expansion interval has been restricted to $[-\pi, \pi]$. To solve many physical problems, it is necessary to develop a Fourier series that will be valid over a wider interval. To obtain an expansion valid in the interval $[-l, l]$, we let

$$f(x) = \frac{a_0}{2} + \sum_{n=1}^{\infty} a_n \cos \phi x + b_n \sin \phi x \tag{7.7}$$

and determine ϕ such that $f(x) = f(x + 2l)$. In this case, $\phi = n\pi/l$; hence, we obtain

$$f(x) = \frac{a_0}{2} + \sum_{n=1}^{\infty} \left[a_n \cos \frac{n\pi x}{l} + b_n \sin \frac{n\pi x}{l} \right] \tag{7.8}$$

for $-l \leq x \leq l$ if $f(x)$ satisfies the Dirichlet conditions in this interval. The corresponding Euler coefficients are

$$a_n = \frac{1}{l} \int_{-l}^{l} f(x) \cos \left(\frac{n\pi x}{l} \right) dx \qquad (n = 0, 1, 2, \ldots) \tag{7.9}$$

and

$$b_n = \frac{1}{l} \int_{-l}^{l} f(x) \sin \left(\frac{n\pi x}{l} \right) dx \qquad (n = 1, 2, 3, \ldots). \tag{7.10}$$

7.1.3 Fourier Integral

We now extend the interval so that the expansion involves all values of x. For the interval $[-l, l]$, we have

$$f(x) = \frac{1}{2l} \int_{-l}^{l} f(t)\, dt + \frac{1}{l} \sum_{n=1}^{\infty} \left[\int_{-l}^{l} f(t) \left\{ \cos \left(\frac{n\pi x}{l} \right) \cos \left(\frac{n\pi t}{l} \right) \right. \right.$$
$$\left. \left. + \sin \left(\frac{n\pi x}{l} \right) \sin \left(\frac{n\pi t}{l} \right) \right\} dt \right]$$

or

$$f(x) = \frac{1}{2l} \int_{-l}^{l} f(t)\, dt + \frac{1}{l} \sum_{n=1}^{\infty} \int_{-l}^{l} f(t) \cos \left[\frac{n\pi}{l}(x - t) \right] dt. \tag{7.11}$$

If we set $u = n\pi/l$ where $1/l = \Delta u/\pi$ since $\Delta n = 1$, Eq. (7.11) becomes

$$f(x) = \frac{1}{2l}\int_{-l}^{l} f(t)\,dt + \frac{1}{\pi}\sum_{u=\pi/l}^{\infty} \Delta u \int_{-l}^{l} f(t)\cos\left[\frac{n\pi}{l}(x-t)\right]dt$$

$$= \frac{1}{\pi}\int_{0}^{\infty} du \int_{-\infty}^{\infty} f(t)\cos[u(x-t)]\,dt \tag{7.12}$$

since

$$\frac{1}{2l}\int_{-l}^{l} f(t)\,dt \longrightarrow 0 \quad \text{and} \quad \sum_{u=\pi/l}^{\infty} \Delta u \longrightarrow \int_{0}^{\infty} du \qquad (\text{as } l \to \infty).$$

Equation (7.12) is called the **Fourier integral formula.**

7.1.4 Complex Form of the Fourier Series

The complex form of the Fourier series is obtained by expressing $\cos(n\pi x/l)$ and $\sin(n\pi x/l)$ in exponential form, that is,

$$f(x) = \frac{a_0}{2} + \sum_{n=1}^{\infty}\left[a_n \cos\left(\frac{n\pi x}{l}\right) + b_n \sin\left(\frac{n\pi x}{l}\right)\right]$$

$$= \frac{a_0}{2} + \sum_{n=1}^{\infty} a_n\left[\frac{e^{in\pi x/l} + e^{-in\pi x/l}}{2}\right] + \sum_{n=1}^{\infty} b_n\left[\frac{e^{in\pi x/l} - e^{-in\pi x/l}}{2i}\right]$$

or

$$f(x) = \sum_{n=-\infty}^{\infty} c_n e^{in\pi x/l} \qquad [-l, l] \tag{7.13}$$

where

$$c_0 = \tfrac{1}{2}a_0 \tag{7.14}$$

$$c_n = \tfrac{1}{2}(a_n - ib_n) \tag{7.15}$$

$$c_{-n} = \tfrac{1}{2}(a_n + ib_n). \tag{7.16}$$

Equation (7.13) is the **complex form of the Fourier series.** On multiplying both sides of Eq. (7.13) by $e^{-im\pi x/l}$ and integrating with respect to x, we obtain

$$\int_{-l}^{l} f(x)e^{-im\pi x/l}\,dx = \sum_{n=-\infty}^{\infty} c_n \int_{-l}^{l} e^{i\pi x(n-m)/l}\,dx$$

$$= \sum_{n=-\infty}^{\infty} c_n 2l\,\delta_{mn}$$

$$= 2lc_m. \tag{7.17}$$

Therefore the c_n coefficient in Eq. (7.13) is given by

$$c_n = \frac{1}{2l}\int_{-l}^{l} f(x)e^{-in\pi x/l}\,dx. \tag{7.18}$$

Before investigating other mathematical properties of the Fourier series, we consider several examples so that a working knowledge of the Fourier series can be achieved. In these examples, it is assumed that (1) the developed Fourier expansions converge uniformly to the original functions, and (2) there is a steady increase in the accuracy of the developed expansion as the number of terms included is increased. A proof of this convergence is given in Section 7.3.

EXAMPLE 7.1 Expand $f_1(x) = x^2$ for $-\pi \leq x \leq \pi$ in a Fourier series.

Solution The graphical representation of $f_1(x)$ in $[-\pi, \pi]$ and its periodic extension outside of $[-\pi, \pi]$ are shown in Fig. 7.1. The expansion coefficients are given by

$$
\begin{aligned}
a_0 &= \frac{1}{\pi}\int_{-\pi}^{\pi} f_1(x)\,dx \\
&= \frac{1}{\pi}\int_{-\pi}^{\pi} x^2\,dx \\
&= \tfrac{2}{3}\pi^2
\end{aligned}
$$

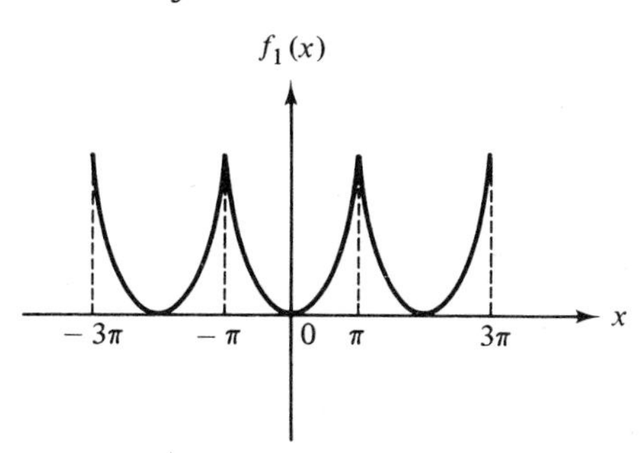

Figure 7.1

$$
\begin{aligned}
a_{n>0} &= \frac{1}{\pi}\int_{-\pi}^{\pi} f_1(x)\cos nx\,dx \\
&= \frac{2}{\pi}\int_{0}^{\pi} x^2\cos nx\,dx \qquad \text{(by symmetry)} \\
&= \frac{2}{\pi n^3}\int_{0}^{n\pi} y^2\cos y\,dy \\
&= \frac{2}{\pi n^3}\Big[2y\cos y + (y^2-2)\sin y\Big]_0^{n\pi} \\
&= \frac{2}{\pi n^3}[2n\pi(-1)^n] \\
&= (-1)^n\frac{4}{n^2}
\end{aligned}
$$

$$b_{n>0} = \frac{1}{\pi}\int_{-\pi}^{\pi} f_1(x) \sin nx \, dx$$

$$= \frac{1}{\pi}\int_{-\pi}^{\pi} x^2 \sin nx \, dx = 0 \qquad \text{(by symmetry).}$$

The required Fourier expansion is

$$f_1(x) = x^2 = \frac{a_0}{2} + \sum_{n=1}^{\infty} [a_n \cos nx + b_n \sin nx]$$

$$= \frac{\pi^2}{3} + 4\sum_{n=1}^{\infty} \frac{(-1)^n}{n^2} \cos nx.$$

EXAMPLE 7.2 *Sawtooth Wave* Expand $f_2(x) = x$ for $-\pi \leq x \leq \pi$ in a Fourier series.

Solution The graphical representation of $f_2(x)$ in $[-\pi, \pi]$ and its periodic extension outside of $[-\pi, \pi]$ are shown in Fig. 7.2. The expansion coefficients are given by

$$a_0 = \frac{1}{\pi}\int_{-\pi}^{\pi} f_2(x) \, dx = \frac{1}{\pi}\int_{-\pi}^{\pi} x \, dx = 0$$

$$a_{n>0} = \frac{1}{\pi}\int_{-\pi}^{\pi} f_2(x) \cos nx \, dx = \frac{1}{\pi}\int_{-\pi}^{\pi} x \cos nx \, dx = 0$$

and

$$b_{n>0} = \frac{1}{\pi}\int_{-\pi}^{\pi} f_2(x) \sin nx$$

$$= \frac{1}{\pi}\int_{-\pi}^{\pi} x \sin nx \, dx$$

$$= -\frac{2}{n} \cos n\pi$$

$$= \frac{2}{n}(-1)^{n+1}.$$

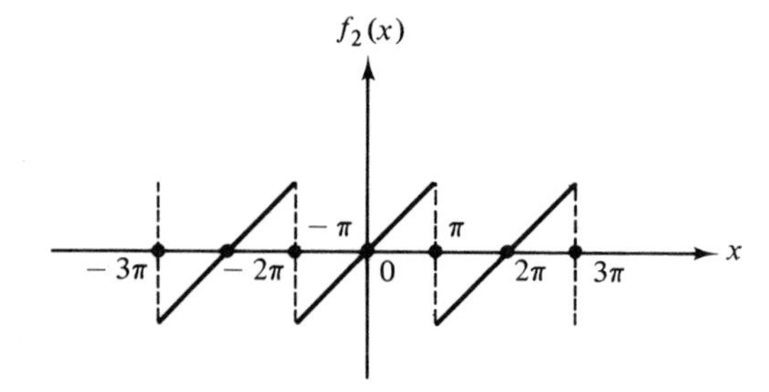

Figure 7.2

The required Fourier expansion is

$$f_2(x) = x = \frac{a_0}{2} + \sum_{n=1}^{\infty} [a_n \cos nx + b_n \sin nx]$$
$$= 2 \sum_{n=1}^{\infty} \frac{(-1)^{n+1} \sin nx}{n}.$$

EXAMPLE 7.3 *Square Wave* Expand in a Fourier series:

$$f_3(x) = \begin{cases} 0 & (-\pi \leq x < 0) \\ h & (0 \leq x \leq \pi). \end{cases}$$

Solution The graphical representation of $f_3(x)$ in $[-\pi, \pi]$ and its periodic extension outside of $[-\pi, \pi]$ are shown in Fig. 7.3. The expansion coefficients are

$$a_0 = \frac{1}{\pi} \int_{-\pi}^{0} f_3(x)\, dx + \frac{1}{\pi} \int_{0}^{\pi} f_3(x)\, dx$$
$$= \frac{1}{\pi} \int_{0}^{\pi} h\, dx$$
$$= h$$
$$a_{n>0} = \frac{1}{\pi} \int_{-\pi}^{0} f_3(x) \cos nx\, dx + \frac{1}{\pi} \int_{0}^{\pi} f_3(x) \cos nx\, dx$$
$$= \frac{h}{\pi} \int_{0}^{\pi} \cos nx\, dx = 0$$

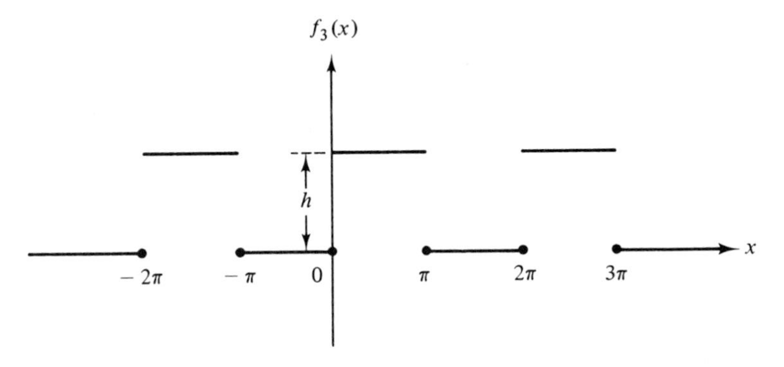

Figure 7.3

and

$$\begin{aligned}
b_{n>0} &= \frac{1}{\pi}\int_{-\pi}^{0} f_3(x)\sin nx\,dx + \frac{1}{\pi}\int_{0}^{\pi} f_3(x)\sin nx\,dx \\
&= \frac{h}{\pi}\int_{0}^{\pi} \sin nx\,dx \\
&= \frac{h}{\pi}\left[-\frac{\cos nx}{n}\right]_0^{\pi} \\
&= \begin{cases} 0 & \text{(for } n \text{ even)} \\ \dfrac{2h}{n\pi} & \text{(for } n \text{ odd).} \end{cases}
\end{aligned}$$

The required Fourier series is

$$\begin{aligned}
f_3(x) &= \frac{a_0}{2} + \sum_{n=1}^{\infty} [a_n \cos nx + b_n \sin nx] \\
&= \frac{h}{2} + \frac{2h}{\pi}\sum_{\substack{n=1 \\ \text{odd}}}^{\infty} \frac{\sin nx}{n}.
\end{aligned}$$

EXAMPLE 7.4 *Full-Wave Rectifier* Expand in a Fourier series:

$$f_4(x) = \begin{cases} \sin x & (0 \le x \le \pi) \\ -\sin x & (-\pi \le x \le 0). \end{cases}$$

Solution The graphical representation of $f_4(x)$ in $[-\pi, \pi]$ and its periodic extension outside of $[-\pi, \pi]$ are shown in Fig. 7.4.

The expansion coefficients are

$$a_0 = \frac{1}{\pi}\int_{-\pi}^{0} f_4(x)\,dx + \frac{1}{\pi}\int_{0}^{\pi} f_4(x)\,dx$$

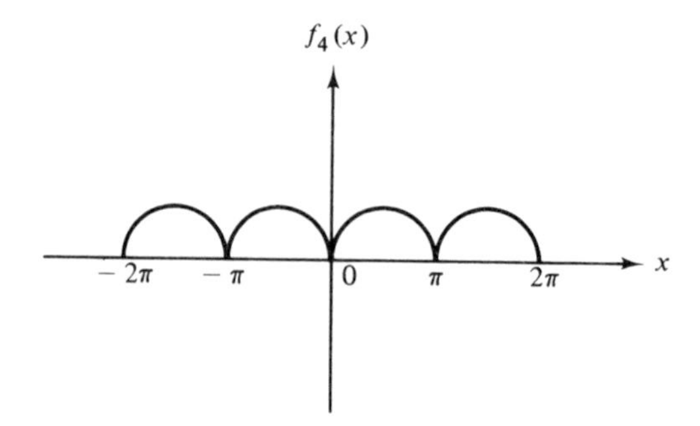

Figure 7.4

$$= -\frac{1}{\pi}\int_{-\pi}^{0} \sin x\, dx + \frac{1}{\pi}\int_{0}^{\pi} \sin x\, dx$$

$$= \frac{2}{\pi}\int_{0}^{\pi} \sin x\, dx = \frac{4}{\pi}$$

$$a_{n>0} = \frac{1}{\pi}\int_{-\pi}^{0} -\sin x \cos nx\, dx + \frac{1}{\pi}\int_{0}^{\pi} \sin x \cos nx\, dx$$

$$= \frac{2}{\pi}\int_{0}^{\pi} \sin x \cos nx\, dx$$

$$= \begin{cases} -\dfrac{4}{\pi}\dfrac{1}{n^2 - 1} & \text{(for } n \text{ even)} \\ 0 & \text{(for } n \text{ odd)} \end{cases}$$

and

$$b_{n>0} = \frac{2}{\pi}\int_{0}^{\pi} \sin x \sin nx\, dx = 0.$$

The required Fourier series is

$$f_4(x) = \frac{a_0}{2} + \sum_{n=1}^{\infty} [a_n \cos nx + b_n \sin nx]$$

$$= \frac{2}{\pi} - \frac{4}{\pi}\sum_{\substack{n=0 \\ \text{even}}}^{\infty} \frac{\cos nx}{n^2 - 1}.$$

EXAMPLE 7.5 Find the particular solution of the heat conduction equation, $u_{xx} = (1/\sigma)u_t$, for $u_x(0, t) = u_x(1, t) = 0$ and $u_t(x, 0) = f(x)$.

Solution By use of the method of separation of variables (Section 5.3.4 of Chapter 5) where we assume $u(x, t) = X(x)T(t)$, we obtain

$$\frac{X''}{X} = \frac{1}{\sigma}\frac{\dot{T}}{T} = \lambda$$

or

$$X'' - \lambda X = 0 \quad \text{where} \quad X(x) = c_1 e^{-\sqrt{\lambda}x} + c_2 e^{\sqrt{\lambda}x}$$

and

$$\dot{T} = \sigma\lambda T \quad \text{where} \quad T(t) = c_3 e^{\sigma\lambda t}.$$

The general solution is

$$u(x, t) = (c_1 e^{-\sqrt{\lambda}x} + c_2 e^{\sqrt{\lambda}x})c_3 e^{\sigma\lambda t}.$$

By use of the condition $u_x(0, t) = 0$, where

$$u_x(x, t) = (-\sqrt{\lambda}\, c_1 e^{-\sqrt{\lambda} x} + \sqrt{\lambda}\, c_2 e^{\sqrt{\lambda} x}) c_3 e^{\sigma\lambda t}$$

we see that $c_1 = c_2$. Hence the solution becomes

$$u(x, t) = c_1(e^{\sqrt{\lambda} x} + e^{-\sqrt{\lambda} x}) c_3 e^{\sigma\lambda t}.$$

From the second condition on space, $u_x(1, t) = 0$, where

$$u_x(1, t) = c_1\sqrt{\lambda}\,(e^{\sqrt{\lambda}} - e^{-\sqrt{\lambda}}) c_3 e^{\sigma\lambda t} = 0$$

we find that $\lambda < 0$ since $c_1 \neq 0$ and $\lambda \neq 0$ for a nontrivial solution. For $\lambda < 0$, we have

$$u(x, t) = (A \cos \sqrt{\lambda}\, x + B \sin \sqrt{\lambda}\, x)(De^{-\lambda\sigma t})$$
$$u_x(0, t) = 0 = (B\sqrt{\lambda})De^{-\lambda\sigma t}; \quad B = 0$$

and

$$u_x(1, t) = 0 = (-A\sqrt{\lambda} \sin \sqrt{\lambda})De^{-\lambda\sigma t}.$$

Since $A \neq 0$ for a nontrivial solution, we take

$$\sin \sqrt{\lambda} = 0 \quad \text{or} \quad \sqrt{\lambda} = n\pi \qquad \text{(eigenvalues)} \qquad n = 0, 1, \ldots.$$

The solution now becomes

$$u(x, t) = \sum_{n=0}^{\infty} u_n(x, t) = \sum_{n=0}^{\infty} [(A'_n \cos n\pi x)e^{-n^2\pi^2\sigma t}].$$

The condition on time, $u_t(x, 0) = f(x)$, leads to the specific form for A'_n. Here we have

$$\begin{aligned} u_t(x, 0) &= f(x) \\ &= \sum_{n=0}^{\infty} (-n^2\pi^2\sigma)(A'_n \cos n\pi x) \\ &= \sum_{n=0}^{\infty} a_n \cos \frac{n\pi x}{l} \qquad ; l = 1 \end{aligned}$$

where

$$\begin{aligned} a_n &= \frac{1}{l} \int_{-l}^{l} f(x) \cos \frac{n\pi x}{l}\, dx \\ &= -n^2\pi^2\sigma A'_n \end{aligned}$$

or

$$A'_n = -\frac{1}{n^2\pi^2\sigma}\int_{-1}^{1} f(x)\cos n\pi x\,dx.$$

The above development assumes that $f(x)$ satisfies the Dirichlet conditions.

7.2 GENERALIZED FOURIER SERIES AND THE DIRAC DELTA FUNCTION

If the set of functions $\{\psi_n(x)\}$ is the orthonormal basis for a vector space E_n, and $F(x)$ is an arbitrary function in E_n, then (see Chapter 2)

$$F(x) = \sum_{n=-\infty}^{\infty} c_n\psi_n(x) \tag{7.19}$$

where the c_n are called **expansion coefficients** and

$$\int_{-\infty}^{\infty} \psi_{n'}^*(x)\psi_n(x)\,dx = \delta_{nn'} \tag{7.20}$$

The orthonormality property expressed in Eq. (7.20) will now be used to develop an expression for c_n in Eq. (7.19). On multiplying Eq. (7.19) by $\psi_{n'}^*(x)$ and integrating over the range of x, we obtain

$$\begin{aligned}\int_{-\infty}^{\infty} \psi_{n'}^*(x)F(x)\,dx &= \sum_{n=-\infty}^{\infty} c_n\int_{-\infty}^{\infty}\psi_{n'}^*(x)\psi_n(x)\,dx\\ &= \sum_{n=-\infty}^{\infty} c_n\delta_{nn'} = c_{n'}.\end{aligned} \tag{7.21}$$

Substituting c_n from Eq. (7.21) into Eq. (7.19), we find that

$$F(x) = \sum_{n=-\infty}^{\infty}\int_{-\infty}^{\infty}\psi_n^*(x')F(x')\psi_n(x)\,dx'. \tag{7.22}$$

The expansion in Eq. (7.22) is called a **generalized Fourier series.** Interchanging the sum with the integral in Eq. (7.22), we may write

$$\begin{aligned}F(x) &= \int_{-\infty}^{\infty} F(x')\left[\sum_{n=-\infty}^{\infty}\psi_n^*(x')\psi_n(x)\right]dx'\\ &= \int_{-\infty}^{\infty} F(x')\delta(x-x')\,dx'\end{aligned} \tag{7.23}$$

where

$$\delta(x-x') = \sum_{n=-\infty}^{\infty}\psi_n^*(x')\psi_n(x). \tag{7.24}$$

The $\delta(x - x')$ quantity is the one-dimensional **Dirac delta function**, and its most important property is expressed by Eq. (7.23). The following are the various notations for the Dirac δ-function used in the literature: $\delta(x - x')$, $\delta(x)$, and $\delta(x, x')$. The Dirac δ-function is defined such that

$$\int_{-\infty}^{\infty} \delta(x - x')\, dx' = 1 \tag{7.25}$$

and

$$\delta(x - x') = \begin{cases} 0 & (\text{for } x - x' \neq 0) \\ \infty & (\text{for } x - x' = 0). \end{cases} \tag{7.26}$$

The properties of the δ-function listed in Eqs. (7.25) and (7.26) make it clear that the Dirac δ-function is not a function in a rigorous mathematical sense since the integral (if it exists) of a function that is zero everywhere except at one point must vanish.†

We will develop some other useful relations involving the Dirac δ-function since its concept is extremely useful in analyzing physical properties of systems involving impulsive-type forces. Consider the case of harmonic functions

$$\psi_n(x) = \frac{1}{(2l)^{1/2}} e^{in\pi x/l}. \tag{7.27}$$

On substituting Eq. (7.27) into Eq. (7.24), we obtain

$$\begin{aligned} \delta(x - x') &= \frac{1}{2l} \sum_{n=-\infty}^{\infty} e^{in\pi(x-x')/l} \\ &= \frac{1}{2\pi} \sum_{\substack{n=-\infty \\ =kl/\pi}}^{\infty} \Delta k e^{ik(x-x')} \end{aligned} \tag{7.28}$$

where $k = \pi n/l$ and $1/2l = \Delta k/2\pi$ since $\Delta n = 1$. If $l \to \infty$ $(\Delta k \to 0)$ in Eq. (7.28), the sum changes into an integral, that is,

$$\delta(x - x') = \frac{1}{2\pi} \int_{-\infty}^{\infty} e^{ik(x-x')}\, dk \tag{7.29}$$

or

$$\begin{aligned} \delta(x - x') &= \frac{1}{2\pi} \lim_{l\to\infty} \int_{-l}^{l} e^{ik(x-x')}\, dk \\ &= \lim_{l\to\infty} \frac{\sin[l(x - x')]}{\pi(x - x')}. \end{aligned} \tag{7.30}$$

†A rigorous mathematical treatment of the Dirac δ-function in terms of the theory of distributions is given in M. J. Lighthill, *Introduction to Fourier Analysis and Generalized Functions*, Cambridge University Press, New York, 1958.

Equations (7.29) and (7.30) are two widely used representations for the Dirac δ-function. In three dimensions, we write

$$\delta(\mathbf{r}) = \delta(x)\delta(y)\delta(z) = \frac{1}{(2\pi)^3}\int_{-\infty}^{\infty} e^{i\mathbf{k}\cdot\mathbf{r}}\, d^3k \tag{7.31}$$

where

$$\delta(\mathbf{r}) = 0 \qquad \mathbf{r} \neq 0 \tag{7.32}$$

and

$$\int_{-\infty}^{\infty} \delta(\mathbf{r})\, d^3r = 1. \tag{7.33}$$

7.3 SUMMATION OF THE FOURIER SERIES

Thus far, it has been necessary to accept the fact that the trigonometric expansion (Fourier series) of a function, where the coefficients of the expansion are determined by use of the Euler's formulas, does indeed converge to the function in question. In this section, it is shown that

$$\lim_{p\to\infty} s_p(x) \longrightarrow \frac{1}{2}[f(x+0) + f(x-0)];$$

Dirichlet's theorem [see Eq. (7.4)] is valid.

Consider the partial sum of a Fourier series

$$\begin{aligned} s_p(x) &= \frac{a_0}{2} + \sum_{n=1}^{p} (a_n \cos nx + b_n \sin nx) \\ &= \frac{1}{2\pi}\int_0^{2\pi} f(t)\, dt + \frac{1}{\pi}\sum_{n=1}^{p}\left[\cos nx \int_0^{2\pi} f(t)\cos nt\, dt \right. \\ &\quad \left. + \sin nx \int_0^{2\pi} f(t) \sin nt\, dt\right] \end{aligned}$$

or

$$s_p(x) = \frac{1}{\pi}\int_0^{2\pi}\left[\frac{1}{2} + \sum_{n=1}^{p} \cos n(x-t)\right] f(t)\, dt. \tag{7.34}$$

It can be shown (Problem 7.20) that

$$\frac{1}{2} + \sum_{n=1}^{p} \cos nx = \frac{\sin (p+\frac{1}{2})x}{2 \sin x/2}. \tag{7.35}$$

On substituting Eq. (7.35) into Eq. (7.34), we obtain the **Dirichlet integral**

$$s_p(x) = \frac{1}{2\pi}\int_0^{2\pi} \frac{\sin[(p+\frac{1}{2})(x-t)]f(t)}{\sin[(x-t)/2]}\,dt$$
$$= \frac{1}{2\pi}\int_{-\pi+x}^{\pi+x} \frac{\sin[(p+\frac{1}{2})(x-t)]f(t)}{\sin[(x-t)/2]}\,dt. \tag{7.36}$$

The last step is valid since the integrand is periodic. We now make the following change of variable:

$$\theta = -\frac{(x-t)}{2}. \tag{7.37}$$

On substituting Eq. (7.37) into Eq. (7.36), we find that

$$s_p(x) = \frac{1}{\pi}\int_{-\pi/2}^{\pi/2} \frac{\sin[(2p+1)\theta]f(x+2\theta)}{\sin\theta}\,d\theta. \tag{7.38}$$

The path of integration will now be bisected so that

$$s_p(x) = \frac{1}{\pi}\int_0^{\pi/2} \frac{\sin[(2p+1)\theta]f(x+2\theta)}{\sin\theta}\,d\theta$$
$$+ \frac{1}{\pi}\int_0^{\pi/2} \frac{\sin[(2p+1)\theta]f(x-2\theta)}{\sin\theta}\,d\theta$$

or

$$s_p(x) = \frac{1}{\pi}\int_0^{\pi/2} \frac{\sin[(2p+1)\theta]}{\sin\theta}\{f(x+2\theta)+f(x-2\theta)\}\,d\theta. \tag{7.39}$$

The limit of $s_p(x)$ as $p \to \infty$ is given by

$$\lim_{p\to\infty} s_p(x) = \lim_{p\to\infty}\int_0^{\pi/2} \frac{\sin[(2p+1)\theta]}{\pi\sin\theta}\{f(x+2\theta)+f(x-2\theta)\}\,d\theta$$
$$\approx \int_0^{\pi/2} \delta(\theta)\{f(x+2\theta)+f(x-2\theta)\}\,d\theta$$

or

$$\lim_{p\to\infty} s_p(x) = \frac{1}{2}\int_{-\infty}^{\infty} \delta(\theta)[f(x+2\theta)+f(x-2\theta)]\,d\theta. \tag{7.40}$$

The last step is valid since $\delta(\theta)$ is an even function. Hence we obtain

$$\lim_{p\to\infty} s_p(x) = \frac{1}{2}[f(x+0)+f(x-0)] \tag{7.41}$$

where $f(x+0)$ and $f(x-0)$ represent the limits from the right and left of the point x, respectively. If $f(x)$ is continuous at the point x, then

$$\lim_{p \to \infty} s_p(x) \longrightarrow f(x).$$

7.4 THE GIBBS PHENOMENON

The partial sums of the Fourier series of a function, $f(x)$, approach $f(x)$ uniformly in every interval that does not contain a discontinuity of $f(x)$. In the immediate vicinity of a jump discontinuity, the convergence of the Fourier series is not uniform since $f(x) \to \frac{1}{2}[f(x+0)+f(x-0)]$ (see Section 7.3). Here the partial sums move progressively closer to the function as the number of terms is increased, but the approximating curves (partial sums) overshoot, by about 18%, the function at the jump discontinuity. This behavior is known as the **Gibbs** (1899) **phenomenon**, although the behavior had apparently been investigated earlier by Wilbraham (1848) and du Bois-Raymond (1874).

In 1906, Bôcher greatly extended Gibbs' results by considering the following specific series

$$f(x) = \frac{\pi - x}{2} \qquad (0 \leq x \leq 2\pi) \tag{7.42}$$

where the Fourier expansion is

$$s(x) = \lim_{p \to \infty} s_p(x) = \sum_{n=1}^{\infty} \frac{\sin nx}{n}. \tag{7.43}$$

The partial sum is given by

$$\begin{aligned} s_p(x) &= \sum_{n=1}^{p} \frac{\sin nx}{n} \\ &= \sin x + \frac{1}{2}\sin 2x + \cdots + \frac{1}{p}\sin px \\ &= \int_0^x (\cos u + \cos 2u + \cdots + \cos pu)\, du \\ &= \int_0^x \left(\sum_{n=1}^{p} \cos nu\right) du \\ &= \frac{1}{2}\int_0^x \frac{\sin[(p+\frac{1}{2})u]}{\sin u/2}\, du - \frac{x}{2} \end{aligned} \tag{7.44}$$

since

$$\sum_{n=1}^{p} \cos nx = \frac{\sin(p+\frac{1}{2})x}{2\sin x/2} - \frac{1}{2}. \tag{7.45}$$

The remainder, $R_p(x)$, is

$$\begin{aligned} R_p(x) &= \sum_{n=p+1}^{\infty} \frac{\sin nx}{n} \\ &= f(x) - s_p(x) \\ &= \frac{\pi - x}{2} - s_p(x) \\ &= \frac{\pi}{2} - \frac{1}{2}\int_0^x \frac{\sin[(p+\frac{1}{2})u]}{\sin u/2}\,du \end{aligned} \tag{7.46}$$

or

$$R_p(x) = \frac{\pi}{2} - \int_0^{(p+1/2)x} \frac{\sin u}{u}\,du + \rho_p(x) \tag{7.47}$$

where

$$\rho_p(x) = \int_0^x \left\{\frac{\sin u/2 - u/2}{u \sin u/2}\right\} \sin\left[\left(p + \frac{1}{2}\right)u\right]du. \tag{7.48}$$

On differentiating $R_p(x)$ [Eq. (7.46)] with respect to x, we find that $R_p(x)$ has maxima or minima at

$$x_k = \frac{2\pi k}{2p+1} \qquad (k = 0, 1, 2, \ldots). \tag{7.49}$$

The value of $R_p(x)$ at x_k is

$$R_p(x_k) = \frac{\pi}{2} - \int_0^{\pi k} \frac{\sin u}{u}\,du + \rho_p\left(\frac{2\pi k}{2p+1}\right). \tag{7.50}$$

As $p \to \infty$ for fixed k, $\rho_p \to 0$. Hence the remainder, the deviation of the approximation from $(\pi - x)/2$ at x_k which approaches the point of the discontinuity at $x = 0$ (end point), tends to the limit

$$\lim_{p\to\infty} R_p(x_k) \longrightarrow \frac{\pi}{2} - \int_0^{k\pi} \frac{\sin u}{u}\,du. \tag{7.51}$$

For $k = 1$, we find that

$$\begin{aligned} \lim_{p\to\infty} R_p(x_1) &\longrightarrow \frac{\pi}{2} - \int_0^{\pi} \frac{\sin u}{u}\,du \\ &= \frac{\pi}{2} - \frac{\pi}{2}(1.179) \\ &= -0.281. \end{aligned} \tag{7.52}$$

Hence the Fourier expansion, as indicated by the negative sign in Eq. (7.52), overshoots the curve for $f(x)$ by about 18% at $x = 0$, a jump discontinuity.

7.5 SUMMARY OF SOME PROPERTIES OF FOURIER SERIES

1. The Fourier series can be used to represent discontinuous functions where all orders of derivatives need not exist. (This is not true for the Taylor expansion.)
2. The Fourier series is useful in expanding periodic functions since outside of the closed interval in question there exists a periodic extension of the function.
3. The Fourier expansion of an oscillating function gives all modes of oscillation (fundamental and all overtones). This kind of representation is extremely useful in physics.
4. Since $\lim s_p(x) \to \frac{1}{2}[f(x+0) + f(x-0)]$ as $p \to \infty$, the Fourier series will not be uniformly convergent at all points if it represents a discontinuous function. In the vicinity of a discontinuity, the Fourier representation overshoots the function (see Section 7.4).
5. Term-by-term integration of a convergent Fourier series is always valid, and it may be valid even if the series is not convergent. However, term-by-term differentiation of a Fourier series must be investigated for the series in question; it is not valid in most cases.

7.6 PROBLEMS

7.1 Classify the following functions as even, odd, or neither:

(a) $\sin x$ (b) $\cos x$

(c) x^2 (d) $x \sin x$

(e) e^x (f) $x^3 \cos nx$

(g) $g(x^2)$ (h) $xg(x^2)$

(i) $\log \left[\frac{1+x}{1-x}\right]$

7.2 Show that the period of each term in the trigonometric series is 2π.

7.3 If $f(x)$ is periodic with period T, show that

$$\int_a^{a+T} f(x)dx = \int_b^{b+T} f(x)dx$$

where a and b are arbitrary.

7.4 By use of the orthogonality property of sines and cosines,

$$\int_{-\pi}^{\pi} \sin mx \sin nx \, dx = \pi\delta_{mn}$$

$$\int_{-\pi}^{\pi} \cos mx \cos nx \, dx = \pi\delta_{mn}$$

$$\int_{-\pi}^{\pi} \sin mx \cos nx \, dx = 0 \qquad (\text{all } n, m > 0)$$

derive the required expressions for the coefficients (Euler's formulas) in the trigonometric series.

7.5 Suppose the function $f(x)$ is represented by a finite Fourier series,

$$\frac{a_0}{2} + \sum_{n=1}^{p} [a_n \cos nx + b_n \sin nx].$$

A measure of the accuracy of this series representation is given by

$$\Delta_p = \int_0^{2\pi} \left[f(x) - \frac{a_0}{2} - \sum_{n=1}^{p} (a_n \cos nx + b_n \sin nx)\right]^2 dx.$$

By minimizing Δ_p ($\partial\Delta_p/\partial a_n = 0$ and $\partial\Delta_p/\partial b_n = 0$), solve for a_n and b_n.

7.6 Solve Example 7.2 by use of the complex form of the Fourier series.

7.7 Develop the Fourier expansion for $f(x) = x$ in the interval $(0\ 2l)$.

7.8 Develop the Fourier sine expansion for $f(x) = x$ in the half-interval $(0, l)$.

7.9 Develop the Fourier expansion for $f(x) = |x|$ for the interval $-l \leq x \leq l$.

7.10 From the result in Example 7.1, show that

$$\frac{\pi^2}{6} = \sum_{n=1}^{\infty} \frac{1}{n^2}.$$

Note that

$$\zeta(2) = \sum_{n=1}^{\infty} \frac{1}{n^2}, \ \left\{\zeta(k) = \sum_{n=1}^{\infty} \frac{1}{n^k}\right\},$$

is called the **Riemann zeta function.**

7.11 Develop the Fourier expansion for

$$f(t) = \begin{cases} 0 & (-\pi \leq \omega t \leq 0) \\ \sin \omega t & (0 \leq \omega t \leq \pi). \end{cases}$$

7.12 A **triangular wave** may be represented by

$$f(x) = \begin{cases} x & (0 < x < \pi) \\ -x & (-\pi < x < 0). \end{cases}$$

Develop the Fourier expansion for $f(x)$.

7.13 Find the Fourier series for $f(x)$ if

$$f(x) = \begin{cases} \pi & \left(-\pi < x \leq \frac{\pi}{2}\right) \\ 0 & \left(\frac{\pi}{2} < x \leq \pi\right). \end{cases}$$

Problems 7.14 and 7.15 deal with a *wave equation.* An elastic string stretched taut on the x-axis with ends at $x = 0$ and $x = L$. If this string is displaced a small distance and released, it vibrates in such a way that $y(x, t)$ is the solution of the wave equation

$$\frac{\partial^2 y(x, t)}{\partial x^2} = \frac{1}{v^2} \frac{\partial^2 y(x, t)}{\partial t^2}$$

where v is the speed of the wave.

7.14 (a) Solve the wave equation subject to the following initial conditions: $y(x, 0) = 0$ and

$$y_t(x, t)|_{t=0} = 2v_0 \sum_{n=1}^{\infty} \sin\left(\frac{n\pi a}{L}\right) \sin\left(\frac{n\pi x}{L}\right).$$

(b) Calculate the transverse speed $y_t(x, t)$ of the string.

7.15 Assume a Fourier expansion of the form

$$y(x, t) = \sum_{n=1}^{\infty} b_n(t) \sin\left(\frac{n\pi x}{L}\right)$$

for $y(x, t)$ with the following initial conditions: $y(x, 0) = f(x)$ and $y_t(x, 0) = g(x)$. Show that

$$b_n(t) = A_n \cos\left(\frac{n\pi vt}{L}\right) + B_n \sin\left(\frac{n\pi vt}{L}\right)$$

where

$$A_n = \frac{2}{L} \int_0^L f(x) \sin\left(\frac{n\pi x}{L}\right) dx$$

and

$$B_n = \frac{2}{n\pi v} \int_0^L g(x) \sin\left(\frac{n\pi x}{L}\right) dx.$$

7.16 Consider the heat conduction, $u_{xx} = (1/\sigma)u_t$ in a rod of length l. Find $u(x, t)$ for $t > 0$ where $u(0, t) = T_1$, $u(l, t) = T_2$, and $u(x, 0) = f(x)$. Assume that T_1 and T_2 are constant.

7.17 The boundary and initial conditions for a stretched string of length l are $u(0, t) = u(l, t) = u(x, 0) = 0$ and $u_t(x, 0) = kx(l - x)$, where k is a constant. Find $u(x, t)$.

7.18 Assuming that the Fourier expansion of $f(x)$ is uniformly convergent, show that

$$\frac{1}{\pi}\int_{-\pi}^{\pi} [f(x)]^2\, dx = \frac{a_0^2}{2} + \sum_{n=1}^{\infty} (a_n^2 + b_n^2) \qquad \textbf{(Parseval relation).}$$

7.19 Show that the term-by-term integration of a convergent Fourier series results in a Fourier series that converges more rapidly than the original Fourier series.

7.20 Show that

$$\frac{1}{2} + \sum_{n=1}^{p} \cos nx = \frac{\sin (p + \frac{1}{2})x}{2 \sin (x/2)}.$$

7.21 Show that $f(x)$ in Example 7.2 is convergent and that $f'(x)$ is not convergent.

7.22 Show that the following relations involving the Dirac delta function are valid:

(a) $\delta(x) = \delta(-x)$

(b) $x\,\delta(x) = 0$

(c) $\delta'(x) = -\delta'(-x)$

(d) $x\,\delta'(x) = -\delta(x)$

(e) $\delta(ax) = \dfrac{\delta(x)}{a}.$

8

Fourier Transforms

8.1 INTRODUCTION

When the method of separation of variables is applied to certain partial differential equations of mathematical physics, integrals of the form

$$F(\alpha) = \int_a^b f(x)K(\alpha, x)\,dx \tag{8.1}$$

often occur. The function $F(\alpha)$ is said to be the **integral transform** of $f(x)$ by the kernel $K(\alpha, x)$. The **kernels** associated with Fourier, Laplace, Fourier-Bessel (Hankel), and Mellin transforms are

$$F(\alpha) = \frac{1}{\sqrt{2\pi}}\int_{-\infty}^{\infty} f(x)e^{i\alpha x}\,dx \qquad \text{(Fourier transform)} \tag{8.2}$$

$$F(\alpha) = \int_0^{\infty} f(x)e^{-\alpha x}\,dx \qquad \text{(Laplace transform)} \tag{8.3}$$

$$F(\alpha) = \int_0^{\infty} f(x)xJ_n(\alpha x)\,dx \qquad \text{(Fourier-Bessel transform)} \tag{8.4}$$

$$F(\alpha) = \int_0^{\infty} f(x)x^{\alpha-1}\,dx \qquad \text{(Mellin transform).} \tag{8.5}$$

The Fourier and Laplace transforms are the most often used in mathematical physics.

8.2 THEORY OF FOURIER TRANSFORMS

8.2.1 Formal Development of the Complex Fourier Transform

The form of the complex Fourier series (see Chapter 7) is

$$f(x) = \sum_{n=-\infty}^{\infty} c_n e^{in\pi x/l} \qquad (-l \le x \le l) \tag{8.6}$$

where

$$c_n = \frac{1}{2l}\int_{-l}^{l} f(x)e^{-in\pi x/l}\,dx. \tag{8.7}$$

To make the transition $l \to \infty$, we introduce a new variable which is defined by

$$k = \frac{n\pi}{l} \tag{8.8}$$

where $(\Delta k/\pi)l = 1$ since $\Delta n = 1$. Hence we may write Eqs. (8.6) and (8.7) in the following forms:

$$f(x) = \sum_{n=kl/\pi=-\infty}^{\infty} C_l(k)e^{ikx}\,\Delta k \tag{8.9}$$

and

$$C_l(k) = \frac{1}{2\pi}\int_{-l}^{l} f(x)e^{-ikx}\,dx \tag{8.10}$$

where $C_l(k) = lc_n/\pi$. If we let $l \to \infty$, we obtain

$$C(k) = \frac{1}{2\pi}\int_{-\infty}^{\infty} f(x)e^{-ikx}\,dx \tag{8.11}$$

and

$$f(x) = \int_{-\infty}^{\infty} C(k)e^{ikx}\,dk. \tag{8.12}$$

There are several ways of defining Fourier transforms, but the differences among the various forms are not significant. To put Eqs. (8.11) and (8.12) in the modern form (the form most often used by authors), we let $F(k) = \sqrt{2\pi}C(-k)$ and obtain

$$F(k) = \frac{1}{\sqrt{2\pi}} \int_{-\infty}^{\infty} f(x)e^{ikx}\, dx \tag{8.13}$$

and

$$f(x) = \frac{1}{\sqrt{2\pi}} \int_{-\infty}^{\infty} F(k)e^{-ikx}\, dk. \tag{8.14}$$

Equations (8.13) and (8.14) are called the **Fourier transform pair.** If $f(x)$ satisfies the Dirichlet conditions and the integral $\int_{-\infty}^{\infty} |f(x)|\, dx$ is finite, then $F(k)$ exists for all k and is called the Fourier transform of $f(x)$. The function $f(x)$ in Eq. (8.14) is called the Fourier transform (inverse transform) of $F(k)$. As a physical requirement, it is usually assumed that $f(x) \to 0$ as $x \to \pm\infty$. Naturally, it is required that kx be dimensionless since it is the argument of an exponential quantity. In physics, k (or k_x) is the magnitude (x-component) of the wave vector with dimension of inverse length, and x is a distance. The combination of ωt, where ω is circular frequency with dimension of inverse time and t is time, is also used.

A Fourier transform pair equivalent to Eqs. (8.13) and (8.14), where the signs of the exponential terms are interchanged, is widely used in quantum mechanics.

The above development of the Fourier transform is purely formal. For a rigorous treatment of the subject, the many books on Fourier or integral transforms and operational mathematics are highly recommended.†

8.2.2 Cosine and Sine Transforms

The **cosine** and **sine transform pairs** are defined, respectively, by

$$\begin{aligned} F_c(k) &= \sqrt{\frac{2}{\pi}} \int_0^{\infty} f(x) \cos kx\, dx \\ f(x) &= \sqrt{\frac{2}{\pi}} \int_0^{\infty} F_c(k) \cos kx\, dk \end{aligned} \tag{8.15}$$

and

$$\begin{aligned} F_s(k) &= \sqrt{\frac{2}{\pi}} \int_0^{\infty} f(x) \sin kx\, dx \\ f(x) &= \sqrt{\frac{2}{\pi}} \int_0^{\infty} F_s(k) \sin kx\, dk. \end{aligned} \tag{8.16}$$

†For example, see E. C. Titchmarsh, *Introduction to the Theory of Fourier Integrals*, Oxford University Press, New York, 1937.

Note that the Fourier transform of $f(x)$ [Eq. (8.13)] may be written as

$$\begin{aligned} F(k) &= \frac{1}{\sqrt{2\pi}} \int_{-\infty}^{\infty} f(x) e^{ikx}\,dx \\ &= \frac{1}{\sqrt{2\pi}} \int_{-\infty}^{\infty} f(x)[\cos kx + i \sin kx]\,dx. \end{aligned} \tag{8.17}$$

If $f(x)$ is an even function of x [$f(x) = f(-x)$], then we see from Eq. (8.17) that the cosine transform is equal to the Fourier transform. If, however, $f(x)$ is an odd function of x [$f(x) = -f(-x)$], then we see from Eq. (8.17) that the sine transform is equal to the Fourier transform provided $-iF(k)$ is replaced by $F_s(k)$.

The one-dimensional Fourier transforms of a function of two and three independent variables are given, respectively, by

$$\begin{aligned} F(k, y) &= \frac{1}{\sqrt{2\pi}} \int_{-\infty}^{\infty} f(x, y) e^{ikx}\,dx \\ f(x, y) &= \frac{1}{\sqrt{2\pi}} \int_{-\infty}^{\infty} F(k, y)^{-ikx}\,dk \end{aligned} \tag{8.18}$$

and

$$\begin{aligned} F(k, y, z) &= \frac{1}{\sqrt{2\pi}} \int_{-\infty}^{\infty} f(x, y, z) e^{ikx}\,dx \\ f(x, y, z) &= \frac{1}{\sqrt{2\pi}} \int_{-\infty}^{\infty} F(k, y, z) e^{-ikx}\,dk. \end{aligned} \tag{8.19}$$

The corresponding cosine and sine transform pairs are

$$\begin{aligned} F_c(k, y) &= \sqrt{\frac{2}{\pi}} \int_0^{\infty} f(x, y) \cos kx\,dx \\ f(x, y) &= \sqrt{\frac{2}{\pi}} \int_0^{\infty} F_c(k, y) \cos kx\,dk \end{aligned} \tag{8.20}$$

and

$$\begin{aligned} F_s(k, y) &= \sqrt{\frac{2}{\pi}} \int_0^{\infty} f(x, y) \sin kx\,dx \\ f(x, y) &= \sqrt{\frac{2}{\pi}} \int_0^{\infty} F_s(k, y) \sin kx\,dk. \end{aligned} \tag{8.21}$$

Equations similar to Eqs. (8.20) and (8.21) are valid for $F_c(k, y, z)$ and $F_s(k, y, z)$.

8.2.3 Multiple-Dimensional Fourier Transforms

The one-dimensional transform theory may be extended in a natural way to the cases of two- and three-dimensional Fourier transforms. In equation form, the two- and three-dimensional Fourier transform pairs are, respectively,

$$F(\alpha, \beta) = \frac{1}{2\pi}\int_{-\infty}^{\infty}\int_{-\infty}^{\infty} f(x, y)e^{i(\alpha x+\beta y)}\,dx\,dy$$
$$f(x, y) = \frac{1}{2\pi}\int_{-\infty}^{\infty}\int_{-\infty}^{\infty} F(\alpha, \beta)e^{-i(\alpha x+\beta y)}\,d\alpha\,d\beta \tag{8.22}$$

and

$$F(\alpha, \beta, \gamma) = \frac{1}{(2\pi)^{3/2}}\int_{-\infty}^{\infty}\int_{-\infty}^{\infty}\int_{-\infty}^{\infty} f(x, y, z)e^{i(\alpha x+\beta y+\gamma z)}\,dx\,dy\,dz$$
$$f(x, y, z) = \frac{1}{(2\pi)^{3/2}}\int_{-\infty}^{\infty}\int_{-\infty}^{\infty}\int_{-\infty}^{\infty} F(\alpha, \beta, \gamma)e^{-i(\alpha x+\beta y+\gamma z)}\,d\alpha\,d\beta\,d\gamma. \tag{8.23}$$

The generalization to the case of an arbitrary number of dimensions is straightforward.

8.2.4 The Transforms of Derivatives

Fourier transforms, cosine transforms, and sine transforms can often be used to transform a differential equation (ordinary or partial) which describes a complicated physical problem into a simpler equation (algebraic or ordinary differential) that can be easily solved. The required solution of the original differential equation is then obtained by finding the inverse transform of the solution of the simpler equation (in transform space). In order to use the transform method to solve first- and second-order differential equations, the transforms of first- and second-order derivatives are needed.

We now develop the transforms of first- and second-order derivatives. The Fourier transforms of first- and second-order derivatives will be represented by $F^{(1)}(k)$ and $F^{(2)}(k)$, respectively. That is to say,

$$F^{(1)}(k) = \frac{1}{\sqrt{2\pi}}\int_{-\infty}^{\infty} \frac{df(x)}{dx}e^{ikx}\,dx. \tag{8.24}$$

On integrating Eq. (8.24) by parts, we obtain

$$\begin{aligned} F^{(1)}(k) &= \frac{1}{\sqrt{2\pi}}\left\{ f(x)e^{ikx}\Big|_{-\infty}^{\infty} - ik\int_{-\infty}^{\infty} f(x)e^{ikx}\,dx\right\} \\ &= -\frac{ik}{\sqrt{2\pi}}\int_{-\infty}^{\infty} f(x)e^{ikx}\,dx \\ &= -ikF(k) \end{aligned} \tag{8.25}$$

where $f(x) \to 0$ as $x \to \pm\infty$. Similarly, we find that

$$F^{(2)}(k) = \frac{1}{\sqrt{2\pi}} \int_{-\infty}^{\infty} \frac{d^2 f(x)}{dx^2} e^{ikx}\, dx$$
$$= -k^2 F(k) \tag{8.26}$$

where $f(x) \to 0$ and $f'(x) \to 0$ as $x \to \pm\infty$.

The corresponding relations for $F^{(1)}(k, y)$ and $F^{(2)}(k, y)$ are

$$F^{(1)}(k, y) = \frac{1}{\sqrt{2\pi}} \int_{-\infty}^{\infty} \frac{\partial f(x, y)}{\partial x} e^{ikx}\, dx$$
$$= -ikF(k, y) \tag{8.27}$$

for $f(x) \to 0$ as $x \to \pm\infty$, and

$$F^{(2)}(k, y) = \frac{1}{\sqrt{2\pi}} \int_{-\infty}^{\infty} \frac{\partial^2 f(x, y)}{\partial x^2} e^{ikx}\, dx$$
$$= -k^2 F(k, y) \tag{8.28}$$

for $f(x) \to 0$ and $\partial f(x, y)/\partial x \to 0$ as $x \to \pm\infty$.

The cosine and sine transforms of first- and second-order derivatives are:

$$F_c^{(1)}(k) = \sqrt{\frac{2}{\pi}} \int_0^{\infty} \frac{df(x)}{dx} \cos kx\, dx$$
$$= kF_s(k) - \sqrt{\frac{2}{\pi}} f(0) \tag{8.29}$$

for $f(x) \to 0$ as $x \to \infty$,

$$F_c^{(2)}(k) = \sqrt{\frac{2}{\pi}} \int_0^{\infty} \frac{d^2 f(x)}{dx^2} \cos kx\, dx$$
$$= -k^2 F_c(k) - \sqrt{\frac{2}{\pi}} f'(0) \tag{8.30}$$

for $f(x) \to 0$ and $f'(x) \to 0$ as $x \to \infty$, and

$$F_s^{(1)}(k) = \sqrt{\frac{2}{\pi}} \int_0^{\infty} \frac{df(x)}{dx} \sin kx\, dx$$
$$= -kF_c(k) \tag{8.31}$$

for $f(x) \to 0$ as $x \to \infty$,

$$F_s^{(2)}(k) = \sqrt{\frac{2}{\pi}} \int_0^\infty \frac{d^2 f(x)}{dx^2} \sin kx \, dx$$
$$= \sqrt{\frac{2}{\pi}}\, k f(0) - k^2 F_s(k) \tag{8.32}$$

for $f(x) \to 0$ and $f'(x) \to 0$ as $x \to \infty$.

The choice of using the cosine or sine transform is dictated by the given boundary conditions at the lower limit. The use of the cosine transform to remove (transform) a first-order derivative term in a differential equation requires a knowledge of $f(0)$, but a knowledge of $f(0)$ is not required when the sine transform is used to remove a first-order derivative term. A knowledge of $f'(0)$ is required to remove a second-order derivative term in a differential equation if the cosine transform is used. However, only a knowledge of $f(0)$ is needed to successfully use a sine transform to remove a second-order derivative term in a differential equation.

The relations for $F_c^{(1)}(k, y)$, $F_c^{(2)}(k, y)$, $F_s^{(1)}(k, y)$, and $F_s^{(2)}(k, y)$ are:

$$F_c^{(1)}(k, y) = \sqrt{\frac{2}{\pi}} \int_0^\infty \frac{\partial f(x, y)}{\partial x} \cos kx \, dx$$
$$= kF_s(k, y) - \sqrt{\frac{2}{\pi}}\, f(0, y) \tag{8.33}$$

for $f(x, y) \to 0$ as $x \to \infty$,

$$F_c^{(2)}(k, y) = \sqrt{\frac{2}{\pi}} \int_0^\infty \frac{\partial^2 f(x, y)}{\partial x^2} \cos kx \, dx$$
$$= -k^2 F_c(k, y) - \sqrt{\frac{2}{\pi}}\, f'(0, y) \tag{8.34}$$

for $f(x, y) \to 0$ and $\partial f(x, y)/\partial x \to 0$ as $x \to \infty$,

$$F_s^{(1)}(k, y) = \sqrt{\frac{2}{\pi}} \int_0^\infty \frac{\partial f(x, y)}{\partial x} \sin kx \, dx$$
$$= -kF_c(k, y) \tag{8.35}$$

for $f(x, y) \to 0$ as $x \to \infty$, and

$$F_s^{(2)}(k, y) = \sqrt{\frac{2}{\pi}} \int_0^\infty \frac{\partial^2 f(x, y)}{\partial x^2} \sin kx \, dx$$
$$= \sqrt{\frac{2}{\pi}}\, k f(0, y) - k^2 F_s(k, y) \tag{8.36}$$

for $f(x, y) \to 0$ and $\partial f(x, y)/\partial x \to 0$ as $x \to \infty$.

8.2.5 The Convolution Theorem

In linear response theory, the general equation for the one-dimensional transform, Eq. (8.1), takes the form

$$F(\alpha) = \int_a^b f(x)K(\alpha, x)\, dx$$

$$= \int_{-\infty}^{\infty} K(\alpha - x)f(x)\, dx \tag{8.37}$$

where $K(\alpha - x)$ is called the **response** of the linear system, $f(x)$ is the input (signal) to the linear system, and $F(\alpha)$ is the output (signal). If $K(\alpha - x) = \delta(\alpha - x)$, then

$$F(\alpha) = \int_{-\infty}^{\infty} \delta(\alpha - x)f(x)\, dx$$

which is consistent with the properties of the Dirac delta function. In this latter case, $K(\alpha - x)$ is called the **impulse response** of the system. When Eq. (8.37) is written in the form

$$F(x) = f * g = \frac{1}{\sqrt{2\pi}} \int_{-\infty}^{\infty} f(x - \xi)g(\xi)\, d\xi \tag{8.38}$$

it is called the one-dimensional **convolution (faltung—folding)** integral of two integrable functions $f(x)$ and $g(x)$. The corresponding two-dimensional form is

$$f * g = \frac{1}{2\pi} \int_{-\infty}^{\infty}\int_{-\infty}^{\infty} f(x - \xi; x - \beta)g(\xi, \beta)\, d\xi\, d\beta.$$

Let $F(k)$ and $G(k)$ be the Fourier transforms of $f(x)$ and $g(x)$, respectively. For these functions, the convolution integral becomes

$$\frac{1}{\sqrt{2\pi}} \int_{-\infty}^{\infty} f(x - \xi)g(\xi)\, d\xi = \frac{1}{\sqrt{2\pi}} \int_{-\infty}^{\infty} g(\xi)\left\{\frac{1}{\sqrt{2\pi}} \int_{-\infty}^{\infty} F(k)e^{-ik(x-\xi)}\, dk\right\} d\xi$$

$$= \frac{1}{2\pi} \int_{-\infty}^{\infty} F(k)e^{-ikx}\, dk \int_{-\infty}^{\infty} g(\xi)e^{ik\xi}\, d\xi$$

$$= \frac{1}{\sqrt{2\pi}} \int_{-\infty}^{\infty} F(k)G(k)e^{-ikx}\, dk. \tag{8.39}$$

In obtaining Eq. (8.39), we have tacitly assumed that the process of interchanging the order of integration is valid. The result in Eq. (8.39) is known as the **convolution theorem for Fourier transforms.** It means that the Fourier transform (inverse transform) of the product $F(k)G(k)$, the right-hand side of Eq. (8.39), is the convolution of the original functions, $f * g$. For an illustration of the use of the convolution theorem, see Examples 8.5 and 8.6.

8.2.6 Parseval's Relation

The integral of the product of two functions $f(x)$ and $g^*(x)$, the complex conjugate of $g(x)$, may be written as

$$\begin{aligned}\int_{-\infty}^{\infty} f(x)g^*(x)\,dx &= \int_{-\infty}^{\infty}\left[\frac{1}{\sqrt{2\pi}}\int_{-\infty}^{\infty} F(k)e^{-ikx}\,dk\right] \\ &\qquad\times\left[\frac{1}{\sqrt{2\pi}}\int_{-\infty}^{\infty} G^*(k')e^{ik'x}\,dk'\right]dx \\ &= \int_{-\infty}^{\infty} dk \int_{-\infty}^{\infty} dk'\,F(k)\,G^*(k')\left[\frac{1}{2\pi}\int_{-\infty}^{\infty} e^{ix(k'-k)}\,dx\right] \\ &= \int_{-\infty}^{\infty} dk\,F(k)\int_{-\infty}^{\infty} dk'\,G^*(k')\,\delta(k'-k) \\ &= \int_{-\infty}^{\infty} F(k)\,G^*(k)\,dk \end{aligned} \tag{8.40}$$

The relation in Eq. (8.40) is known as **Parseval's relation** (or **Parseval's formula**) and is widely used in optics, electromagnetism, and quantum mechanics (see Example 8.8).

EXAMPLE 8.1 Find the Fourier transform, $F(k)$, of the Gaussian distribution function, $f(x) = Ne^{-\alpha x^2}$, where N and α are constants.

Solution By use of Eq. (8.13), we find that

$$\begin{aligned} F(k) &= \frac{1}{\sqrt{2\pi}}\int_{-\infty}^{\infty} f(x)e^{ikx}\,dx \\ &= \frac{N}{\sqrt{2\pi}}\int_{-\infty}^{\infty} e^{ikx-\alpha x^2}\,dx \\ &= \frac{N}{\sqrt{2\pi}}\left\{\int_{-\infty}^{\infty} e^{-\alpha x^2}\cos kx\,dx + i\int_{-\infty}^{\infty} e^{-\alpha x^2}\sin kx\,dx\right\} \\ &= \frac{N}{\sqrt{2\pi}}e^{-k^2/4\alpha} \end{aligned}$$

where the second integral is zero since it is the integral of an odd function from $-\infty$ to ∞ and

$$\int_{-\infty}^{\infty} e^{-\alpha x^2}\cos kx\,dx = \sqrt{\frac{\pi}{\alpha}}e^{-k^2/4\alpha},$$

Note that both $f(x)$ and $F(k)$ are Gaussian distribution functions with peaks at the origin. The **standard deviation**, width, is defined as the range of the variable x (or k) for which the function $f(x)$ [or $F(k)$] drops by a factor of $e^{-1/2} = 0.606$ of its maximum value. For $f(x) = Ne^{-\alpha x^2}$, the standard deviation is given by

$$\sigma_x = \frac{\Delta x}{2} = \frac{1}{\sqrt{2\alpha}}.$$

For

$$F(k) = \frac{N}{\sqrt{2\alpha}} e^{-k^2/4\alpha}$$

the standard deviation is given by

$$\sigma_k = \frac{\Delta k}{2} = \sqrt{2\alpha}.$$

Note that $\Delta x\, \Delta k = (2/\sqrt{2\alpha})(2\sqrt{2\alpha}) = 4$. If $\alpha \to 0$ (small), then $\Delta x \to \infty$ and $\Delta k \to 0$. For $\alpha \to \infty$ (large), $\Delta x \to 0$ and $\Delta k \to \infty$ (see Fig. 8.1).

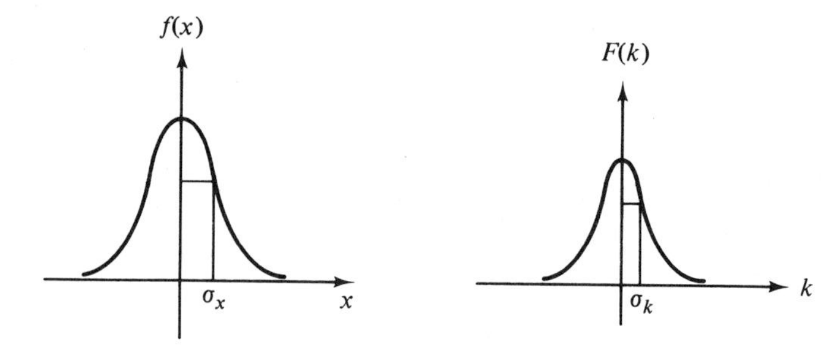

Figure 8.1

EXAMPLE 8.2 Find the Fourier transform for the box function $f(x)$ where

$$f(x) = \begin{cases} 1 & (-a \leq x \leq a) \\ 0 & (|x| > a). \end{cases}$$

Solution By use of Eq. (8.13), we obtain

$$\begin{aligned} F(k) &= \frac{1}{\sqrt{2\pi}} \int_{-\infty}^{\infty} f(x) e^{ikx}\, dx \\ &= \int_{-a}^{a} e^{ikx}\, dx \end{aligned}$$

$$= \frac{1}{\sqrt{2\pi}} \left[\frac{e^{ikx}}{ik} \right]_{-a}^{a}$$

$$= \sqrt{\frac{2}{\pi}} \frac{\sin ka}{k}.$$

A sketch of $f(x)$ and $F(k)$ is given in Fig. 8.2.

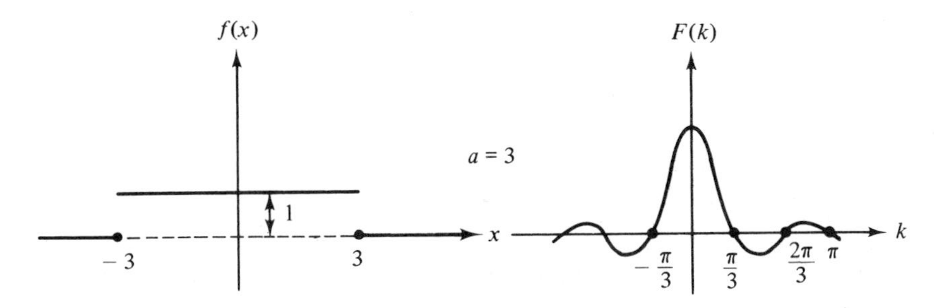

Figure 8.2

EXAMPLE 8.3 Find the cosine and sine transforms of $f(x) = e^{-x}$.

Solution Using Eq. (8.15), we find that

$$F_c(k) = \sqrt{\frac{2}{\pi}} \int_0^\infty f(x) \cos kx \, dx$$

$$= \sqrt{\frac{2}{\pi}} \int_0^\infty e^{-x} \cos kx \, dx$$

$$= \sqrt{\frac{2}{\pi}} \left(\frac{1}{1 + k^2} \right).$$

The corresponding sine transform is obtained by means of Eq. (8.16). We obtain

$$F_s(k) = \sqrt{\frac{2}{\pi}} \int_0^\infty f(x) \sin kx \, dx$$

$$= \sqrt{\frac{2}{\pi}} \int_0^\infty e^{-x} \sin kx \, dx$$

$$= \sqrt{\frac{2}{\pi}} \left(\frac{k}{1 + k^2} \right).$$

EXAMPLE 8.4 By the Fourier transform method, develop the formal solution of the differential equation which characterizes the motion of a damped harmonic oscillator.

Solution The required equation is

$$\ddot{x}(t) + 2\alpha\dot{x}(t) + \omega_0^2 x(t) = f(t)$$

where $f(t)$ is any function such that

$$f(t) = \frac{1}{\sqrt{2\pi}} \int_{-\infty}^{\infty} F(\omega) e^{-i\omega t} \, d\omega$$

and

$$F(\omega) = \frac{1}{\sqrt{2\pi}} \int_{-\infty}^{\infty} f(t) e^{i\omega t} \, dt.$$

On taking the Fourier transform of each term in the differential equation for the damped harmonic oscillator, we obtain

$$\frac{1}{\sqrt{2\pi}} \int_{-\infty}^{\infty} \frac{d^2x}{dt^2} e^{i\omega t} \, dt + \frac{2\alpha}{\sqrt{2\pi}} \int_{-\infty}^{\infty} \frac{dx}{dt} e^{i\omega t} \, dt + \frac{\omega_0^2}{\sqrt{2\pi}} \int_{-\infty}^{\infty} x(t) e^{i\omega t} \, dt$$
$$= \frac{1}{\sqrt{2\pi}} \int_{-\infty}^{\infty} f(t) e^{i\omega t} \, dt.$$

If we use Eqs. (8.25) and (8.26) in the above equation, it reduces to

$$-\omega^2 X(\omega) - 2\alpha i\omega X(\omega) + \omega_0^2 X(\omega) = F(\omega)$$

or

$$X(\omega) = \frac{F(\omega)}{\omega_0^2 - \omega^2 - 2\alpha i\omega}$$

where

$$X(\omega) = \frac{1}{\sqrt{2\pi}} \int_{-\infty}^{\infty} x(t) e^{i\omega t} \, dt.$$

The solution of the original equation is obtained if we take the Fourier transform (inverse transform) of $X(\omega)$. It is

$$x(t) = \frac{1}{\sqrt{2\pi}} \int_{-\infty}^{\infty} \frac{F(\omega) e^{-i\omega t} \, d\omega}{\omega_0^2 - \omega^2 - 2\alpha i\omega}.$$

This integral may be evaluated by use of the calculus of residues method developed in Chapter 4.

EXAMPLE 8.5 *Wave Equation* Consider the motion of a very long (infinite) elastic string (see Chapter 5). The one-dimensional equation for transverse waves in this string is

$$\frac{\partial^2 \varphi(x, t)}{\partial x^2} = \frac{1}{v^2} \frac{\partial^2 \varphi(x, t)}{\partial t^2}$$

where

$$\varphi(x, t) \longrightarrow 0 \quad \text{and} \quad \frac{\partial \varphi}{\partial x} \longrightarrow 0 \qquad (\text{as } x \longrightarrow \pm\infty)$$

$$\varphi(x, 0) = f(x) \quad \text{and} \quad \left.\frac{\partial \varphi}{\partial t}\right|_{t=0} = 0.$$

Solution Let

$$\Phi(k, t) = \frac{1}{\sqrt{2\pi}} \int_{-\infty}^{\infty} \varphi(x, t) e^{ikx}\, dx.$$

On taking the Fourier transform of the wave equation, we obtain

$$\frac{1}{\sqrt{2\pi}} \int_{-\infty}^{\infty} \frac{\partial^2 \varphi}{\partial x^2} e^{ikx}\, dx = \frac{1}{v^2} \frac{1}{\sqrt{2\pi}} \frac{\partial^2}{\partial t^2} \int_{-\infty}^{\infty} \varphi(x, t) e^{ikx}\, dx.$$

By use of Eq. (8.28), we see that the above equation reduces to

$$-k^2 \Phi(k, t) = \frac{1}{v^2} \frac{\partial^2 \Phi(k, t)}{\partial t^2}$$

where $\varphi(x, t) \to 0$ and $\partial\varphi/\partial x \to 0$ for $x \to \pm\infty$. In other words, there is no wave motion at $\pm\infty$ (the ends of a very long string). The solution of the above ordinary differential equation (since the derivative with respect to only one independent variable appears) is

$$\Phi(k, t) = c_1 e^{ikvt} + c_2 e^{-ikvt}.$$

The given initial conditions will now be used to evaluate the constants c_1 and c_2. From $\varphi(x, 0) = f(x)$, we obtain

$$\begin{aligned} \Phi(k, 0) &= \frac{1}{\sqrt{2\pi}} \int_{-\infty}^{\infty} \varphi(x, 0) e^{-ikx}\, dx \\ &= \frac{1}{\sqrt{2\pi}} \int_{-\infty}^{\infty} f(x) e^{-ikx}\, dx \\ &= F(k) \\ &= c_1 + c_2. \end{aligned}$$

The second initial condition gives

$$\begin{aligned} \dot{\Phi}(k, 0) &= \frac{1}{\sqrt{2\pi}} \int_{-\infty}^{\infty} \left(\frac{\partial \varphi}{\partial t}\right)_{t=0} e^{-ikx}\, dx \\ &= 0 \\ &= ikvc_1 - ikvc_2. \end{aligned}$$

Therefore the solution in k-space is

$$\Phi(k, t) = \frac{F(k)}{2} e^{ikvt} + \frac{F(k)}{2} e^{-ikvt}.$$

The required expression for $\varphi(x, t)$, in this simple case, may be obtained directly by taking the Fourier transform of $\Phi(k, t)$. We have

$$\begin{aligned}\varphi(x, t) &= \frac{1}{\sqrt{2\pi}} \int_{-\infty}^{\infty} \Phi(k, t) e^{-ikx}\, dk \\ &= \frac{1}{\sqrt{2\pi}} \left\{ \frac{1}{2} \int_{-\infty}^{\infty} (F(k)e^{ikvt})e^{-ikx}\, dk + \frac{1}{2} \int_{-\infty}^{\infty} (F(k)e^{-ikvt})e^{-ikx}\, dk \right\}\end{aligned}$$

or

$$\varphi(x, t) = \tfrac{1}{2} f(x - vt) + \tfrac{1}{2} f(x + vt).$$

The above procedure for finding $\varphi(x, t)$ from $\Phi(k, t)$ leads to the desired result only in relatively simple cases. Use of the convolution theorem in the expression for the inverse transform is the general procedure for finding $\varphi(x, t)$ from $\Phi(k, t)$, that is,

$$\begin{aligned}\varphi(x, t) &= \frac{1}{\sqrt{2\pi}} \int_{-\infty}^{\infty} \Phi(k, t) e^{-ikx}\, dk \\ &= \frac{1}{2\sqrt{2\pi}} \left\{ \int_{-\infty}^{\infty} (F(k)e^{ikvt})e^{-ikx}\, dk + \int_{-\infty}^{\infty} (F(k)e^{-ikvt})e^{-ikx}\, dk \right\} \\ &= \frac{1}{2\sqrt{2\pi}} \left\{ \int_{-\infty}^{\infty} f(\xi) g_+(x - \xi)\, d\xi + \int_{-\infty}^{\infty} f(\xi) g_-(x - \xi)\, d\xi \right\}.\end{aligned}$$

Here we need the original forms (in x-space) of e^{ikvt} and e^{-ikvt} to obtain $g_+(x - \xi)$ and $g_-(x - \xi)$. To obtain these forms, we take the inverse transforms of e^{ikvt} and e^{-ikvt}. For e^{ikvt}, we find

$$\begin{aligned}g_+(x) &= \frac{1}{\sqrt{2\pi}} \int_{-\infty}^{\infty} (e^{ikvt}) e^{-ikx}\, dk \\ &= \frac{1}{\sqrt{2\pi}} \int_{-\infty}^{\infty} e^{-ik(x - vt)}\, dk \\ &= \sqrt{2\pi}\, \delta(x - vt).\end{aligned}$$

Similarly, we find that

$$g_-(x) = \sqrt{2\pi}\, \delta(x + vt).$$

The required expression for $\varphi(x, t)$ in terms of the convolution integral now becomes

$$\begin{aligned}\varphi(x, t) &= \frac{1}{2\sqrt{2\pi}}\left\{\int_{-\infty}^{\infty} f(\xi)g_+(x-\xi)\,d\xi + \int_{-\infty}^{\infty} f(\xi)g_-(x-\xi)\,d\xi\right\}\\ &= \frac{1}{2}\left\{\int_{-\infty}^{\infty} f(\xi)\,\delta(x-vt-\xi)\,d\xi + \int_{-\infty}^{\infty} f(\xi)\,\delta(x+vt-\xi)\,d\xi\right\}\\ &= \frac{1}{2}f(x-vt) + \frac{1}{2}f(x+vt).\end{aligned}$$

This result is the same as that obtained by use of the direct method.

The solution of the wave equation subject to the indicated conditions is therefore

$$\varphi(x, t) = \tfrac{1}{2}f(x-vt) + \tfrac{1}{2}f(x+vt).$$

Let us examine the behavior of $\varphi_+(x, t) = f(x - vt)$ at time $t + \Delta t$ and position $x + v\,\Delta t$. For this case, we have

$$\begin{aligned}\varphi_+(x+v\Delta t, t+\Delta t) &= f[x + v\Delta t - v(t+\Delta t)]\\ &= f(x-vt).\end{aligned}$$

Hence φ_+ represents a wave motion (a signal) traveling in the positive x-direction with speed v which repeats itself at $t + \Delta t$ and $x + \Delta x$. Similarly, $\varphi_-(x, t) = f(x + vt)$ represents a signal traveling in the negative x-direction which repeats itself at $t + \Delta t$ and $x - \Delta x$.

Example 8.6 *Heat Equation* The temperature distribution in a very long heat-conducting rod is described by the one-dimensional heat equation,

$$\frac{\partial^2 T(x, t)}{\partial x^2} = \frac{1}{\sigma}\frac{\partial T(x, t)}{\partial t} \qquad [T(x, 0) = f(x)].$$

Solution Here it is assumed that $T(x, t) \to 0$ and $\partial T/\partial x \to 0$ as $x \to \pm\infty$, and the Fourier transform of $f(x)$ exists. Let

$$T(k, t) = \frac{1}{\sqrt{2\pi}}\int_{-\infty}^{\infty} T(x, t)e^{ikx}\,dx.$$

Taking the Fourier transform of the heat equation, we obtain

$$\frac{1}{\sqrt{2\pi}}\int_{-\infty}^{\infty}\frac{\partial^2 T}{\partial x^2}e^{ikx}\,dx = \frac{1}{\sigma}\frac{1}{\sqrt{2\pi}}\frac{\partial}{\partial t}\int_{-\infty}^{\infty} T(x, t)e^{ikx}\,dx.$$

By use of Eq. (8.28), the above equation reduces to

$$-k^2 T(k, t) = \frac{1}{\sigma} \frac{\partial T(k, t)}{\partial t}$$

where $T(x, t) \to 0$ and $\partial T/\partial x \to 0$ as $x \to \pm\infty$ (the temperature is zero at the ends of the very long rod). The solution of the above equation is

$$T(k, t) = Ae^{-\sigma k^2 t}.$$

Since

$$\begin{aligned} T(k, 0) &= A \\ &= \frac{1}{\sqrt{2\pi}} \int_{-\infty}^{\infty} f(x) e^{ikx}\, dx \\ &= F(k) \end{aligned}$$

the solution in k-space becomes

$$T(k, t) = F(k) e^{-\sigma k^2 t}.$$

The object is to develop an expression for $T(x, t)$. Here we need the inverse transform of $F(k)e^{-\sigma k^2 t}$ which can be obtained by use of the convolution theorem, that is,

$$\begin{aligned} T(x, t) &= \frac{1}{\sqrt{2\pi}} \int_{-\infty}^{\infty} (F(k) e^{-\sigma k^2 t}) e^{-ikx}\, dk \\ &= \frac{1}{\sqrt{2\pi}} \int_{-\infty}^{\infty} f(\xi) g(x - \xi)\, d\xi. \end{aligned}$$

We need the original form (in x-space) of the second function, $g(x - \xi)$. It is just the Fourier transform of $e^{-\sigma k^2 t}$ which becomes

$$g(x) = \frac{1}{\sqrt{2\pi}} \int_{-\infty}^{\infty} (e^{-\sigma k^2 t}) e^{-ikx}\, dk.$$

If $u^2 = 2k^2 \sigma t$ where $dk = du/\sqrt{2\sigma t}$, then

$$\begin{aligned} g(x) &= \frac{1}{2\sqrt{\sigma t \pi}} \int_{-\infty}^{\infty} (e^{-u^2/2}) e^{-i(u/\sqrt{2\sigma t})x}\, du \\ &= \frac{1}{\sqrt{2\sigma t}} e^{-x^2/4\sigma t}. \end{aligned}$$

The expression for $T(x, t)$ in terms of the convolution integral now becomes

$$T(x, t) = \frac{1}{\sqrt{2\pi}}\left(\frac{1}{\sqrt{2\sigma t}}\right)\int_{-\infty}^{\infty} f(\xi)e^{-(x-\xi)^2/4\sigma t}\, d\xi$$

$$= \frac{1}{2\sqrt{\pi\sigma t}}\int_{-\infty}^{\infty} f(\xi)e^{-(x-\xi)^2/4\sigma t}\, d\xi.$$

The specific form of $f(x)$, the initial temperature distribution, must of course be given before the above integral can be evaluated.

Example 8.7. By use of the three-dimensional Fourier transform method, solve Poisson's equation for the electrostatic potential function,

$$\nabla^2\phi(\mathbf{r}) = -\frac{\rho(\mathbf{r})}{\epsilon}.$$

Solution In three-dimensions, we write

$$\Phi(\mathbf{k}) = \frac{1}{(2\pi)^{3/2}}\int_{-\infty}^{\infty} \phi(\mathbf{r})e^{i\mathbf{k}\cdot\mathbf{r}}\, d^3r$$

where $d^3r = dx\, dy\, dz$ (triple integral). The Fourier transform of $\Phi(\mathbf{k})$ is given by

$$\phi(\mathbf{r}) = \frac{1}{(2\pi)^{3/2}}\int_{-\infty}^{\infty} \Phi(\mathbf{k})e^{-i\mathbf{k}\cdot\mathbf{r}}\, d^3k.$$

On taking the Fourier transform of both sides of Poisson's equation, we obtain

$$-k^2\Phi(\mathbf{k}) = -\frac{P(\mathbf{k})}{\epsilon}$$

or

$$\Phi(\mathbf{k}) = \frac{P(\mathbf{k})}{k^2\epsilon}$$

where $\phi(\mathbf{r}) \to 0$ and $\partial\phi/\partial r \to$ for $r \to \pm\infty$. The expression for $P(\mathbf{k})$ is

$$P(\mathbf{k}) = \frac{1}{(2\pi)^{3/2}}\int_{-\infty}^{\infty} \rho(\mathbf{r}')e^{i\mathbf{k}\cdot\mathbf{r}'}\, d^3r'.$$

The required solution of the original differential equation, $\phi(\mathbf{r})$, is obtained by taking the inverse transform of $P(\mathbf{k})/k^2\epsilon$; we have

$$\begin{aligned}\phi(\mathbf{r}) &= \frac{1}{(2\pi)^{3/2}}\int_{-\infty}^{\infty}\frac{P(\mathbf{k})}{k^2\epsilon}e^{-i\mathbf{k}\cdot\mathbf{r}}\,d^3k\\ &= \frac{1}{(2\pi)^3}\int_{-\infty}^{\infty}\int_{-\infty}^{\infty}\frac{\rho(\mathbf{r}')e^{-i\mathbf{k}\cdot(\mathbf{r}-\mathbf{r}')}}{k^2\epsilon}\,d^3k\,d^3r'\\ &= \frac{1}{(2\pi)^3\epsilon}\int_{-\infty}^{\infty}\rho(\mathbf{r}')G(\mathbf{r}-\mathbf{r}')\,d^3r'\end{aligned}$$

where

$$G(\mathbf{r}-\mathbf{r}') = \int_{-\infty}^{\infty}\frac{e^{-i\mathbf{k}\cdot(\mathbf{r}-\mathbf{r}')}}{k^2}\,d^3k.$$

If spherical polar coordinates are chosen for the variables in k-space, we have

$$\begin{aligned}d^3k &= k^2\sin\theta\,dk\,d\theta\,d\phi\\ &= -k^2\,d(\cos\theta)\,d\phi\,dk.\end{aligned}$$

The polar axis is the direction of $\mathbf{r}-\mathbf{r}'$. The expression for $G(\mathbf{r}-\mathbf{r}')$ becomes

$$\begin{aligned}G(\mathbf{r}-\mathbf{r}') &= -\int_{+1}^{-1}\int_0^{2\pi}\int_0^{\infty}\frac{k^2e^{-ik|\mathbf{r}-\mathbf{r}'|\cos\theta}\,d(\cos\theta)\,d\phi\,dk}{k^2}\\ &= 2\pi\int_0^{\infty}\left[\frac{e^{-ik|\mathbf{r}-\mathbf{r}'|}-e^{ik|\mathbf{r}-\mathbf{r}'|}}{-ik|\mathbf{r}-\mathbf{r}'|}\right]dk\\ &= 4\pi\int_0^{\infty}\frac{\sin k|\mathbf{r}-\mathbf{r}'|}{k|\mathbf{r}-\mathbf{r}'|}\,dk\\ &= \frac{2\pi^2}{|\mathbf{r}-\mathbf{r}'|}.\end{aligned}$$

The final form for $\phi(\mathbf{r})$ is

$$\begin{aligned}\phi(\mathbf{r}) &= \left(\frac{1}{(2\pi)^3\epsilon}\right)(2\pi^2)\int_{-\infty}^{\infty}\frac{\rho(\mathbf{r}')\,d^3r'}{|\mathbf{r}-\mathbf{r}'|}\\ &= \frac{1}{4\pi\epsilon}\int_{-\infty}^{\infty}\frac{\rho(\mathbf{r}')\,d^3r'}{|\mathbf{r}-\mathbf{r}'|}.\end{aligned}$$

In this chapter, we have discussed Fourier transforms in which the ranges of integration have been $(0, \infty)$ for the cosine and sine transforms and $(-\infty, \infty)$ for the complex Fourier transforms. These transform methods may be extended to cases of finite intervals; however, the use of finite transforms does not lead to new results.

We conclude this chapter with a detailed discussion of the role of Fourier transforms in the foundation of quantum (wave) mechanics.

8.3 THE WAVE PACKET IN QUANTUM MECHANICS

8.3.1 Origin of the Problem: Quantization of Energy

Prior to 1900, physicists assumed that the laws governing the macroworld, together with the appropriate statistical considerations, were valid in the microworld. This approach always led to an unsatisfactory theory of black-body radiation (Wien's and Rayleigh-Jeans' radiation laws).

The development of quantum theory had its origin in the inability of classical physics (mechanics, electromagnetism, optics, and thermodynamics) to account for the experimentally observed energy distribution (energy vs. frequency or wavelength) in the continuous spectrum of black-body radiation. In short, one is required to explain on an atomistic basis the color of light that an object emits when it is heated to a certain temperature. A correct theory of black-body radiation was developed by Planck (1858–1947) in 1900. This theory requires that the radiated energy be quantized, that is, $E = h\nu$, where E is the energy of the radiation, ν is the frequency of the radiation, and h is Planck's constant. However, Planck attributed the quantization phenomenon to the radiating object.

In 1905, Einstein (1879–1955) further developed the quantization of energy concept by (1) assuming that the quantization phenomenon was a property of the radiation itself, (2) assuming that the quantization process applied to both absorption and emission of radiation, and (3) using the quantization concept to develop a correct theory of the photoelectric effect.

Bohr (1885–1962), in 1913, used a mixture of classical physics (mechanics and electromagnetism) and quantization of energy concept to formulate a satisfactory theory for the observed spectrum of the hydrogen atom. Bohr's theory is based on the following three postulates:

1. The electron in the hydrogen atom moves about the nucleus (the proton) in certain circular orbits (stationary states) without radiating energy.
2. The allowed stationary states are such that

$$\begin{aligned} L &= mvr \\ &= n\hbar \qquad (n = 1, 2, 3, \ldots) \end{aligned}$$

where L is the angular momentum of the electron, r is the radius of the orbit, m is the mass of the electron, $\hbar = h/2\pi$, v is the speed of the electron, and n is the principal quantum number.

3. When the electron makes a transition from a state of energy E_i to a state of energy E_f, where $E_i > E_f$, electromagnetic radiation (photons) is emitted from the hydrogen atom. The frequency of this radiation is given by

$$\nu = \frac{E_i - E_f}{h}$$

Bohr's theory was generalized by Wilson and Sommerfeld and applied to other atoms with limited success. By 1924, it was clear that a new theory was needed to explain the basic properties of atoms and molecules in a systematic manner.

8.3.2 The Development of a New Quantum Theory

In an historic paper (1925) "On a Quantum Theoretical Interpretation of Kinematical and Mechanical Relations" which led to the development of matrix mechanics, Heisenberg (1901–) introduced a system of mechanics in which classical concepts of mechanics were drastically revised. Heisenberg assumed that atomic theory should emphasize the observable quantities rather than the shapes of electronic orbits (Bohr's theory). This theory was rapidly developed by means of matrix algebra.

Parallel to the advancement of matrix mechanics, Schrödinger initiated (1926) a new line of study which evolved into wave mechanics. Wave mechanics was inspired by de Broglie's (1892–) wave theory of matter, $\lambda = h/p$, where p is the momentum of a particle and λ is the wavelength of its associated wave (matter wave). Schrödinger (1887–1961) introduced an equation of motion, the Schrödinger wave equation, for matter waves and proved that wave mechanics was mathematically equivalent to matrix mechanics. However, the physical meaning of wave mechanics was not clear at first. Schrödinger first considered the de Broglie wave as a physical entity (the particle, electron, is actually a wave). This interpretation soon led to difficulty since a wave may be partially reflected and partially transmitted at a boundary, but an electron cannot be split into two parts for transmission and reflection. This difficulty was removed by Born (1882–1970) who proposed a statistical interpretation of the de Broglie waves which is now generally accepted. The new theory based on the statistical interpretation was very rapidly developed into a general coherent system of mechanics (called quantum mechanics).

8.3.3 A Wave Equation for Particles: The Wave Packet

In developing a wave equation for particles, Schrödinger knew that (1) Hamilton had established an analogy between the Newtonian mechanics of a particle and geometrical (ray) optics (Hamiltonian mechanics), and (2) equations of wave optics reduced to those of geometrical optics if the wavelength in the former is equal to zero. Schrödinger postulated that classical Newtonian mechanics was the limiting case of a more general wave mechanics and proceeded to obtain a wave equation for particles.

The conditions for the existence of a wave equation for particles will now be established. We begin with the fundamental principle of classical mechanics. For a particle of mass m moving in a force field described by a potential $V(x, y, z)$, we may write

$$\delta \int_A^B p\, ds = 0 \tag{8.41}$$

where $p\, ds$ is called an **element of action** with MKS units of Joule·second. (See Appendix B in Chapter 9 for a discussion of the calculus of variations.) Equation (8.41) is referred to as the **principle of least action** and can be derived from Hamilton's more general principle of **stationary action** which is given by

$$\delta \int_A^B L\, dt = 0 \tag{8.42}$$

where the **Lagrangian** L is defined by

$$L = (\text{kinetic energy}) - (\text{potential energy}). \tag{8.43}$$

An element of **Hamiltonian action** is given by $L\, dt$. To go from Hamilton's principle of stationary action to the principle of least action, the total energy, E, of the system must be assumed to be constant. The energy of the particle is given by

$$E = \frac{p^2}{2m} + V(x, y, z). \tag{8.44}$$

The corresponding momentum is

$$p = \sqrt{2m[E - V(x, y, z)]}. \tag{8.45}$$

In geometrical optics, the basic principle is **Fermat's** (1601–1665) **principle of least time,**

$$\delta \int_A^B \frac{ds}{u} = 0. \tag{8.46}$$

For wave motion, we may write

$$u = \lambda \nu \tag{8.47}$$

where u is the phase velocity (speed) of a monochromatic wave of frequency ν and wavelength λ. On substituting Eq. (8.47) into Eq. (8.46), we get

$$\delta \int_A^B \frac{ds}{\lambda} = 0. \tag{8.48}$$

Note the striking similarity between Eqs. (8.41) and (8.48). If we let

$$p = \frac{H}{\lambda} \tag{8.49}$$

then Eq. (8.41) is identical to Eq. (8.48). Here H is a pure constant of proportionality.

The rays of a monochromatic wave represented by $\delta \int_A^B (ds/\lambda)$ coincide exactly with the trajectories of the particle with fixed energy, E, determined by $\delta \int_A^B p\, ds$ because of Eq. (8.49). Thus far, the representation is purely geometrical and not kinematical. That is to say, the shapes of the trajectories of the particle are identical with the shapes of the rays, but the time rates at which these trajectories or rays are traversed are not yet related to each other.

In order to establish a complete description of the motion of a particle by the motion of a wave, we must (1) find a suitable wave representation of a single particle, and (2) establish the kinematical equivalence of a ray and a particle trajectory. A localized wave whose amplitude is zero everywhere except in a small region in space, a **wave packet**, will satisfy condition (1). We will indicate later how condition (2) can be satisfied.

A monochromatic plane wave in one dimension (the three-dimensional case is straightforward) may be represented by

$$\varphi_k(x, t) = \varphi(k)e^{i(kx-\omega t)} \tag{8.50}$$

where k is the x-component of the propagation vector, $\mathbf{k}$, for $|\mathbf{k}| = 2\pi/\lambda$ and $\omega = \omega(k) = 2\pi\nu$. The superposition of a group of plane waves of nearly the same wavelength and frequency that interfere destructively everywhere except in a small region results in a wave packet. In the one-dimensional case, such a wave packet may be represented in Fourier analysis by

$$\psi(x, t) = \frac{1}{\sqrt{2\pi}} \int_{-\infty}^{\infty} F(k)e^{i(kx-\omega t)} dk. \tag{8.51}$$

A maximum (constructive interference) will occur when $(kx - \omega t) = 0$ since the sum over the oscillating exponential function for different k values would result, on the average, in a flat pattern (destructive interference).

We assume that the form of $\psi(0, 0)$ is (see Fig. 8.3)

$$\psi(0, 0) = \frac{1}{\sqrt{2\pi}} \int_{k_0-(\Delta k/2)}^{k_0+(\Delta k/2)} F(k)\, dk. \tag{8.52}$$

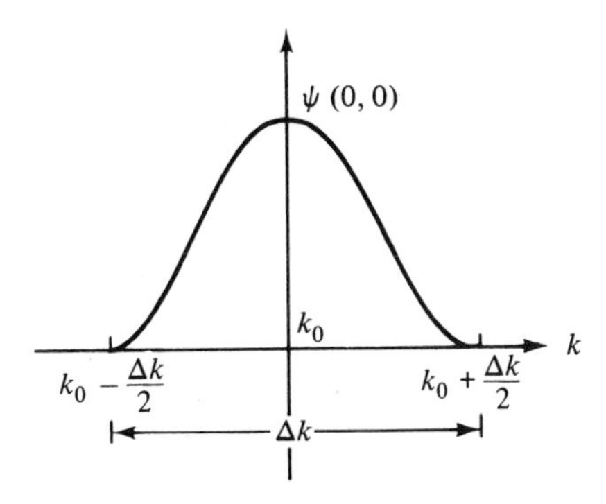

Figure 8.3

At a later time Δt and distance Δx, the form of ψ is

$$\begin{aligned}\psi(\Delta x, \Delta t) &= \frac{1}{\sqrt{2\pi}} \int_{k_0-(\Delta k/2)}^{k_0+(\Delta k/2)} F(k) e^{i(k\Delta x - \omega \Delta t)}\, dk \\ &= \frac{1}{\sqrt{2\pi}} e^{i(k_0\Delta x - \omega_0 \Delta t)} \int_{k_0-(\Delta k/2)}^{k_0+(\Delta k/2)} F(k) e^{i[(k-k_0)\Delta x - (\omega-\omega_0)\Delta t]}\, dk.\end{aligned} \tag{8.53}$$

For a maximum, we require that

$$i[(k - k_0)\, \Delta x - (\omega - \omega_0)\, \Delta t] = 0$$

or

$$\frac{\Delta x}{\Delta t} = \frac{\omega - \omega_0}{k - k_0}. \tag{8.54}$$

Expanding $\omega(k)$ in a Taylor series about k_0, we get

$$\omega = \omega_0 + (k - k_0)\left(\frac{d\omega}{dk}\right)_{\omega=\omega_0} + (k - k_0)^2 \left(\frac{d^2\omega}{dk^2}\right)_{\omega=\omega_0} + \cdots. \tag{8.55}$$

Neglecting $d^2\omega/dk^2$ and higher-order terms in the above expansion, we find that Eq. (8.54) becomes

$$\frac{\Delta x}{\Delta t} = \left(\frac{d\omega}{dk}\right)_{\omega=\omega_0}$$

or

$$v_{\substack{\text{group}\\ \text{velocity}}} = \lim_{\Delta t \to 0} \left(\frac{\Delta x}{\Delta t}\right) = \frac{d\omega}{dk}. \tag{8.56}$$

The establishment of the kinematical equivalence of a ray and a particle trajectory, condition (2) above, can be accomplished by requiring that the group velocity of the wave packet equal the velocity of the particle. In other words, we require that

$$v_g = v_p. \tag{8.57}$$

The velocity of the particle, v_p, is given by

$$\begin{aligned} v_p &= \frac{dE}{dp} \\ &= \frac{d(E/H)}{d(1/\lambda)} \end{aligned} \tag{8.58}$$

since

$$E = \frac{p^2}{2m} + V$$

and

$$\begin{aligned} p &= 2m[E - V] \\ &= \frac{H}{\lambda}. \end{aligned}$$

The group velocity of the wave, v_g, is given by

$$\begin{aligned} v_g &= \frac{d\omega}{dk} \\ &= \frac{d\omega}{d(2\pi/\lambda)} \\ &= \frac{d(\omega/2\pi)}{d(1/\lambda)}. \end{aligned} \tag{8.59}$$

Since we require that $v_p = v_g$, we must have

$$\begin{aligned} \frac{E}{H} &= \frac{\omega}{2\pi} \\ &= \nu \end{aligned}$$

or

$$E = H\nu. \tag{8.60}$$

We have established the fact that it is reasonable to consider describing the motion of a particle by use of a localized wave, wave packet, if we require that $E =$ constant ν. This is just Planck's quantization of energy condition where $H = h$.

Note that an alternate form (change of sign of the exponential term) of the Fourier transform is used. This form is the most often used form in quantum mechanics. Here the corresponding transform pairs are

$$\psi(k_x, t) = \frac{1}{\sqrt{2\pi}} \int_{-\infty}^{\infty} \psi(x, t) e^{-ik_x x} \, dx \tag{8.61}$$

$$= \frac{1}{\sqrt{2\pi}} \int_{-\infty}^{\infty} \phi(x) e^{-i(k_x x - \omega t)} \, dx \tag{8.62}$$

or

$$\psi(p_x, t) = \frac{1}{\sqrt{2\pi\hbar}} \int_{-\infty}^{\infty} \psi(x, t) e^{-ip_x x/\hbar} \, dx \tag{8.63}$$

$$= \frac{1}{\sqrt{2\pi\hbar}} \int_{-\infty}^{\infty} \phi(x) e^{-i(p_x x - Et)/\hbar} \, dx \tag{8.64}$$

and

$$\psi(x, t) = \frac{1}{\sqrt{2\pi}} \int_{-\infty}^{\infty} \psi(k_x, t) e^{ik_x x} \, dk_x \tag{8.65}$$

$$= \frac{1}{\sqrt{2\pi}} \int_{-\infty}^{\infty} \phi(k_x) e^{i(k_x x - \omega t)} \, dk_x \tag{8.66}$$

or

$$\psi(x, t) = \frac{1}{\sqrt{2\pi\hbar}} \int_{-\infty}^{\infty} \psi(p_x, t) e^{ip_x x/\hbar} \, dp_x \tag{8.67}$$

$$= \frac{1}{\sqrt{2\pi\hbar}} \int_{-\infty}^{\infty} \phi(p_x) e^{i(p_x x - Et)/\hbar} \, dp_x \tag{8.68}$$

where $p_x = \hbar k_x$ ($\mathbf{p} = \hbar\mathbf{k}$) and E is kinetic energy.

EXAMPLE 8.8 By use of Parseval's formula, show that

$$\int_{-\infty}^{\infty} \phi^*(p_x)\phi(p_x) \, dp_x = 1$$

if

$$\int_{-\infty}^{\infty} \psi^*(x)\psi(x) \, dx = 1.$$

Solution Here we write

$$\psi(x) = \frac{1}{\sqrt{2\pi\hbar}} \int_{-\infty}^{\infty} \phi(p_x) e^{ip_x x/\hbar} \, dp_x.$$

Parseval's formula,

$$\int_{-\infty}^{\infty} f(x)g^*(x)\,dx = \int_{-\infty}^{\infty} F(k)G^*(k)\,dk$$

when applied to ψ and ϕ, leads to

$$\int_{-\infty}^{\infty} \psi^*(x)\psi(x)\,dx = \int_{-\infty}^{\infty} \phi^*(p_x)\phi(p_x)\,dp_x = 1$$

since

$$\int_{-\infty}^{\infty} \psi^*(x)\psi(x)\,dx = 1.$$

8.4 PROBLEMS

8.1 Calculate the Laplace transform of $f(x) = e^{-x}$.

8.2 Calculate the Mellin transform of $f(x) = e^{-x}$.

8.3 (a) Calculate the Fourier transform (inverse transform) of

$$F(k) = \frac{N}{\sqrt{2\alpha}} e^{-k^2/4\alpha}.$$

(b) Sketch an appropriate diagram for $f(x)$ and $F(k)$.

8.4 Calculate the cosine and sine transforms of e^{-bx}, where b is a positive integer.

8.5 Calculate the cosine transform of $f(x) = e^{-x^2/2}$.

8.6 (a) Calculate the Fourier transform of $f(x)$ where

$$f(x) = \begin{cases} 1 & (\text{for } -a \le x \le a) \\ 0 & (\text{for } |x| > a). \end{cases}$$

(b) Sketch an appropriate diagram for $f(x)$ and $F(k)$.

8.7 (a) Calculate the Fourier transform, $F(\omega)$, of $f(t)$ where

$$f(t) = \begin{cases} e^{-i\omega_0 t} & \left(\text{for } -\frac{\tau_0}{2} \le t \le \frac{\tau_0}{2}\right) \\ 0 & \left(\text{for } |t| > \frac{\tau_0}{2}\right). \end{cases}$$

(b) Sketch $f(t)$, $F(\omega)$, and $|F(\omega)|^2$. The power spectrum is given by $|F(\omega)|^2$.

8.8 Show that the Fourier transform of

$$\psi(x) = \begin{cases} \sum_{n=-\infty}^{\infty} c_n e^{inx} & (\text{for } 0 \le x \le 2\pi) \\ 0 & (\text{otherwise}) \end{cases}$$

does not vanish outside of any finite interval.

8.9 Show that the convolution of $f(x)$ and $g(x)$ is commutative, $f * g = g * f$.

8.10 Calculate the convolution, $f * g$, of $f(x) = g(x) = e^{-|x|}$.

8.11 (a) By use of the sine transform method, solve the one-dimensional heat conduction equation,

$$\frac{\partial^2 T(x, t)}{\partial x^2} = \frac{1}{\sigma} \frac{\partial T(x, t)}{\partial t} \qquad (x > 0)$$

for $T(x, 0) = 0$ and $T(0, t) = T_0$; $T(x, t) \to 0$ and $\partial T(x, t)/\partial x \to 0$ as $x \to \infty$.

(b) For the solution, show that $T(x, t) \to T_0$ as $t \to \infty$.

8.12 By use of the cosine transform method, solve the one-dimensional heat conduction equation,

$$\frac{\partial^2 T(x, t)}{\partial x^2} = \frac{1}{\sigma} \frac{\partial T(x, t)}{\partial t} \qquad (x > 0)$$

for

$$T(x, 0) = 0 \quad \text{and} \quad \left.\frac{\partial T(x, t)}{\partial x}\right|_{x=0} = Q_0$$

$$T(x, t) \longrightarrow 0 \quad \text{and} \quad \frac{\partial T(x, t)}{\partial x} \longrightarrow 0 \qquad (\text{as } x \longrightarrow \infty).$$

8.13 Consider the steady flow of heat in a metal plate. In this case, the form of the heat equation is

$$\frac{\partial^2 T(x, y)}{\partial x^2} + \frac{\partial^2 T(x, y)}{\partial y^2} = 0.$$

By use of the Fourier transform method, find $T(x, y)$ for $T(x, 0) = \varphi(x)$; $T(x, y) \to 0$ and $\partial T(x, y)/\partial x \to 0$ as $x \to \pm\infty$.

8.14 By use of the Fourier transform method, solve the one-dimensional heat conduction equation,

$$\frac{\partial^2 T(x, t)}{\partial x^2} = \frac{1}{\sigma} \frac{\partial T(x, t)}{\partial t}$$

for $T(x, t) \to 0$ and $\partial T(x, t)/\partial x \to 0$ as $x \to \pm\infty$; $T(x, 0) = e^{-a^2x^2}$, where a is a positive constant.

8.15 By use of the Fourier transform method, solve the infinite string problem when $u(x, t) \to 0$ and $\partial u(x, t)/\partial x \to 0$ as $x \to \pm\infty$;

$$u(x, 0) = f(x) \quad \text{and} \quad \left.\frac{\partial u(x, t)}{\partial t}\right|_{t=0} = g(x).$$

8.16 (a) By use of the Fourier transform method, solve the one-dimensional Schrödinger wave equation for a free particle,

$$i\hbar \frac{\partial \psi(x, t)}{\partial t} = -\frac{\hbar^2}{2m} \frac{\partial^2 \psi(x, t)}{\partial x^2}.$$

(b) Write the solution of Part (a) in the form

$$\psi(x, t) = \frac{1}{\sqrt{2\pi\hbar}} \int_{-\infty}^{\infty} \phi(p_x)\, e^{i(p_x x - p_x^2 t/2m)/\hbar}\, dp_x$$

where $\psi(p_x, 0) = \phi(p_x)$.

(c) For $\psi(x, 0) = e^{-x^2/2\alpha^2}$ calculate $\phi(p_x)$.

(d) Obtain the explicit expression for $\psi(x, t)$. The real quantity $\psi^*\psi = |\psi|^2$ has physical significance in quantum mechanics.

(e) Obtain the explicit expression for $|\psi|^2$ and show that the width of this Gaussian form increases with time, that is, the wave packet spreads in x-space as time increases.

9

Tensor Analysis

9.1 INTRODUCTION

It is a fundamental postulate of physics that the laws of nature be expressed by equations that are valid for all coordinate systems (**reference frames**). Alternately, we say that the laws of nature are **covariant**, which means that they have the same forms in all coordinate systems. A systematic method of investigating the behavior of quantities that undergo a coordinate transformation is the subject matter of tensor analysis.

The chief purpose of this chapter is the development, in an introductory manner, of the essentials of tensor algebra and tensor calculus that are needed for an understanding of a tensorial presentation of mechanics, electromagnetic theory, and special and general relativity. (Tensor analysis is indispensable in a mathematical treatment of general relativity.)

In developing the mathematical subject of absolute differential geometry, Gauss, Riemann, and Christoffel (1829–1900) introduced the concept of a tensor. The subject of absolute tensor calculus (tensor analysis) was introduced and developed by Ricci (1853–1925) and his student Levi-Civita (1872–1941). Einstein made extensive use of tensor calculus in his formulation of the general theory of relativity.

A **tensor** consists of a set of quantities called **components** whose properties are independent of the coordinate system used to describe them. The components of a tensor in two different coordinate systems are related by the characteristic transformation laws discussed below.

9.1.1 Notations

A collection of indices (subscripts and/or superscripts) is used to make the mathematical development of tensor analysis compact. The superscripts, contravariant indices, are used to denote the contravariant components of a tensor, $T^{ij\cdots}$. The subscripts, covariant indices, are used to represent the covariant components of a tensor, $T_{ij\ldots}$. The components of a mixed tensor are specified by indicating both subscripts and superscripts, $T^{ij\cdots}_{lmn\cdots}$. Throughout this chapter, we will use this notation (letters plus indices) to denote the components of a tensor or the tensor itself.

9.1.2 The Rank and Number of Components of a Tensor

The **rank (order)** of a tensor is the number (without counting an index which appears once as a subscript and once as a superscript) of indices in the symbol representing a tensor (or the components of a tensor). For example,

A is a tensor of rank zero (scalar)
B^i is a contravariant tensor of rank one (vector)
C_k is a covariant tensor of rank one (vector)
D_{ij} is a covariant tensor of rank two
T^{il}_{jki} is a mixed tensor of rank three.

In an n-dimensional space, the number of components of a tensor of rank r is n^r.

9.2 TRANSFORMATION OF COORDINATES IN LINEAR SPACES

Consider an ordered set of n mutually independent real variables, $x^1, x^2, \ldots, x^n = \{x^i\}$, called the **coordinates of a point**. The collection of all such points corresponding to all the sets of values of $\{x^i\}$ forms an n-**dimensional linear space** which we specify by V_n. The set of n equations

$$\bar{x}^i = \phi^i(x^1, x^2, \ldots, x^n) \qquad (i = 1, 2, \ldots, n) \tag{9.1}$$

defines a new coordinate system specified by the mutually independent variables, $\bar{x}^1, \bar{x}^2, \ldots, \bar{x}^n$. The ϕ^i are assumed to be single-valued real functions of the coordinates and possess continuous partial derivatives. For convenience and clarity of notation, the symbol $\bar{x}^i$ is used to represent the new coordinate system instead of the traditional symbol x'^i.

On differentiating Eq. (9.1) with respect to x^j, we obtain the following representation for an infinitesimal displacement in the original coordinate system, x^j, in terms of the new coordinate system, $\bar{x}^i$:

$$\begin{aligned} d\bar{x}^i &= \sum_{j=1}^{n} \frac{\partial \phi^i}{\partial x^j} dx^j \\ &= \sum_{j=1}^{n} \frac{\partial \bar{x}^i}{\partial x^j} dx^j \\ &= \frac{\partial \bar{x}^i}{\partial x^j} dx^j. \end{aligned} \tag{9.2}$$

The **Einstein summation convention**—when the same letter, index, appears once as a superscript and once as a subscript, it is understood that a summation is to occur on this index—is used in Eq. (9.2). Note that the j index in the denominator of Eq. (9.2) is considered as a subscript, and a summation over j is understood.

If

$$\det\left|\frac{\partial \bar{x}^i}{\partial x^j}\right| = \begin{vmatrix} \dfrac{\partial \bar{x}^1}{\partial x^1} & \cdots & \dfrac{\partial \bar{x}^1}{\partial x^n} \\ \vdots & & \vdots \\ \dfrac{\partial \bar{x}^n}{\partial x^1} & \cdots & \dfrac{\partial \bar{x}^n}{\partial x^n} \end{vmatrix} \neq 0 \tag{9.3}$$

then the inverse transformation in Eq. (9.2) exists (see Section 2.8 of Chapter 2). The determinant in Eq. (9.3) is called the **Jacobian** (Jacobi, 1804–1851) of the transformation.

EXAMPLE 9.1 Find the Jacobian of the transformation for rotation in two dimensions (see Section 2.9 of Chapter 2).

Solution In this case, we have

$$\bar{x}^1 = x^1 \cos\theta + x^2 \sin\theta \quad \text{and} \quad \bar{x}^2 = x^2 \cos\theta - x^1 \sin\theta.$$

The Jacobian of the transformation is given by

$$\det\left|\frac{\partial \bar{x}^i}{\partial x^j}\right| = \begin{vmatrix} \dfrac{\partial \bar{x}^1}{\partial x^1} & \dfrac{\partial \bar{x}^1}{\partial x^2} \\ \dfrac{\partial \bar{x}^2}{\partial x^1} & \dfrac{\partial \bar{x}^2}{\partial x^2} \end{vmatrix} = \begin{vmatrix} \cos\theta & \sin\theta \\ -\sin\theta & \cos\theta \end{vmatrix}$$

which is just the determinant of the rotation matrix.

9.3 CONTRAVARIANT AND COVARIANT TENSORS

In this section, we discuss the characteristic transformation laws for the components of tensors.

9.3.1 First-Rank Tensors (Vectors)

A set of components, A^j, which transforms according to Eq. (9.2),

$$\bar{A}^i = \frac{\partial \bar{x}^i}{\partial x^j} A^j \tag{9.4}$$

forms the components of a contravariant tensor of rank one. In Eq. (9.4), $A^j = A^j(x^j)$ and $\bar{A}^i = \bar{A}^i(\bar{x}^i)$, where x^j and $\bar{x}^i$ refer to the old and new coordinate systems, respectively. Equation (9.4) may be solved for A^j if it is multiplied by $\partial x^k/\partial \bar{x}^i$ and summed over i. In this case, we obtain

$$\begin{aligned}\frac{\partial x^k}{\partial \bar{x}^i} \bar{A}^i &= \frac{\partial x^k}{\partial \bar{x}^i} \frac{\partial \bar{x}^i}{\partial x^j} A^j \\ &= \frac{\partial x^k}{\partial x^j} A^j \\ &= \delta^k_j A^j \\ &= A^k\end{aligned}$$

or

$$A^j = \frac{\partial x^j}{\partial \bar{x}^i} \bar{A}^i \tag{9.5}$$

where the Kronecker delta is given by

$$\delta^k_j = \begin{cases} 1 & (j = k) \\ 0 & (j \neq k). \end{cases} \tag{9.6}$$

A shortcut for obtaining Eq. (9.5) from Eq. (9.4) is to remove the bars from originally barred symbols and place bars on originally unbarred symbols.

A set of quantities B_k is called the components of a covariant tensor of rank one if

$$\begin{aligned}A^k B_k &= \text{invariant (a scalar)} \\ &= \bar{A}^k \bar{B}_k\end{aligned} \tag{9.7}$$

for an arbitrary contravariant tensor with components A^k. If A^k from Eq. (9.5) is substituted into Eq. (9.7), the result is

$$\bar{A}^k \bar{B}_k = A^k B_k$$
$$= \frac{\partial x^k}{\partial \bar{x}^i} \bar{A}^i B_k$$
$$= \frac{\partial x^\alpha}{\partial \bar{x}^k} \bar{A}^k B_\alpha$$

or

$$\bar{A}^k \left[\bar{B}_k - \frac{\partial x^\alpha}{\partial \bar{x}^k} B_\alpha \right] = 0$$

or

$$\bar{B}_k = \frac{\partial x^\alpha}{\partial \bar{x}^k} B_\alpha \tag{9.8}$$

since the $\bar{A}^k$ are arbitrary. Equation (9.8) is the transformation law for the components of a covariant tensor of rank one.

A mnemonic scheme for remembering the transformation laws for contravariant and covariant components with reference to the placement of $\partial \bar{x}^i$ is "CO BELOW" for covariant components.

9.3.2 Higher-Rank Tensors

If the n^2 quantities, A^{kl}, transform according to

$$\bar{A}^{ij} = \frac{\partial \bar{x}^i}{\partial x^k} \frac{\partial \bar{x}^j}{\partial x^l} A^{kl} \tag{9.9}$$

then A^{kl} are the components of a second-rank contravariant tensor. If the n^2 quantities, A_{kl}, transform according to

$$\bar{A}_{ij} = \frac{\partial x^k}{\partial \bar{x}^i} \frac{\partial x^l}{\partial \bar{x}^j} A_{kl} \tag{9.10}$$

then the A_{kl} are the components of a second-rank covariant tensor. Similarly,

$$\bar{A}^i_j = \frac{\partial \bar{x}^i}{\partial x^k} \frac{\partial x^l}{\partial \bar{x}^j} A^k_l \tag{9.11}$$

means that A^k_l are the components of a second-rank mixed tensor.

A set of n^{s+p} quantities,

$$A^{t_1 \cdots t_s}_{g_1 \cdots g_p}$$

is said to be the components of a mixed tensor whose rank is $(s + p)$ if they transform according to

$$\bar{A}^{u_1 \cdots u_s}_{r_1 \cdots r_p} = \frac{\partial \bar{x}^{u_1}}{\partial x^{t_1}} \cdots \frac{\partial \bar{x}^{u_s}}{\partial x^{t_s}} \frac{\partial x^{g_1}}{\partial \bar{x}^{r_1}} \cdots \frac{\partial x^{g_p}}{\partial \bar{x}^{r_p}} A^{t_1 \cdots t_s}_{g_1 \cdots g_p}. \tag{9.12}$$

If the components of a tensor are zero in one coordinate system, it follows from the characteristic transformation law, Eq. (9.12), that they are zero in all coordinate systems. This property is extremely important when we discuss physical quantities.

9.3.3 Symmetric and Antisymmetric Tensors

If

$$A^k_{rs} = A^k_{sr} \qquad (A^{rs}_k = A^{sr}_k)$$

or (9.13)

$$B^k_{rs} = -B^k_{sr} \qquad (B^{rs}_k = -B^{sr}_k)$$

then A^k_{rs} (or A^{rs}_k) and B^k_{rs} (or B^{rs}_k) are said to be **symmetric** and **antisymmetric** (skew-symmetric), respectively, in the r and s indices. The symmetric (or antisymmetric) property is conserved under a transformation of coordinates and may be extended to higher-rank tensors for any two contravariant or two covariant indices.

9.3.4 Polar and Axial Vectors

We are now at a stage of development where a technical discussion of vectors is appropriate.

It is important to understand that components of vectors as used in Chapter 1 referred to the projections of vectors on the three coordinate axes in a right-hand Cartesian system. For clarity, we will call such components **ordinary components** of vectors. By contrast, contravariant and covariant components of vectors as developed in this chapter are not restricted to a specific coordinate system. It will be shown in Section 9.5.2 that ordinary, contravariant, and covariant components of vectors are identical in Cartesian coordinates.

Vectors whose components are completely specified by

$$\bar{A}^i = \frac{\partial \bar{x}^i}{\partial x^j} A^j$$

or

$$\bar{A}_i = \frac{\partial x^j}{\partial \bar{x}^i} A_j$$

are called **polar** (ordinary, true, or proper) vectors, and they are employed to represent quantities such as displacements and mechanical forces. In Chapter 1, we used a directed line segment to represent these vectors.

In Cartesian coordinates, we may obtain a vector **T** from the cross product of two polar vectors **A** and **B**;

$$\mathbf{T} = \mathbf{A} \times \mathbf{B}.$$

The components of **T** may be written as

$$\begin{aligned} T_{ij} &= A_i B_j - A_j B_i \\ &= -T_{ji}. \end{aligned}$$

That is to say, the components of a cross product of two polar vectors transform like the components of a second-rank antisymmetric tensor. Vectors like **T** are called **axial** (or pseudo) vectors and are used to represent directions of rotation such as angular velocity and torque (moment). Geometrically, axial vectors correspond to areas.

9.4 TENSOR ALGEBRA

9.4.1 Addition (Subtraction)

As is clear from the basic transformation law in Eq. (9.12), two tensors of the same type (having the same number of covariant and contravariant indices) can be added (or subtracted) to produce a single tensor. For example,

$$C^{j_1 \cdots j_s}_{i_1 \cdots i_p} = A^{j_1 \cdots j_s}_{i_1 \cdots i_p} \pm B^{j_1 \cdots j_s}_{i_1 \cdots i_p}. \tag{9.14}$$

9.4.2 Multiplication (Outer Product)

The **outer product** of two tensors with components $A^{j_1 \cdots j_s}_{i_1 \cdots i_p}$ and $B^{k_1 \cdots k_\mu}_{l_1 \cdots l_\nu}$ is defined by

$$C^{j_1 \cdots j_s k_1 \cdots k_\mu}_{i_1 \cdots i_p l_1 \cdots l_\nu} = A^{j_1 \cdots j_s}_{i_1 \cdots i_p} B^{k_1 \cdots k_\mu}_{l_1 \cdots l_\nu}. \tag{9.15}$$

The division of a tensor of rank greater than zero by another tensor of rank greater than zero is not uniquely defined.

9.4.3 Contraction

The operation of **contraction** is the process by which the number of covariant and contravariant indices of a mixed tensor is reduced by one. For example, consider the contraction of a mixed tensor with components A^{ij}_{klm},

$$B^{j}_{lm} = A^{ij}_{ilm}. \tag{9.16}$$

Here the components B^{j}_{lm} are obtained from the components A^{ij}_{klm} by contracting the indices i and k ($i = k$). A summation on i is understood in Eq. (9.16). In general, the contraction operation enables a mixed tensor of rank $r - 2$ to be obtained from a mixed tensor of rank r. Any index of the contravariant set and any index of the covariant set may be used to form the components of the new tensor.

9.4.4 Inner Product

The process of combining outer multiplication and contraction to produce a new tensor is called **inner multiplication**, and the resulting tensor is called the **inner product** of the two tensors involved. For example, we may write

$$C^{ij}_{mnt} = A^{ij}_{k} B^{k}_{mnt} \tag{9.17}$$

for the multiplication (outer product) and contraction of two tensors with components A^{ij}_{k} and B^{l}_{mnt}.

9.4.5 Quotient Law

The **quotient law** is a simple indirect test which can be used to ascertain whether a set of quantities forms the components of a tensor. (The direct method is to determine if the quantities satisfy the appropriate transformation equation.)

QUOTIENT LAW If the product (outer or inner) of $A^{i_1 \cdots i_s}_{g_1 \cdots g_k}$ with the components of an arbitrary tensor yields a nonzero tensor, then the quantities $A^{i_1 \cdots i_s}_{g_1 \cdots g_k}$ are the components of a tensor. To illustrate the quotient law, consider the specific inner product

$$C^{pk} = A^{ijk} B^{p}_{ij}. \tag{9.18}$$

The quotient law states that the A^{ijk} quantities form the components of a contravariant tensor if the B^{p}_{ij} are the components of an arbitrary tensor and the C^{pk} are the components of a tensor.

In the $\bar{x}^i$ coordinate system, Eq. (9.18) becomes

$$\bar{C}^{pk} = \bar{A}^{ijk} \bar{B}^{p}_{ij} \tag{9.19}$$

or

$$\frac{\partial \bar{x}^p}{\partial x^l} \frac{\partial \bar{x}^k}{\partial x^m} C^{lm} = \bar{A}^{ijk} \frac{\partial \bar{x}^p}{\partial x^r} \frac{\partial x^s}{\partial \bar{x}^i} \frac{\partial x^t}{\partial \bar{x}^j} B^{r}_{st}. \tag{9.20}$$

Substituting Eq. (9.18) into Eq. (9.20) and making an appropriate change of indices, we obtain

$$\frac{\partial \bar{x}^p}{\partial x^l}\left[\bar{A}^{ijk}\frac{\partial x^s}{\partial \bar{x}^i}\frac{\partial x^t}{\partial \bar{x}^j} - \frac{\partial \bar{x}^k}{\partial x^m}A^{stm}\right]B^l_{st} = 0. \tag{9.21}$$

Multiplying Eq. (9.21) by $\partial x^a/\partial \bar{x}^p$ and summing over p, we get

$$\left[\bar{A}^{ijk}\frac{\partial x^s}{\partial \bar{x}^i}\frac{\partial x^t}{\partial \bar{x}^j} - \frac{\partial \bar{x}^k}{\partial x^m}A^{stm}\right]B^a_{st} = 0. \tag{9.22}$$

Since Eq. (9.22) is valid for arbitrary B^a_{st}, we require that

$$\bar{A}^{ijk}\frac{\partial x^s}{\partial \bar{x}^i}\frac{\partial x^t}{\partial \bar{x}^j} = \frac{\partial \bar{x}^k}{\partial x^m}A^{stm}$$

or

$$\bar{A}^{rgk} = \frac{\partial \bar{x}^r}{\partial x^s}\frac{\partial \bar{x}^g}{\partial x^t}\frac{\partial \bar{x}^k}{\partial x^m}A^{stm}. \tag{9.23}$$

From Eq. (9.23), we see that the A^{stm} form the components of a third-rank contravariant tensor.

For an arbitrary tensor with components A_ν, we may form the inner product

$$\delta^\nu_\mu A_\nu = A_\mu. \tag{9.24}$$

By use of the fundamental transformation law for the components of a tensor, we see that the quantities $\delta^\nu_\mu A_\nu$ transform like the components of a tensor; hence A_μ are the components of a tensor and δ^ν_μ are the components of a second-rank mixed tensor by use of the quotient law. The transformation equation for δ^ν_μ is

$$\begin{aligned}\bar{\delta}^\nu_\mu &= \frac{\partial \bar{x}^\nu}{\partial x^\alpha}\frac{\partial x^\beta}{\partial \bar{x}^\mu}\delta^\alpha_\beta \\ &= \frac{\partial \bar{x}^\nu}{\partial x^\alpha}\frac{\partial x^\alpha}{\partial \bar{x}^\mu} \\ &= \frac{\partial \bar{x}^\nu}{\partial \bar{x}^\mu} \\ &= \delta^\nu_\mu.\end{aligned} \tag{9.25}$$

Therefore the components of δ^ν_μ have the same values in all coordinate systems.

9.5 THE LINE ELEMENT

9.5.1 The Fundamental Metric Tensor

The generalized form of the element of length (arc length or interval), ds, between two coordinate points x^i and $x^i + dx^i$ as defined by Riemann is given by the following quadratic differential form:

$$ds^2 = g_{ij}\, dx^i\, dx^j \tag{9.26}$$

where it is assumed that (1) the g_{ij} are functions of the x^i, (2) $g_{ij} = g_{ji}$ (symmetric), and (3) det $|g_{ij}| \neq 0$. In this case, the space is called a **Riemannian space**, and the quadratic form $g_{ij}\, dx^i\, dx^j$ is called the **metric.**

If rectangular Cartesian coordinates are introduced in a Euclidean space, we have $g_{ij} = \delta_{ij}$; hence

$$\begin{aligned} ds^2 &= g_{ij}\, dx^i\, dx^j \\ &= \delta_{ij}\, dx^i\, dx^j \\ &= \sum_i^3 (dx^i)^2 = dx^2 + dy^2 + dz^2 \end{aligned}$$

which is the familar form of ds^2.

It is required that ds be independent of the system of coordinates. Hence we may write

$$\begin{aligned} \bar{g}_{\mu\nu}\, d\bar{x}^\mu\, d\bar{x}^\nu &= g_{\mu\nu}\, dx^\mu\, dx^\nu \\ &= g_{\mu\nu} \frac{\partial x^\mu}{\partial \bar{x}^\alpha}\, d\bar{x}^\alpha \frac{\partial x^\nu}{\partial \bar{x}^\beta}\, d\bar{x}^\beta \\ &= g_{\mu\nu} \frac{\partial x^\mu}{\partial \bar{x}^\alpha} \frac{\partial x^\nu}{\partial \bar{x}^\beta}\, d\bar{x}^\alpha\, d\bar{x}^\beta \end{aligned}$$

or

$$\left[\bar{g}_{\alpha\beta} - g_{\mu\nu} \frac{\partial x^\mu}{\partial \bar{x}^\alpha} \frac{\partial x^\nu}{\partial \bar{x}^\beta}\right] d\bar{x}^\alpha\, d\bar{x}^\beta = 0.$$

Since the differentials in the above equation are arbitrary, we require that

$$\bar{g}_{\alpha\beta} = \frac{\partial x^\mu}{\partial \bar{x}^\alpha} \frac{\partial x^\nu}{\partial \bar{x}^\beta}\, g_{\mu\nu}. \tag{9.27}$$

Hence the components of $g_{\mu\nu}$ transform like the components of a second-rank covariant tensor, and $g_{\mu\nu}$ are called the components of the **fundamental** (or metric) **tensor**. (We say that $g_{\mu\nu}$ is the fundamental metric tensor.)

9.5.2 Associate Tensors

The **tensor associate** to an arbitrary tensor A^j is the result of the inner product of A^j and the fundamental tensor, g_{ij}, that is,

$$A_i = g_{ij}A^j \tag{9.28}$$

where A_i, which is a new form (covariant) of the old tensor A^j, is called the associate to A^j. The process in Eq. (9.28) is called **lowering** the superscript.

We now show that A_i in Eq. (9.28) is just a new form of A^j. Note that

$$\begin{aligned}
A_i &= g_{ij}A^j \\
&= \left(\frac{\partial \bar{x}^\alpha}{\partial x^i}\frac{\partial \bar{x}^\beta}{\partial x^j}\bar{g}_{\alpha\beta}\right)\left(\frac{\partial x^j}{\partial \bar{x}^l}\bar{A}^l\right) \\
&= \frac{\partial \bar{x}^\alpha}{\partial x^i}\frac{\partial \bar{x}^\beta}{\partial \bar{x}^l}\bar{g}_{\alpha\beta}\bar{A}^l \\
&= \frac{\partial \bar{x}^\alpha}{\partial x^i}\delta^\beta_l\bar{g}_{\alpha\beta}\bar{A}^l \\
&= \frac{\partial \bar{x}^\alpha}{\partial x^i}\bar{g}_{\alpha\beta}\bar{A}^\beta \\
&= \frac{\partial \bar{x}^\alpha}{\partial x^i}\bar{A}_\alpha
\end{aligned}$$

or

$$\bar{A}_i = \frac{\partial x^\alpha}{\partial \bar{x}^i}A_\alpha$$

which is the transformation law for a covariant vector. Hence assuming the A_i in Eq. (9.28) is just a new form for A^i is consistent with the required transformation law for the components of a tensor.

In Cartesian coordinates, $g_{ij} = 0$ for $i \neq j$ and $g_{11} = g_{22} = g_{33} = 1$. Therefore the contravariant components of a vector (tensor of rank one) are identical to the covariant components as can be seen from Eq. (9.28).

If Eq. (9.28) is solved for A^j, we obtain

$$A^j = g^{ij}A_i \tag{9.29}$$

where

$$g^{ij} = \frac{(g_{ij})^{CT}}{g} \qquad (\text{for } g = \det|g_{ij}|). \tag{9.30}$$

Here $(g_{ij})^{CT}$ means the cofactor transpose of the matrix g_{ij}. On applying the quotient law in Eq. (9.29), we see that the quantities g^{ij} form the components of a second-rank contravariant tensor. The process in Eq. (9.29) is known as

raising the subscript. The tensor A^j is the associate to A_i, and g^{ij} is called the **reciprocal tensor** to g_{ij}.

Multiplying Eq. (9.29) by g_{kj} and summing over j, we get

$$g_{kj}A^j = A_k$$
$$= g_{kj}g^{ij}A_i$$

or

$$g_{kj}g^{ij} = \delta^i_k. \tag{9.31}$$

The process of lowering and raising indices may be performed on higher-rank tensors. For example,

$$g_{ri}A^{ijk}_{lmn} = A^{\cdot jk}_{rlmn}. \tag{9.32}$$

The dot notation is introduced to indicate which index has been lowered (or raised).

9.6 TENSOR CALCULUS

Consider the invariant $\bar{\phi}(\bar{x}) = \phi(x)$; differentiating both sides of this equation with respect to $\bar{x}^i$, we obtain

$$\frac{\partial\bar{\phi}}{\partial\bar{x}^i} = \frac{\partial x^j}{\partial\bar{x}^i}\frac{\partial\phi}{\partial x^j}. \tag{9.33}$$

In Eq. (9.33), we see that the $\partial\phi/\partial x^j$ are the components of a first-rank covariant tensor. The above differentiation enabled us to generate a tensor of rank one from a tensor of rank zero. If we try to extend this process to obtain a tensor of rank two by differentiating a tensor of rank one

$$A^\beta = \frac{\partial x^\beta}{\partial\bar{x}^n}\bar{A}^n \tag{9.34}$$

we find that

$$\frac{\partial A^\beta}{\partial x^\alpha} = \frac{\partial x^\beta}{\partial\bar{x}^n}\frac{\partial\bar{x}^m}{\partial x^\alpha}\frac{\partial\bar{A}^n}{\partial\bar{x}^m} + \frac{\partial^2 x^\beta}{\partial x^\alpha\partial\bar{x}^n}\bar{A}^n. \tag{9.35}$$

In this latter case, we see that the presence of the second term in Eq. (9.35) means that $\partial\bar{A}^n/\partial\bar{x}^m$ do not form the components of a tensor.

We now develop a scheme for a new kind of derivative, the **covariant derivative**, which enables us to obtain tensors from the differentiation of other tensors by expressing the second term in Eq. (9.35) in terms of the Christoffel symbols. The process of developing a scheme such that the derivative of a tensor always leads to a tensor is the chief aim of tensor calculus. We begin this process by first discussing the two Christoffel symbols since they are directly involved in this development.

9.6.1 Christoffel Symbols

The **Christoffel symbols** of the first and second kinds are defined, respectively, by

$$[ij, k] = \frac{1}{2}\left(\frac{\partial g_{ik}}{\partial x^j} + \frac{\partial g_{jk}}{\partial x^i} - \frac{\partial g_{ij}}{\partial x^k}\right) \tag{9.36}$$

and

$$\left\{ {l \atop ij} \right\} = g^{lk}[ij, k] \tag{9.37}$$

where $i, j,$ and k are regarded as subscripts and l is to be considered as a superscript when the summation convention is used. The symbol $\mathbf{\Gamma}^l_{ij}$ is used by many authors to denote the Christoffel symbol of the second kind; when this notation is used, one must be aware of the fact that $\mathbf{\Gamma}^l_{ij}$ is not always a tensor as suggested by the symbol.

From the above definitions, we see that the two Christoffel symbols are symmetric with respect to the i and j indices. Hence we may write

$$[ij, k] = [ji, k] \tag{9.38}$$

and

$$\left\{ {l \atop ij} \right\} = \left\{ {l \atop ji} \right\}. \tag{9.39}$$

Also note that

$$\begin{aligned} g_{lm}\left\{ {l \atop ij} \right\} &= g_{lm}g^{lk}[ij, k] \\ &= \delta^k_m\,[ij, k] \\ &= [ij, m]. \end{aligned} \tag{9.40}$$

In the second step above, we made use of inner multiplication.

9.6.2 Covariant Differentiation of Tensors

By use of Eq. (A9.7), the second term in Eq. (9.35) may be written as

$$\bar{A}^n \frac{\partial \bar{x}^s}{\partial x^\alpha}\frac{\partial^2 x^\beta}{\partial \bar{x}^s\,\partial \bar{x}^n} = \bar{A}^n \frac{\partial \bar{x}^s}{\partial x^\alpha}\left(\left\{ {p \atop sn} \right\}\frac{\partial x^\beta}{\partial \bar{x}^p} - \left\{ {\beta \atop ij} \right\}\frac{\partial x^i}{\partial \bar{x}^s}\frac{\partial x^j}{\partial \bar{x}^n}\right). \tag{9.41}$$

Substituting Eq. (9.41) into Eq. (9.35), we find that

$$\frac{\partial A^\beta}{\partial x^\alpha} = \frac{\partial x^\beta}{\partial \bar{x}^n}\frac{\partial \bar{x}^m}{\partial x^\alpha}\frac{\partial \bar{A}^n}{\partial \bar{x}^m} + \bar{A}^n \frac{\partial \bar{x}^s}{\partial x^\alpha}\left(\overline{\left\{\begin{matrix} p \\ sn \end{matrix}\right\}}\frac{\partial x^\beta}{\partial \bar{x}^p} - \left\{\begin{matrix} \beta \\ ij \end{matrix}\right\}\frac{\partial x^i}{\partial \bar{x}^s}\frac{\partial x^j}{\partial \bar{x}^n}\right)$$

or

$$\frac{\partial A^\beta}{\partial x^\alpha} + \bar{A}^n \frac{\partial \bar{x}^s}{\partial x^\alpha}\frac{\partial x^i}{\partial \bar{x}^s}\frac{\partial x^j}{\partial \bar{x}^n}\left\{\begin{matrix} \beta \\ ij \end{matrix}\right\} = \frac{\partial x^\beta}{\partial \bar{x}^n}\frac{\partial \bar{x}^m}{\partial x^\alpha}\frac{\partial \bar{A}^n}{\partial \bar{x}^m} + \bar{A}^n \frac{\partial \bar{x}^s}{\partial x^\alpha}\frac{\partial x^\beta}{\partial \bar{x}^p}\overline{\left\{\begin{matrix} p \\ sn \end{matrix}\right\}}.$$

The above equation reduces to

$$\frac{\partial A^\beta}{\partial x^\alpha} + \left\{\begin{matrix} \beta \\ \alpha j \end{matrix}\right\} A^j = \left(\frac{\partial \bar{A}^n}{\partial \bar{x}^m} + \overline{\left\{\begin{matrix} n \\ mp \end{matrix}\right\}} \bar{A}^p\right)\frac{\partial x^\beta}{\partial \bar{x}^n}\frac{\partial \bar{x}^m}{\partial x^\alpha}. \tag{9.42}$$

Introducing the notation $A^\beta_{,\alpha}$ which is defined as

$$A^\beta_{,\alpha} = \frac{\partial A^\beta}{\partial x^\alpha} + \left\{\begin{matrix} \beta \\ \alpha j \end{matrix}\right\} A^j \tag{9.43}$$

we see that Eq. (9.42) becomes

$$A^\beta_{,\alpha} = \frac{\partial x^\beta}{\partial \bar{x}^n}\frac{\partial \bar{x}^m}{\partial x^\alpha}\bar{A}^n_{,m}. \tag{9.44}$$

Equation (9.44) has the same form as the basic transformation law for the components of a tensor, Eq. (9.12). Hence $A^\beta_{,\alpha}$ is a second-rank mixed tensor, and it is called the **covariant derivative** of A^β with respect to x^α.

By use of a similar procedure, we find that the **covariant derivative** of B_β with respect to x^α is given by

$$\bar{B}_{\beta,\alpha} = \frac{\partial x^i}{\partial \bar{x}^\beta}\frac{\partial x^k}{\partial \bar{x}^\alpha}B_{i,k} \tag{9.45}$$

where

$$B_{i,k} = \frac{\partial B_i}{\partial x^k} - \left\{\begin{matrix} j \\ ik \end{matrix}\right\} B_j. \tag{9.46}$$

The above relations for the covariant derivatives of a tensor may be extended in a natural way to the case of a general mixed tensor. Consider, for example, the tensor $T^{j_1\cdots j_s}_{i_1\cdots i_p}$; here we have

$$\begin{aligned} T^{j_1\cdots j_s}_{i_1\cdots i_p,l} = \frac{\partial A^{j_1\cdots j_s}_{i_1\cdots i_p}}{\partial x^l} &- \left\{\begin{matrix} \alpha \\ i_1 l \end{matrix}\right\} T^{j_1\cdots j_s}_{\alpha i_2\cdots i_p} - \left\{\begin{matrix} \alpha \\ i_2 l \end{matrix}\right\} T^{j_1\cdots j_s}_{i_1\alpha i_3\cdots i_p} - \cdots - \left\{\begin{matrix} \alpha \\ i_p l \end{matrix}\right\} T^{j_1\cdots j_s}_{i_1\cdots \alpha} \\ &+ \left\{\begin{matrix} j_1 \\ \alpha l \end{matrix}\right\} T^{\alpha j_2\cdots j_s}_{i_1\cdots i_p} + \left\{\begin{matrix} j_2 \\ \alpha l \end{matrix}\right\} T^{j_1\alpha j_3\cdots j_s}_{i_1\cdots i_p} + \cdots + \left\{\begin{matrix} j_s \\ \alpha l \end{matrix}\right\} T^{j_1\cdots \alpha}_{i_1\cdots i_p}. \end{aligned} \tag{9.47}$$

The covariant derivative of a tensor of rank zero (a scalar) is defined to be the same as the ordinary derivative, that is,

$$A_{,l} = \frac{\partial A}{\partial x^l}. \tag{9.48}$$

EXAMPLE 9.2 By use of the electromagnetic field tensor, develop the tensor form of Maxwell's equations. (Consider the case for a vacuum.)

Solution In rationalized MKS units, Maxwell's equations (the foundation of electromagnetic theory) in free space are

$$\nabla \cdot \mathbf{D} = \rho \qquad \nabla \times \mathbf{E} = -\frac{\partial \mathbf{B}}{\partial t}$$

$$\nabla \cdot \mathbf{B} = 0 \qquad \nabla \times \mathbf{H} = \mathbf{J} + \frac{\partial \mathbf{D}}{\partial t}$$

where $\mathbf{B} = \mu_0 \mathbf{H}$ and $\mathbf{D} = \epsilon_0 \mathbf{E}$; the quantities $\mu_0 = 4\pi \times 10^{-7}\ kg \cdot m/C^2$ and $\epsilon_0 = 8.854 \times 10^{-12}\ C^2/N \cdot m^2$ are the permeability and permittivity of the vacuum, respectively; the quantities ρ and $\mathbf{J}$ are the charge density and current density, respectively; the quantities $\mathbf{D}$ and $\mathbf{B}$ are the electric displacement and magnetic induction, respectively; the quantities $\mathbf{E}$ and $\mathbf{H}$ are the electric field intensity and magnetic field intensity, respectively.

If the **Lorentz condition**

$$\nabla \cdot \mathbf{A} + \frac{1}{c^2}\frac{\partial \phi}{\partial t} = 0$$

is satisfied, the basic equations for the vector and scalar potentials are

$$\Box^2 \mathbf{A} = -\mu_0 \mathbf{J}$$

and

$$\Box^2 \phi = -\frac{\rho}{\epsilon_0}$$

where the **d'Alembertian operator**, $\Box^2$, is defined by

$$\Box^2 = \nabla^2 - \frac{1}{c^2}\frac{\partial^2}{\partial t^2}.$$

The definitions of the **vector** and **scalar potentials** are, respectively,

$$\mathbf{B} = \nabla \times \mathbf{A} \quad \text{and} \quad \mathbf{E} = -\nabla \phi - \frac{\partial \mathbf{A}}{\partial t}.$$

The components of the four-vectors (four-dimensional orthogonal coordinate system—Minkowski (1864–1909) space) for **position**, **current density**, and **potential** are respectively taken to be

$$x^{\nu} = (x, y, z, ict)$$
$$J_{\lambda} = (J_x, J_y, J_z, ic\rho)$$

and

$$A_{\mu} = \left(A_x, A_y, A_z, i\frac{\phi}{c}\right).$$

The **electromagnetic field tensor** is defined by

$$F_{\mu\nu} = \frac{\partial A_{\nu}}{\partial x^{\mu}} - \frac{\partial A_{\mu}}{\partial x^{\nu}} \qquad (\mu, \nu = 1, 2, 3, 4) \tag{9.49}$$

where

$$\begin{aligned}
F_{11} &= \frac{\partial A_1}{\partial x^1} - \frac{\partial A_1}{\partial x^1} \\
&= F_{22} = F_{33} = F_{44} = 0 \\
F_{12} &= \frac{\partial A_2}{\partial x^1} - \frac{\partial A_1}{\partial x^2} \\
&= \frac{\partial A_y}{\partial x} - \frac{\partial A_x}{\partial y} \\
&= B_z \\
F_{13} &= \frac{\partial A_3}{\partial x^1} - \frac{\partial A_1}{\partial x^3} \\
&= \frac{\partial A_z}{\partial x} - \frac{\partial A_x}{\partial z} \\
&= -B_y
\end{aligned}$$

and

$$\begin{aligned}
F_{14} &= \frac{\partial A_4}{\partial x^1} - \frac{\partial A_1}{\partial x^4} \\
&= \frac{\partial\left(i\frac{\phi}{c}\right)}{\partial x} - \frac{\partial A_x}{\partial(ict)} \\
&= \frac{i}{c}\left[\frac{\partial \phi}{\partial x} + \frac{\partial A_x}{\partial t}\right] \\
&= -\frac{i}{c}E_x.
\end{aligned}$$

The remaining components of $F_{\mu\nu}$ may be obtained in a similar manner; the result in matrix form is

$$F_{\mu\nu} = \begin{pmatrix} 0 & B_z & -B_y & -\frac{i}{c}E_x \\ -B_z & 0 & B_x & -\frac{i}{c}E_y \\ B_y & -B_x & 0 & -\frac{i}{c}E_z \\ \frac{i}{c}E_x & \frac{i}{c}E_y & \frac{i}{c}E_z & 0 \end{pmatrix}.$$

Consider the tensor equation (which is valid for an arbitrary antisymmetric tensor)

$$\frac{\partial F_{\alpha\beta}}{\partial x^\mu} + \frac{\partial F_{\beta\mu}}{\partial x^\alpha} + \frac{\partial F_{\mu\alpha}}{\partial x^\beta} = 0 \qquad (\alpha, \beta, \mu = 1, 2, 3, 4).$$

Since $F_{\mu\nu}$ is an antisymmetric tensor, it can be shown that only four independent equations result from the sixty-four possible equations that are indicated by the above tensor equation. Typical values for α, β, and μ are set #1: 1, 2, 3; set #2: 4, 2, 3; set #3: 4, 3, 1; and set #4: 4, 1, 2. The resulting equations are:

Set 1 $\alpha, \beta, \mu = 1, 2, 3$

$$\frac{\partial F_{12}}{\partial x^3} + \frac{\partial F_{23}}{\partial x^1} + \frac{\partial F_{31}}{\partial x^2} = 0$$

or

$$\frac{\partial B_z}{\partial z} + \frac{\partial B_x}{\partial x} + \frac{\partial B_y}{\partial y} = 0$$

or

$$\nabla \cdot \mathbf{B} = 0 \qquad \text{(Maxwell's second equation)}.$$

Set 2 $\alpha, \beta, \mu = 4, 2, 3$

$$\frac{\partial F_{42}}{\partial x^3} + \frac{\partial F_{23}}{\partial x^4} + \frac{\partial F_{34}}{\partial x^2} = 0$$

or

$$\frac{\partial\left(\frac{i}{c}E_y\right)}{\partial z} + \frac{\partial B_x}{\partial(ict)} + \frac{\partial\left(-\frac{i}{c}E_z\right)}{\partial y} = 0$$

or

$$\frac{\partial E_z}{\partial y} - \frac{\partial E_y}{\partial z} = -\frac{\partial B_x}{\partial t}.$$

Set 3 $\alpha, \beta, \mu = 4, 3, 1$

$$\frac{\partial F_{43}}{\partial x^1} + \frac{\partial F_{31}}{\partial x^4} + \frac{\partial F_{14}}{\partial x^3} = 0$$

or

$$\frac{\partial\left(\frac{i}{c}E_z\right)}{\partial x} + \frac{\partial B_y}{\partial(ict)} + \frac{\partial\left(-\frac{i}{c}E_x\right)}{\partial z} = 0$$

or

$$\frac{\partial E_x}{\partial z} - \frac{\partial E_z}{\partial x} = -\frac{\partial B_y}{\partial t}.$$

Set 4 $\alpha, \beta, \mu = 4, 1, 2$

$$\frac{\partial F_{41}}{\partial x^2} + \frac{\partial F_{12}}{\partial x^4} + \frac{\partial F_{24}}{\partial x^1} = 0$$

or

$$\frac{\partial\left(\frac{i}{c}E_x\right)}{\partial y} + \frac{\partial B_z}{\partial(ict)} + \frac{\partial\left(-\frac{i}{c}E_y\right)}{\partial x} = 0$$

or

$$\frac{\partial E_y}{\partial x} - \frac{\partial E_x}{\partial y} = -\frac{\partial B_z}{\partial t}.$$

On combining the results of sets 2, 3, and 4, we obtain

$$\nabla \times \mathbf{E} = -\frac{\partial \mathbf{B}}{\partial t} \qquad \text{(Maxwell's third equation).}$$

The remaining two Maxwell's equations may be obtained from the following tensor equation:

$$\sum_{\nu=1}^{4} \frac{\partial F_{\mu\nu}}{\partial x^\nu} = \mu_0 J_\mu.$$

9.6.3 The Equation of the Geodesic Line

In three-dimensional Euclidean space, a straight line is the shortest distance between two points (see Example A9B.1 in Appendix B). In this section, we generalize this fundamental concept to Riemannian space.

If the curve $x^\nu = x^\nu(\lambda)$ joins two fixed points $P_1(\lambda_1)$ and $P_2(\lambda_2)$, then the distance along this curve between the two points is given by (see Appendix B to Chapter 9)

$$s = \int_{p_1}^{p_2} ds$$

$$= \int_{\lambda_1}^{\lambda_2} [F(x^\nu, \dot{x}^\nu, \lambda)]^{1/2}\, d\lambda$$

where

$$F = g_{ij}\frac{dx^i}{d\lambda}\frac{dx^j}{d\lambda} \qquad \text{and} \quad \dot{x}^\nu = \frac{dx^\nu}{d\lambda}.$$

Since this distance is independent of the system of coordinates, a curve drawn between P_1 and P_2 such that $\int_{p_1}^{p_2} ds$ is stationary (a minimum) is also independent of the system of coordinates; this curve is called a **geodesic**. We now develop the differential equation of the geodesic in Riemannian space.

Note that

$$\delta s = \delta \int_{\lambda_1}^{\lambda_2} F^{1/2}\, d\lambda = 0.$$

By use of the result derived for the extreme of functionals [Euler's equation, Eq. (A9B.11)], we may write

$$\begin{aligned} 0 &= \frac{\partial F^{1/2}}{\partial x^i} - \frac{d}{d\lambda}\left(\frac{\partial F^{1/2}}{\partial \dot{x}^i}\right) \\ &= \frac{d}{d\lambda}\left[\frac{1}{2F^{1/2}}\frac{\partial F}{\partial \dot{x}^i}\right] - \frac{1}{2F^{1/2}}\frac{\partial F}{\partial x^i} \\ &= \frac{d}{d\lambda}\left(\frac{1}{F^{1/2}}\right)\frac{\partial F}{\partial \dot{x}^i} + \frac{1}{F^{1/2}}\frac{d}{d\lambda}\left(\frac{\partial F}{\partial \dot{x}^i}\right) - \frac{1}{F^{1/2}}\frac{\partial F}{\partial x^i} \\ &= \frac{d}{d\lambda}\left(\frac{\partial F}{\partial \dot{x}^i}\right) - \frac{\partial F}{\partial x^i} - \frac{1}{2F}\frac{\partial F}{\partial \dot{x}^i}\frac{dF}{d\lambda}. \end{aligned} \tag{9.50}$$

Here we may take λ to equal the distance s along the curve in question and $dF/ds = 0$ since it is an arbitrary parameter. In this case, we have

$$\dot{x}^i = \frac{dx^i}{ds}$$

$$F = g_{ij}\frac{dx^i}{ds}\frac{dx^j}{ds}$$

$$\frac{\partial F}{\partial \dot{x}^i} = 2g_{ij}\dot{x}^j$$

and

$$\frac{\partial F}{\partial x^i} = \dot{x}^\mu \dot{x}^\nu \frac{\partial g_{\mu\nu}}{\partial x^i}.$$

Equation (9.50) now becomes

$$0 = \frac{d}{ds}\left(\frac{\partial F}{\partial \dot{x}^i}\right) - \frac{\partial F}{\partial x^i}$$

$$= \frac{d}{ds}(2g_{ij}\dot{x}^j) - \dot{x}^\mu \dot{x}^\nu \frac{\partial g_{\mu\nu}}{\partial x^i}$$

$$= g_{ij}\frac{d^2x^j}{ds^2} + \frac{\partial g_{ij}}{\partial x^k}\frac{dx^k}{ds}\frac{dx^j}{ds} - \frac{1}{2}\frac{\partial g_{\mu\nu}}{\partial x^i}\frac{dx^\mu}{ds}\frac{dx^\nu}{ds}.$$

By use of the Christoffel symbol of the first kind, we may write the above equation in the form

$$0 = g_{ij}\frac{d^2x^j}{ds^2} + [\mu\nu, i]\frac{dx^\mu}{ds}\frac{dx^\nu}{ds}$$

$$= g^{i\alpha}g_{ij}\frac{d^2x^j}{ds^2} + g^{i\alpha}[\mu\nu, i]\frac{dx^\mu}{ds}\frac{dx^\nu}{ds}. \tag{9.51}$$

Inner multiplication with respect to α in Eq. (9.51) leads to

$$\frac{d^2x^\alpha}{ds^2} + \begin{Bmatrix} \alpha \\ \mu\nu \end{Bmatrix}\frac{dx^\mu}{ds}\frac{dx^\nu}{ds} = 0. \tag{9.52}$$

Equation (9.52) is the required equation of the geodesic in Riemannian space.

EXAMPLE 9.3 Find the equation of the geodesic for a three-dimensional Cartesian coordinate system.

Solution For Cartesian coordinates in a three-dimensional Euclidean space, we have

$$g_{\mu\nu} = \begin{cases} 1 & (\mu = \nu) \\ 0 & (\mu \neq \nu) \end{cases}, \qquad \nu = 1, 2, 3 \quad \text{and} \quad \begin{Bmatrix} \alpha \\ \mu\nu \end{Bmatrix} = 0.$$

The equation of the geodesic in this special case becomes

$$\frac{d^2x^\alpha}{ds^2} = 0 \qquad (\alpha = 1, 2, 3).$$

If we consider an observer who is traveling with an object that is moving from P_1 to P_2 (the object is at rest with respect to the observer), $ds = dt$ where dt is called the **proper time**. Since $dv^\alpha/dt = 0$, then $v^\alpha = dx^\alpha/dt$ is a constant. On integrating $dx^\alpha/dt = A$, we obtain

$$x^\alpha = At + B$$

where A and B are arbitrary constants. The above equation is the required equation for the geodesic; it is the equation for a straight line.

9.6.4 The Riemann-Christoffel Tensor

Riemann assumed that the quadratic form $ds^2 = g_{\mu\nu}\, dx^\mu\, dx^\nu$ defined the metrical properties of the world and should be regarded as a physical reality. However, it was Einstein, in his theory of general relativity, who attached physical significance to $g_{\mu\nu}$ by assuming that $g_{\mu\nu}$ formed the components of the gravitational potential. In this theory, Einstein assumed that the phenomena of gravitation are intimately connected with geometry and that the laws by which matter affects measurements are the laws of gravitation. Here it must be noted that the gravitational potential will then have an invariant quadratic differential form, while electromagnetic phenomena are governed by a potential (four-potential), A_μ, which has an invariant linear differential form given by $A_\mu\, dx^\mu$. The existence of these two separate invariant forms is the source of the difficulty involved in developing a theory which unifies gravitational and electromagnetic phenomena.

We begin this section with the covariant derivative of anarbitary covariant vector A_β

$$A_{\mu,\nu} = \frac{\partial A_\mu}{\partial x^\nu} - \begin{Bmatrix} \beta \\ \mu\nu \end{Bmatrix} A_\beta. \tag{9.53}$$

By use of Eq. (9.47), we see that

$$A_{i,jk} = \frac{\partial A_{i,j}}{\partial x^k} - \begin{Bmatrix} \alpha \\ ki \end{Bmatrix} A_{\alpha,j} - \begin{Bmatrix} \alpha \\ kj \end{Bmatrix} A_{i,\alpha}. \tag{9.54}$$

If we substitute $A_{\mu,\nu}$ (with the appropriate change of indices) from Eq. (9.53) into Eq. (9.54), Eq. (9.54) becomes

$$\begin{aligned} A_{i,jk} = {} & \frac{\partial}{\partial x^k}\left(\frac{\partial A_i}{\partial x^j} - \begin{Bmatrix} \beta \\ ij \end{Bmatrix} A_\beta\right) \\ & - \begin{Bmatrix} \alpha \\ ki \end{Bmatrix}\left(\frac{\partial A_\alpha}{\partial x^j} - \begin{Bmatrix} \beta \\ \alpha j \end{Bmatrix} A_\beta\right) \\ & - \begin{Bmatrix} \alpha \\ kj \end{Bmatrix}\left(\frac{\partial A_i}{\partial x^\alpha} - \begin{Bmatrix} \beta \\ i\alpha \end{Bmatrix} A_\beta\right). \end{aligned} \tag{9.55}$$

Similarly, we obtain

$$\begin{aligned} A_{i,kj} = {} & \frac{\partial}{\partial x^j}\left(\frac{\partial A_i}{\partial x^k} - \begin{Bmatrix} \beta \\ ik \end{Bmatrix} A_\beta\right) \\ & - \begin{Bmatrix} \alpha \\ ji \end{Bmatrix}\left(\frac{\partial A_\alpha}{\partial x^k} - \begin{Bmatrix} \beta \\ \alpha k \end{Bmatrix} A_\beta\right) \\ & - \begin{Bmatrix} \alpha \\ kj \end{Bmatrix}\left(\frac{\partial A_i}{\partial x^\alpha} - \begin{Bmatrix} \beta \\ i\alpha \end{Bmatrix} A_\beta\right). \end{aligned} \tag{9.56}$$

Subtracting Eq. (9.56) from Eq. (9.55), we obtain the result

$$A_{i,jk} - A_{i,kj} = \left[\frac{\partial}{\partial x^j}\begin{Bmatrix}\beta \\ ik\end{Bmatrix} - \frac{\partial}{\partial x^k}\begin{Bmatrix}\beta \\ ij\end{Bmatrix} + \begin{Bmatrix}\alpha \\ ki\end{Bmatrix}\begin{Bmatrix}\beta \\ \alpha j\end{Bmatrix} - \begin{Bmatrix}\alpha \\ ji\end{Bmatrix}\begin{Bmatrix}\beta \\ \alpha k\end{Bmatrix}\right]A_\beta. \tag{9.57}$$

Since $A_{i,jk} - A_{i,kj}$ is a tensor and A_β is an arbitrary tensor, the expression in brackets in Eq. (9.57) is a mixed tensor of contravariant rank one and covariant rank three (fourth rank mixed tensor) by use of the quotient law. Equation (9.57) is normally written as

$$A_{i,jk} - A_{i,kj} = R^\beta_{kij}A_\beta \tag{9.58}$$

where the **Riemann-Christoffel tensor,** R^β_{kij}, is given by

$$R^\beta_{kij} = \frac{\partial}{\partial x^j}\begin{Bmatrix}\beta \\ ik\end{Bmatrix} - \frac{\partial}{\partial x^k}\begin{Bmatrix}\beta \\ ij\end{Bmatrix} + \begin{Bmatrix}\alpha \\ ki\end{Bmatrix}\begin{Bmatrix}\beta \\ \alpha j\end{Bmatrix} - \begin{Bmatrix}\alpha \\ ji\end{Bmatrix}\begin{Bmatrix}\beta \\ \alpha k\end{Bmatrix}. \tag{9.59}$$

The Riemann-Christoffel tensor is formed exclusively from the fundamental tensor $g_{\mu\nu}$. If a coordinate system is selected such that $g_{\mu\nu}$ are constants, then all components of the Riemann-Christoffel tensor vanish in this system and all other systems. The **curvature tensor,** $R_{\alpha kij}$, is obtained by use of the inner product of the fundamental metric tensor $g_{\alpha\beta}$ and the Riemann-Christoffel tensor

$$R_{\alpha kij} = g_{\alpha\beta}R^\beta_{kij}. \tag{9.60}$$

Contracting the Riemann-Christoffel tensor, Eq. (9.59), with respect to the β and j indices, we obtain **Ricci's tensor**:

$$R_{ki} = \frac{\partial}{\partial x^\beta}\begin{Bmatrix}\beta \\ ik\end{Bmatrix} - \frac{\partial}{\partial x^k}\begin{Bmatrix}\beta \\ i\beta\end{Bmatrix} + \begin{Bmatrix}\alpha \\ ki\end{Bmatrix}\begin{Bmatrix}\beta \\ \alpha\beta\end{Bmatrix} - \begin{Bmatrix}\alpha \\ \beta i\end{Bmatrix}\begin{Bmatrix}\beta \\ \alpha k\end{Bmatrix}. \tag{9.61}$$

We conclude this chapter by writing down Einstein's set of equations for the general relativistic theory of gravitation; they are

$$G_{\mu\nu} \equiv R_{\mu\nu} - \tfrac{1}{2}g_{\mu\nu}(R - 2\lambda) = \kappa T_{\mu\nu} \tag{9.62}$$

where $G_{\mu\nu}$ is called the **Einstein tensor,** $R_{\mu\nu}$ is Ricci's tensor, $g_{\mu\nu}$ is the fundamental metric tensor, the **curvature scalar** R is given by $R = g^{\mu\nu}R_{\mu\nu}$, λ is the cosmological constant, κ is a universal constant, and $T_{\mu\nu}$ is the energy-momentum tensor (in matter-free space, $T_{\mu\nu} = 0$).

9.7 PROBLEMS

9.1 By use of the summation convention, rewrite

$$d\phi = \frac{\partial \phi}{\partial x}\,dx + \frac{\partial \phi}{\partial y}\,dy + \frac{\partial \phi}{\partial z}\,dz.$$

9.2 How many equations in a four-dimensional space are represented by $R^{\alpha}_{\beta\gamma\rho} = 0$? Explain.

9.3 Show that the symmetric (or antisymmetric) property of a tensor is conserved under a transformation of coordinates.

9.4 Show that $g_{ij} = 0$ for $i \neq j$ is required for orthogonal coordinate systems.

9.5 By use of the Riemannian metric, show that $g_{ij} = g_{ji}$.

9.6 Under what conditions will the Christoffel symbols be tensors?

9.7 Show that the covariant derivative of g_{ij} is zero (Ricci's theorem).

9.8 By use of the definition of the components of the four-vector current density, write the continuity equation.

9.9 Show that the set of tensor equations

$$\sum_{\nu=1}^{4} \frac{\partial F_{\mu\nu}}{\partial x^{\nu}} = \mu_0 J_{\mu}$$

where $F_{\mu\nu}$ and J_{μ} are the components of the electromagnetic field tensor and four-vector current density, respectively, reduces to Maxwell's first and fourth equations.

9.10 Compute Γ^{i}_{jk} in (a) Cartesian coordinates and (b) cylindrical coordinates.

9.11 For the surface of a sphere (a two-dimensional Riemannian space), compute (a) the components of the fundamental metric tensor and (b) the two Christoffel symbols.

9.12 Compute the components of the fundamental metric tensor and ds^2 for the new coordinate system when the transformation from Cartesian coordinates (x, y, z) to spherical coordinates (r, θ, ϕ) is made.

9.13 By use of the definition of the Christoffel symbol of the first kind, show that

$$\frac{\partial g_{ik}}{\partial x^{j}} = [ij, k] + [kj, i].$$

9.14 By use of the definition of the Christoffel symbol of the second kind, show that

$$\frac{\partial g^{mk}}{\partial x^{l}} = -g^{ik}\left\{ {m \atop il} \right\} - g^{jm}\left\{ {k \atop jl} \right\}.$$

9.15 By use of a procedure similar to that used to obtain Eq. (9.44), show that Eq. (9.45) is valid.

9.16 By use of the fundamental transformation law for the components of a tensor, show that the quantities $\delta^{\nu}_{\mu}A_{\nu}$ transform like the components of a tensor.

9.17 The tensor forms for grad ϕ, div **A**, curl **A**, and $\nabla^2\phi$ in terms of components are

$$\text{grad}\,\phi = \frac{\partial \phi}{\partial x^j}$$

$$\text{div}\,\mathbf{A} = \frac{\partial A^i}{\partial x^i} + \Gamma^i_{ik}A^k$$

$$(\text{curl}\,\mathbf{A})_{ji} = \frac{\partial A_j}{\partial x^i} - \frac{\partial A_i}{\partial x^j}.$$

$$\nabla^2\phi = g^{ij}\left(\frac{\partial^2\phi}{\partial x^i\,\partial x^j} - \Gamma^k_{ij}\frac{\partial \phi}{\partial x^k}\right).$$

(a) Show that div **A** is a scalar.
(b) Show that the above expressions yield the appropriate results for Cartesian and spherical coordinates.

APPENDIX A: THE TRANSFORMATION LAWS FOR THE CHRISTOFFEL SYMBOLS

In this appendix, we investigate the transformation laws for $[ij, k]$ and $\left\{ \begin{matrix} l \\ ij \end{matrix} \right\}$.

In terms of new coordinates $\bar{x}^i$, we may write

$$\overline{[lm, n]} = \frac{1}{2}\left(\frac{\partial \bar{g}_{ln}}{\partial \bar{x}^m} + \frac{\partial \bar{g}_{mn}}{\partial \bar{x}^l} - \frac{\partial \bar{g}_{lm}}{\partial \bar{x}^n}\right). \qquad \text{(A9A.1)}$$

We had

$$\bar{g}_{lm} = \frac{\partial x^i}{\partial \bar{x}^l}\frac{\partial x^j}{\partial \bar{x}^m}g_{ij}.$$

On differentiating $\bar{g}_{lm}$ with respect to $\bar{x}^n$, we obtain

$$\frac{\partial \bar{g}_{lm}}{\partial \bar{x}^n} = \frac{\partial}{\partial \bar{x}^n}\left(\frac{\partial x^i}{\partial \bar{x}^l}\right)\frac{\partial x^j}{\partial \bar{x}^m}g_{ij} + \frac{\partial x^i}{\partial \bar{x}^l}\frac{\partial}{\partial \bar{x}^n}\left(\frac{\partial x^j}{\partial \bar{x}^m}\right)g_{ij} + \frac{\partial x^i}{\partial \bar{x}^l}\frac{\partial x^j}{\partial \bar{x}^m}\frac{\partial g_{ij}}{\partial \bar{x}^n}.$$

Since $g_{ij} = g_{ji}$, the above equation becomes

$$\frac{\partial \bar{g}_{lm}}{\partial \bar{x}^n} = \left(\frac{\partial^2 x^i}{\partial \bar{x}^n\, \partial \bar{x}^l}\frac{\partial x^j}{\partial \bar{x}^m} + \frac{\partial x^j}{\partial \bar{x}^l}\frac{\partial^2 x^i}{\partial \bar{x}^n\, \partial \bar{x}^m}\right) g_{ij} + \frac{\partial x^i}{\partial \bar{x}^l}\frac{\partial x^j}{\partial \bar{x}^m}\frac{\partial g_{ij}}{\partial \bar{x}^n}. \tag{A9A.2}$$

Cyclic permutation of l, m, and n ($lmn \to nlm$) in the above equation leads to

$$\frac{\partial \bar{g}_{nl}}{\partial \bar{x}^m} = \left(\frac{\partial^2 x^i}{\partial \bar{x}^m\, \partial \bar{x}^n}\frac{\partial x^j}{\partial \bar{x}^l} + \frac{\partial x^j}{\partial \bar{x}^n}\frac{\partial^2 x^i}{\partial \bar{x}^m\, \partial \bar{x}^l}\right) g_{ij} + \frac{\partial x^i}{\partial \bar{x}^n}\frac{\partial x^j}{\partial \bar{x}^l}\frac{\partial g_{ij}}{\partial \bar{x}^m}. \tag{A9A.3}$$

For $nlm \to mnl$ in Eq. (A9A.3), we obtain

$$\frac{\partial \bar{g}_{mn}}{\partial \bar{x}^l} = \left(\frac{\partial^2 x^i}{\partial \bar{x}^l\, \partial \bar{x}^m}\frac{\partial x^j}{\partial \bar{x}^n} + \frac{\partial x^j}{\partial \bar{x}^m}\frac{\partial^2 x^i}{\partial \bar{x}^l\, \partial \bar{x}^n}\right) g_{ij} + \frac{\partial x^i}{\partial \bar{x}^m}\frac{\partial x^j}{\partial \bar{x}^n}\frac{\partial g_{ij}}{\partial \bar{x}^l}. \tag{A9A.4}$$

Substituting Eqs. (A9A.2), (A9A.3), and (A9A.4) into Eq. (A9A.1), we find that

$$\overline{[lm, n]} = [ij, k]\frac{\partial x^i}{\partial \bar{x}^l}\frac{\partial x^j}{\partial \bar{x}^m}\frac{\partial x^k}{\partial \bar{x}^n} + \frac{\partial x^i}{\partial \bar{x}^n}\frac{\partial^2 x^j}{\partial \bar{x}^l\, \partial \bar{x}^m} g_{ij} \tag{A9A. 5}$$

which shows that $[ij, k]$ transforms like a tensor only if the second term in Eq. (A9A.5) vanishes. When this second term vanishes, the transformation is said to be **affine**. Equation (A9A.5) is the required transformation law for the Christoffel symbol of the first kind.

The transformation law for the Christoffel symbol of the second kind will now be developed. For the contravariant fundamental tensor, we had

$$\bar{g}^{\nu\mu} = \frac{\partial \bar{x}^\nu}{\partial x^r}\frac{\partial \bar{x}^\mu}{\partial x^s} g^{rs}.$$

Inner multiplication of both sides of Eq. (A9A.5) with the corresponding sides of the above equation for $\bar{g}^{\nu\mu}$ leads to

$$\overline{\begin{Bmatrix} k \\ lm \end{Bmatrix}} = \begin{Bmatrix} \alpha \\ ij \end{Bmatrix}\frac{\partial \bar{x}^k}{\partial x^\alpha}\frac{\partial x^i}{\partial \bar{x}^l}\frac{\partial x^j}{\partial \bar{x}^m} + \frac{\partial \bar{x}^k}{\partial x^j}\frac{\partial^2 x^j}{\partial \bar{x}^l\, \partial \bar{x}^m}. \tag{A9A.6}$$

which is the transformation law for the Christoffel symbol of the second kind.

Inner multiplication of Eq. (A9A.6) with $\partial x^r/\partial \bar{x}^p$ yields

$$\overline{\begin{Bmatrix} k \\ lm \end{Bmatrix}} \frac{\partial x^r}{\partial \bar{x}^p} = \begin{Bmatrix} \alpha \\ ij \end{Bmatrix} \frac{\partial x^r}{\partial \bar{x}^p} \frac{\partial \bar{x}^k}{\partial x^\alpha} \frac{\partial x^i}{\partial \bar{x}^l} \frac{\partial x^j}{\partial \bar{x}^m} + \frac{\partial x^r}{\partial \bar{x}^p} \frac{\partial \bar{x}^k}{\partial x^j} \frac{\partial^2 x^j}{\partial \bar{x}^l \, \partial \bar{x}^m}$$

or

$$\frac{\partial^2 x^r}{\partial \bar{x}^l \, \partial \bar{x}^m} = \overline{\begin{Bmatrix} p \\ lm \end{Bmatrix}} \frac{\partial x^r}{\partial \bar{x}^p} - \begin{Bmatrix} r \\ ij \end{Bmatrix} \frac{\partial x^i}{\partial \bar{x}^l} \frac{\partial x^j}{\partial \bar{x}^m}. \qquad \text{(A9A.7)}$$

Equation (A9A.7) is an extremely useful relation for the second partial derivative of x^r in terms of the first derivative and the Christoffel symbol of the second kind.

APPENDIX B: RUDIMENTS OF THE CALCULUS OF VARIATIONS

A9B.1 Maxima and Minima of Functions

For a continuous function, $f(x)$, to have a maximum or a minimum (an extremum) at a certain point x_0, it is necessary that the first derivative, $df(x)/dx$, vanish at x_0. A necessary condition for a continuous function $g(x, y)$ to have an extremum at some point (x_0, y_0) is that

$$\left.\frac{\partial g(x, y)}{\partial x}\right|_{\substack{x=x_0 \\ y=y_0}} \quad \text{and} \quad \left.\frac{\partial g(x, y)}{\partial y}\right|_{\substack{x=x_0 \\ y=y_0}}$$

vanish at this point. This procedure can be extended, in a natural way, to include functions of several variables as follows:

$$df(x^1, x^2, \ldots, x^n) = \frac{\partial f}{\partial x^1} dx^1 + \frac{\partial f}{\partial x^2} dx^2 + \cdots + \frac{\partial f}{\partial x^n} dx^n.$$

Since $df = 0$ at an extremum for arbitrary dx^i $(i = 1, 2, \ldots, n)$, we require that

$$\left.\frac{\partial f}{\partial x^1}\right|_{x^i = x_0{}^i} = \left.\frac{\partial f}{\partial x^2}\right|_{x^i = x_0{}^i} = \cdots = \left.\frac{\partial f}{\partial x^n}\right|_{x^i = x_0{}^i} = 0$$

for all i.

A9B.2 Maxima and Minima of Functionals

Consider the definite integral

$$J = \int_{x_1}^{x_2} F(y, y', x)\, dx \qquad \left(y' = \frac{dy}{dx}\right). \qquad \text{(A9B.1)}$$

Here $F(y, y', x)$ is called a **functional** (a function of a function).

The fundamental problem of the calculus of variations (a problem which occurs frequently in mathematical physics) is that of finding a function $y(x)$ such that the functional J is stationary (an extremum) for small variations in $y(x)$. Here $F(y, y', x)$ is a known function of y, y', and x, but $y(x)$ is unknown, that is, the exact path of integration is not known. In Fig. 9.1, two of the infinite number of paths which can be drawn between (x_1, y_1) and (x_2, y_2) are indicated.

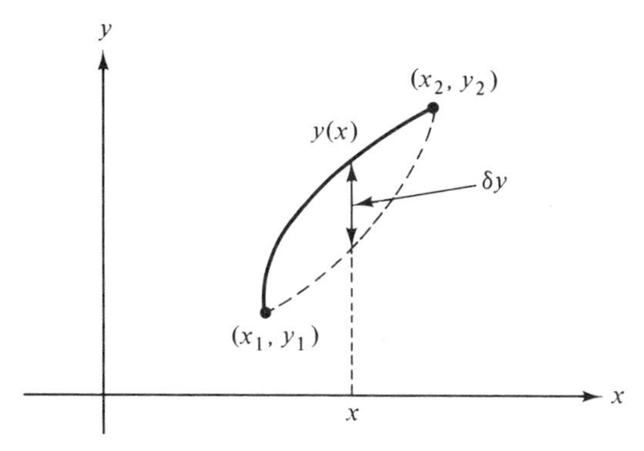

Figure 9.1

The method of ordinary calculus for ordinary functions cannot be used directly to find the extremum of J which involves a functional. To reduce this problem to one that can be solved by use of ordinary calculus, we write

$$y(x, \alpha) = y(x, 0) + \alpha\eta(x) \tag{A9B.2}$$

where $\eta(x_1) = \eta(x_2) = 0$ (all paths must pass through the fixed end points), α is a small scale factor, and $\eta(x)$ is an arbitrary differentiable function of x. Here $y(x, 0)$ is an unknown path that will result in a maximum or a minimum of J. (In most cases of physical interest, the extremum will be a minimum.) The function $y(x, \alpha)$ describes a neighboring path where $\delta y = y(x, \alpha) - y(x, 0) = \alpha\eta(x)$ is the variation of y at some x. In terms of Eq. (A9B.2), J now becomes

$$J(\alpha) = \int_{x_1}^{x_2} F[y(x, \alpha), y'(x, \alpha), x]\, dx. \tag{A9B.3}$$

The condition for an extremum of $J(\alpha)$ is that

$$\left.\frac{dJ}{d\alpha}\right|_{\alpha=0} = 0 \qquad [\text{for all } \eta(x)]. \tag{A9B.4}$$

On differentiating $J(\alpha)$ with respect to α, we obtain

$$\frac{dJ(\alpha)}{d\alpha} = \int_{x_1}^{x_2} \left[\frac{\partial F}{\partial y}\frac{\partial y}{\partial \alpha} + \frac{\partial F}{\partial y'}\frac{\partial y'}{\partial \alpha}\right] dx \tag{A9B.5}$$

since the α-dependence of F is in y and y'. Note that

$$\frac{\partial y(x, \alpha)}{\partial \alpha} = \eta(x) \tag{A9B.6}$$

and

$$\frac{\partial y'(x, \alpha)}{\partial \alpha} = \frac{d\eta(x)}{dx}. \tag{A9B.7}$$

By use of Eqs. (A9B.6) and (A9B.7), Eq. (A9B.5) reduces to

$$\frac{dJ}{d\alpha} = \int_{x_1}^{x_2} \left[\frac{\partial F}{\partial y}\eta(x) + \frac{\partial F}{\partial y'}\frac{d\eta(x)}{dx}\right] dx. \tag{A9B.8}$$

Integrating the second term in Eq. (A9B.8) by parts, we get

$$\begin{aligned}\int_{x_1}^{x_2} \frac{\partial F}{\partial y'}\frac{d\eta(x)}{dx}\,dx &= \eta(x)\frac{\partial F}{\partial y'}\bigg|_{x_1}^{x_2} - \int_{x_1}^{x_2} \eta(x)\frac{d}{dx}\left(\frac{\partial F}{\partial x'}\right) dx \\ &= -\int_{x_1}^{x_2} \eta(x)\frac{d}{dx}\left(\frac{\partial F}{\partial y'}\right) dx \end{aligned} \tag{A9B.9}$$

since $\eta(x_1) = \eta(x_2) = 0$. On substituting Eq. (A9B.9) into Eq. (A9B.8), we find that Eq. (A9B.4) becomes

$$\begin{aligned}\frac{dJ}{d\alpha}\bigg|_{\alpha=0} &= 0 \\ &= \int_{x_1}^{x_2} \left[\frac{\partial F}{\partial y} - \frac{d}{dx}\left(\frac{\partial F}{\partial y'}\right)\right]\eta(x)\,dx. \end{aligned} \tag{A9B.10}$$

Since $\eta(x)$ is an arbitrary function, we may write

$$\frac{\partial F}{\partial y} - \frac{d}{dx}\left(\frac{\partial F}{\partial y'}\right) = 0. \tag{A9B.11}$$

Equation (A9B.11) is known as **Euler's equation** (or Euler-Lagrange equation).

Multiplying Eq. (A9B.10) by α, we obtain

$$\begin{aligned}\alpha\frac{dJ}{d\alpha}\bigg|_{\alpha=0} &= \delta J \\ &= \int_{x_1}^{x_2} \left[\frac{\partial F}{\partial y} - \frac{d}{dx}\left(\frac{\partial F}{\partial y'}\right)\right] \delta y\,dx \end{aligned} \tag{A9B.12}$$

since $\alpha\eta(x) = \delta y$.

The above development may be extended to the case of functionals involving several dependent variables,

$$F(y_1, y_2, \ldots, y_n; y'_1, y'_2, \ldots, y'_n; x).$$

The resulting set of Euler equations is

$$\frac{\partial F}{\partial y_i} - \frac{d}{dx}\left(\frac{\partial F}{\partial y'_i}\right) = 0 \qquad (i = 1, 2, \ldots, n). \tag{A9B.13}$$

The extension to the case of several independent and several dependent variables is also straightforward.

EXAMPLE A9B.1 By use of Euler-Lagrange equation, determine the minimum (shortest) distance between two points (x_1, y_1) and (x_2, y_2) in Cartesian coordinates.

Solution The element of distance is

$$\begin{aligned} ds &= [dx^2 + dy^2]^{1/2} \\ &= \left[1 + \left(\frac{dy}{dx}\right)^2\right]^{1/2} dx. \end{aligned}$$

The distance between the two points is given by

$$\begin{aligned} s &= \int_{x_1, y_1}^{x_2, y_2} ds \\ &= \int_{x_1}^{x_2} \left[1 + \left(\frac{dy}{dx}\right)^2\right]^{1/2} dx. \end{aligned}$$

On comparing the above integrand with the integrand in Eq. (A9B.1), we find that

$$F(y, y', x) = \left[1 + \left(\frac{dy}{dx}\right)^2\right]^{1/2}$$

where

$$\frac{\partial F}{\partial y} - \frac{d}{dx}\left(\frac{\partial F}{\partial y'}\right) = 0.$$

Since

$$\frac{\partial F}{\partial y} = 0 \quad \text{and} \quad \frac{\partial F}{\partial y'} = \frac{y'}{[1 + (y')^2]^{1/2}}$$

the Euler-Lagrange equation,

$$\frac{\partial F}{\partial y} - \frac{d}{dx}\left(\frac{\partial F}{\partial y'}\right) = 0$$

becomes

$$-\frac{d}{dx}\left\{\frac{y'}{[1 + (y')^2]^{1/2}}\right\} = 0$$

or

$$\frac{dy/dx}{[1 + (y')^2]^{1/2}} = c = \text{constant}.$$

This equation may be written as

$$\frac{dy}{dx} = A = \text{constant}$$

or

$$dy = A\, dx$$

where

$$A = \left(\frac{c}{1 - c}\right)^{1/2}.$$

Integrating the above differential equation, we obtain

$$y = Ax + B$$

which is the equation of a straight line for $B =$ constant.

A9B.3 Lagrange's Equations of Classical Mechanics

The fundamental principle of classical mechanics (for a particle or a system of particles) is the Hamilton variational principle which, in mathematical form, is

$$\delta \int_{t_1}^{t_2} L(q_j, \dot{q}_j, t)\, dt = 0 \tag{A9B.14}$$

where t is time and $L\, dt$ is called the **action**. The quantities q_j which completely define the position of the particle are called **generalized coordinates** (they may be selected to match the conditions of the problem to be solved). The corresponding **generalized velocities** are represented by $\dot{q}_j$. The Lagrangian, L, of the system is the difference of the kinetic energy, T, and the potential energy, V. In equation form, the Lagrangian is

$$L = T - V. \tag{A9B.15}$$

Hamilton's variational principle asserts that the actual motion of a particle, under conservative forces, between two given points during a given interval of time (t_1, t_2) is such that Eq. (A9B.14) is valid.

By use of the basic result of the calculus of variations (Euler's equation), Hamilton's principle leads to the following set of equations:

$$\frac{\partial L}{\partial q_j} - \frac{d}{dt}\left(\frac{\partial L}{\partial \dot{q}_j}\right) = 0 \qquad (j = 1, 2, \ldots, n). \tag{A9B.16}$$

This set of equations is called **Lagrange's equations of motion** and is equivalent to Newton's (second law) equations of motion.

EXAMPLE A9B.2 Show that Eq. (A9B.16) reduces to Newton's second law.

Solution Consider a particle with position vector $\mathbf{r}$ and velocity $\dot{\mathbf{r}}$. In this case, we have

$$\begin{aligned} L &= T - V \\ &= \tfrac{1}{2}m\dot{\mathbf{r}}^2 - V(\mathbf{r}). \end{aligned}$$

Equation (A9B.16) now becomes

$$-\nabla V - \frac{d}{dt}(m\dot{\mathbf{r}}) = 0$$

which is Newton's second law since

$$-\nabla V = \mathbf{F} \qquad \text{(force)}$$

and

$$\frac{d}{dt}(m\dot{\mathbf{r}}) = m\mathbf{a} \qquad \text{(mass times acceleration).}$$

The **generalized momentum** is defined by

$$\mathbf{p}_j = \frac{\partial L}{\partial \dot{q}_j}$$

and the **generalized force** is defined by

$$\mathbf{F}_j = \frac{\partial L}{\partial q_j}.$$

A9B.4 Problems

A9B.1 For polar coordinates, show that the path of shortest distance between two points on a Euclidean plane lies along a straight line.

A9B.2 By use of the Euler-Lagrange equation, develop the equation for the trajectory of a (a) free particle and (b) particle moving in a constant gravitational field ($V = mgy$, where y is a Cartesian coordinate) for Cartesian coordinates.

A9B.3 Show that Euler's equation for one independent and one dependent variable can be written as

$$\frac{\partial F}{\partial x} - \frac{d}{dx}\left(F - y'\frac{\partial F}{\partial y'}\right) = 0$$

where

$$F = F(y, y', x) \quad \text{and} \quad y' = \frac{dy}{dx}.$$

A9B.4 By use of the variational technique developed in Appendix A9B, derive the Euler equations for the case of several dependent variables and one independent variable.

General Bibliography

1. ARFKEN, G. B., *Mathematical Methods for Physicists.* New York, Academic Press, Inc., 1970.

2. BOAS, M. L., *Mathematical Methods in the Physical Sciences.* New York, John Wiley & Sons, Inc., 1966.

3. BUTKOV, E., *Mathematical Physics.* Reading, Mass., Addison-Wesley Publishing Company, Inc., 1968.

4. CHISHOLM, J. S. R., and MORRIS, R., *Mathematical Methods in Physics.* Philadelphia, Pa., Saunders, 1965.

5. IRVING, J., and MULLINEUX, N., *Mathematics in Physics and Engineering.* New York, Academic Press, Inc., 1959.

6. KILLINGBECK, J., and COLE, G. H. A., *Mathematical Techniques and Physical Applications.* New York, Academic Press, Inc., 1971.

7. KRAUT, E., *Fundamentals of Mathematical Physics.* New York, McGraw-Hill Book Company, 1967.

8. KREYSZIG, E., *Advanced Engineering Mathematics.* New York, John Wiley & Sons, Inc., 1962.

9. MATHEWS, J., and WALKER, R. L., *Mathematical Methods of Physics.* Reading, Mass., W. A. Benjamin, 1970.

10. PIPES, L. A., and HARVILL, L., *Applied Mathematics for Engineers and Physicists*. New York, McGraw-Hill Book Company, 1970.

11. SOKOLNIKOFF, I. S., and REDHEFFER, R. M., *Mathematics of Physics and Modern Engineering*. New York, McGraw-Hill Book Company, 1966.

12. WYLIE, C. R., *Advanced Engineering Mathematics*. New York, McGraw-Hill Book Company, 1960.

References

CHAPTER 1: VECTOR ANALYSIS

1. MARION, J. B., *Principles of Vector Analysis*. New York, Academic Press, Inc., 1965.
2. PHILLIPS, H. B., *Vector Analysis*. New York, John Wiley & Sons, Inc., 1956.

CHAPTER 2: OPERATOR AND MATRIX ANALYSIS

1. HOHN, F. E., *Elementary Matrix Algebra*. New York, The Macmillan Company, 1958.
2. SCHMEIDLER, W., *Linear Operators in Hilbert Space*. New York, Academic Press, Inc., 1965.

CHAPTERS 3 & 4: FUNCTIONS OF A COMPLEX VARIABLE CALCULUS OF RESIDUES

1. CHURCHILL, R. V., *Complex Variables and Applications*. New York, McGraw-Hill Book Company, 1960.
2. DETTMAN, J. W., *Applied Complex Variables*. New York, The Macmillan Company, 1966.

3. Kyrala, A., *Applied Functions of a Complex Variable.* New York, Wiley-Interscience, 1972.
4. Silverman, R. A., *Complex Analysis with Applications.* Englewood Cliffs, N. J., Prentice-Hall, Inc., 1974.

CHAPTER 5: DIFFERENTIAL EQUATIONS

1. Agnew, R. P., *Differential Equations.* New York, McGraw-Hill Book Company, 1960.
2. Churchill, R. V., *Fourier Series and Boundary Value Problems.* New York, McGraw-Hill Book Company, 1963.
3. Duff, G. F., *Differential Equations of Applied Mathematics.* New York, John Wiley & Sons, Inc., 1966.
4. Jaeger, J. C., *An Introduction to Applied Mathematics.* Oxford, The Clarendon Press, 1956.
5. Myint-u, T., *Partial Differential Equations of Mathematical Physics.* New York, American Elsevier Publishing Company, Inc., 1973.
6. Scarborough, J., *Differential Equations and Applications.* Baltimore, Waverly Press, Inc., 1965.
7. Simmons, G. F., *Differential Equations with Applications and Historical Notes.* New York, McGraw-Hill Book Company, 1972.
8. Weinberger, H. F., *Partial Differential Equations.* New York, Blaisdell Publishing Company, 1965.

CHAPTER 6: SPECIAL FUNCTIONS

1. Churchill, R. V., *Fourier Series and Boundary Value Problems.* New York, McGraw-Hill Book Company, 1963.
2. Hochstadt, H., *Special Functions of Mathematical Physics.* New York, Holt, Rinehart & Winston, 1961.
3. MacRobert, T. M., *Spherical Harmonics.* Oxford, Pergamon Press, 1967.
4. Tranter, C. J., *Bessel Functions with Some Physical Applications.* London, English Universities Press, 1968.

CHAPTER 7: FOURIER SERIES

1. Churchill, R. V., *Fourier Series and Boundary Value Problems.* New York, McGraw-Hill Book Company, 1963.
2. Rogosinski, W., *Fourier Series.* New York, Chelsea Publishing Company, 1959.
3. Sneddon, I. N., *Fourier Series.* New York, Dover Publications, Inc., 1961.
4. Whittaker, E. T., and Watson, G. N., *Modern Analysis.* New York, The Macmillan Company, 1947.

CHAPTER 8: FOURIER TRANSFORMS

1. CHURCHILL, R. V., *Operational Mathematics*. New York, McGraw-Hill Book Company, 1958.
2. FONG, P., *Elementary Quantum Mechanics*. Reading, Mass., Addison-Wesley Publishing Company, Inc., 1962.
3. GASIOROWICZ, S., *Quantum Physics*. New York, John Wiley & Sons, Inc., 1974.
4. JAMMER, M., *The Conceptual Development of Quantum Mechanics*. New York, McGraw-Hill Book Company, 1966.
5. SNEDDON, I. N., *Fourier Transforms*. New York, McGraw-Hill Book Company, 1951.

CHAPTER 9: TENSOR ANALYSIS

1. SOKOLNIKOFF, I. S., *Tensor Analysis*. New York, John Wiley & Sons, Inc., 1960.
2. SPAIN, B., *Tensor Calculus*. New York, Interscience Publishers, Inc., 1960.

Index

Q

R

S

Z